Energy Systems

Series editor

Panos M. Pardalos, Gainesville, USA

More information about this series at http://www.springer.com/series/8368

Sandra D. Eksioglu · Steffen Rebennack
Panos M. Pardalos
Editors

Handbook of Bioenergy

Bioenergy Supply Chain - Models and Applications

Springer

Editors
Sandra D. Eksioglu
Department of Industrial Engineering
Clemson University
Clemson, SC
USA

Panos M. Pardalos
Department of Industrial and Systems Engineering
University of Florida
Gainesville, FL
USA

Steffen Rebennack
Division of Economics and Business
Engineering Hall
Colorado School of Mines
Golden, CO
USA

ISSN 1867-8998 ISSN 1867-9005 (electronic)
Energy Systems
ISBN 978-3-319-20091-0 ISBN 978-3-319-20092-7 (eBook)
DOI 10.1007/978-3-319-20092-7

Library of Congress Control Number: 2015943045

Springer Cham Heidelberg New York Dordrecht London

Printed on acid-free paper

Springer International Publishing AG Switzerland is part of Springer Science+Business Media (www.springer.com)

Preface

In the recent years we have seen an increasing interest in many areas of research related to bioenergy. Bioenergy is renewable energy, and it is produced using biological sources such as, agricultural products and residues, forest products and residues, energy crops and other types of biological waste. The increasing interest in bioenergy has been motivated by its potential to become one of the future energy sources of USA. This in return is expected to reduce the nation's dependency on fossil fuels, and to have a positive impact on the economy, environment and society. Investments in and use of bioenergy have been encouraged by a number of legislations at the federal and state levels. For example, the Renewable Fuel Standard in the Energy Independence and Security Act of 2007 requires an increase of the minimum annual level of renewable fuels used in U.S. transportation fuel from 9 billion gallons in 2008 to 36 in 2022. These targets are expected to be met through a variety of biofuels, among which, grain ethanol and cellulosic ethanol will be major contributors.

Despite the increasing interest, the economic viability of bioenergy and its future have been questioned and challenged for the following reasons. The production of traditional, first-generation biofuels relied on the use of edible products such as, corn and soyabean. This fact stirred the national debate of food versus fuel. Many of the existing conversion technologies are in the pre-commercial stages. Most non-food feedstocks have not reached cultivation status where they may be considered commodity products. Non-food feedstocks are bulky and difficult to transport, thus, the corresponding delivery costs are high. Most of the vehicles we use cannot handle blends with more than 10 % ethanol. Despite the challenges faced today, we believe that investing in bioenergy is very important. The goal of this handbook is to present some of the state-of-the-art developments and tools to help the industry overcome the economical challenges. A number of chapters in this handbook present optimization and simulation models to optimize biomass transportation, and bioenergy supply chain. A few chapters present results from life-cycle cost analysis of different types of biofuels. One of the chapters proposes a bi-objective optimization model that determines the optimum allocation of marginal land and crop land for biomass cultivation and supply chain optimization.

A number of case studies are presented that illustrate the impact of these tools in making bioenergy an economically viable energy option.

Let us provide a short review of the logistical challenges the industry has been facing in the past two decades. The production of traditional, first-generation biofuels (ethanol) relied on the use of corn and soybean. Most ethanol plants were located in the middle of corn fields and within a short radius from their supply. These location decisions were motivated by the high costs of transporting biomass. The limited amount of biomass available within this collection radius did not justify investments on large-scale biorefineries. As a result, traditional biorefineries have low production capacity and have not benefited from the economies of scale associated with high production volumes. The second-generation biofuels utilize agricultural and forest waste and energy crops as feedstock. The logistical efforts with loading, unloading and transporting these types of feedstock are higher since these types of biomass feedstock are bulky, aerobically unstable and have poor flowability properties. The first- and second-generation biofuels are highly acidic, have high moisture content and high oxygen content. Due to these properties, the pipeline system that is currently in place for distribution of fossil fuels cannot be utilized. The next-generation biofuels, referred to as drop-in fuels, are expected to have similar physical properties as fossil fuel, and thus, become interchangeable.

All the types of bioenergy will continue to face biomass feedstock transportation and other logistics challenges. Researchers are currently developing models which support large-scale biomass transportation and consequently large-scale production of bioenergy. Some of these models rely on preprocessing biomass at the farm gate, and using rail and barge as cost-efficient modes for high-volume and long-haul transportation. Some research has shown that locating a bioenergy plant close to an intermodal facility (such as, in-land port or rail station) could result in transportation cost savings. However, such a selection of a facility location may result in higher risk of disruptions due to natural disasters such as flooding. Other research has been focused on using hub-and-spoke distribution networks for biomass. These types of distribution networks have shown to be cost-efficient for airlines and railways. Similar to these industries, the products shipped on the system are widely dispersed geographically. Thus, hub-and-spoke networks could be a valid option for the in-bound supply chain design of bioenergy plants.

Ongoing concerns about bioenergy are focused not only on its economic viability, but also, on the carbon footprint in the course of its life cycle. This is because the steps involved in production and distribution of bioenergy are energy intensive. For example, growing, harvesting and transporting biomass feedstock, converting feedstock into fuel, and distributing fuel require energy. A few studies evaluate costs, energy and carbon footprint during the life cycle of bioenergy. Other studies propose models which optimize costs and emissions of the biofuel supply chain.

This handbook starts off with a survey by Balaman and Selim of existing models to optimize the bioenergy supply chain. This discussion is followed by a number of chapters which present recent developments in this area. The models developed by Chen and Huang and Poudel et al. are deterministic in nature. These models take an integrated view of facility location and transportation decisions in the supply chain.

The chapter by Kaut et al. proposes a stochastic optimization modelling framework. This modelling framework is mindful of supply and demand uncertainties which are often observed in the bioenergy supply chain. The chapter by Cinar et al. presents a two-stage stochastic model which captures uncertainty of costs and biomass supply on a supply chain design problem. A few chapters focus of single objective optimization models. Chapters by Palander et al., Zhang and Osmani present multi-objective optimization models. These models optimize costs and delivery time in the supply chain; and costs and land use. The chapter by Morales-Rincón et al. discusses a GIS-based method to locate bioenergy production facilities. Amundson et al. present a framework which integrates process optimization, supply chain optimization and discrete event simulation capabilities. Their goal is to provide a comprehensive and multidisciplinary tool for bioenergy supply chain design. The tool incorporates risk models in order to capture various uncertainties of this supply chain. Bai and Ouyang provide a comprehensive review of the impacts of bioenergy production on the transportation systems and infrastructure. The chapter by Cinar et al. reviews and classifies the literature on biomass co-firing. Grebner et al. discuss some physical and economic methods for assessing woody biomass availability for bioenergy production. Thomas et al. provide an environmental life-cycle assessment of biofuel production; and Granda-Marulanda et al. provide a life-cycle cost assessment of biofuel production using syngas.

To summarize, this handbook brings together recent advances on a number of topics which are important for the development and long-term sustainability of this industry. The handbook presents a number of models that can be used to optimize the supply chain performance of different types of biomass, such as, forest products and residues, energy crops, etc.; and different types of bioenergy, such as, cellulosic ethanol, biodiesel, electricity, etc. The case studies present problems faced by the industry in Southeast USA (such as, Alabama and Mississippi), or Western USA (such as, California) or Europe (such as, Norway and France).

February 2015

Sandra D. Eksioglu
Steffen Rebennack
Panos M. Pardalos

Contents

Biomass to Energy Supply Chain Network Design: An Overview of Models, Solution Approaches and Applications

Şebnem Yılmaz Balaman and Hasan Selim

Abstract Energy production from biomass is an alternative and additive way to fossil fuel based energy production to reduce the dependency on limited fossil fuel sources and mitigate the harmful environmental impacts of these systems. One of the major challenges in establishing efficient renewable energy systems is the complex supply chain structure in an uncertain decision environment, various decisions to be made and different conflicting criteria/objectives. This study describes the key issues in decision making for biomass to energy supply chains such as decision levels, uncertainty and sustainability concepts. It also provides a comprehensive review and systematic classification of the current literature on decision making approaches for design, management and operation of biomass to energy supply chains. This study allows readers to identify the decision making methods that satisfy the problem specific requirements and offer a clear vision of the advances in the field.

1 Introduction

Design, operation and management issues in renewable energy supply chains are increasingly gaining importance in recent years in parallel with the rising interest in renewable energy sources. Biomass is one of these energy sources, and biomass based energy production has been proposed as a part of the solution to the world wide increasing energy consumption, limited fossil fuel resources, climate change and environmental pollution caused by the use of fossil fuel resources.

All of the activities in biomass to energy supply chain networks should be governed to ensure the efficiency of material, information and cash flows in the networks. The

Ş.Y. Balaman (✉) · H. Selim
Department of Industrial Engineering, Dokuz Eylul University,
Tınaztepe Campus, 35160 Buca, Izmir, Turkey
e-mail: s.yilmaz@deu.edu.tr

S.D. Eksioglu et al. (eds.), *Handbook of Bioenergy*,
Energy Systems, DOI 10.1007/978-3-319-20092-7_1

major activities in a biomass to energy supply chain are cultivation, harvesting, collection, storage and transportation of biomass, biomass to energy conversion, energy distribution and energy consumption. To perform these activities, an efficient supply chain has to be configured including a transportation network design, location and capacity selection for conversion plants and storages, feedstock supply and distribution, process residue handling and operational schedules.

Due to the complex tradeoffs involved in conflicting criteria and objectives, various competing supply chain design decisions that affect supply chain performance cannot be made independently. Given that many mathematical or computational models are being developed to design biomass based supply chains, it is important to develop approaches to identify and incorporate a wide range of sources of uncertainty that can be coupled to these models. Besides the incorporation of inherent uncertainties, considering multiple (usually conflicting) objectives in biomass to energy supply chain design models emphasize on economical, technical, environmental and social issues that have to be considered in real life problems simultaneously.

Considering the fact that the number of researches on the design and optimization of biomass to energy supply chains are continuously increasing, this study attempts to provide an up-to-date and comprehensive survey in this field. This study focuses on key issues, methods and models for design, management and operation of biomass to energy supply chains.

The remainder of this study is organized as follows. Decision levels, uncertainty and sustainability concepts in biomass to energy supply chains are defined in Sects. 2 and 3, respectively. Section 4 presents a literature review of decision-making approaches for biomass to energy supply chain network design with a focus on uncertainty and sustainability concepts. In this regard, 120 articles, which have been published in English speaking peer-reviewed journals from 2000 to 2011, are reviewed. The studies that are reviewed in this work are classified into 4 main categories; review studies, biomass potential analyses, studies with uncertainty concept and studies without uncertainty concept. Studies with/without uncertainty concept are classified into two sub categories according to the number of objectives included by them, namely studies with single objective and studies with multiple objectives. Single/multiple objective studies are then sorted according to the decision-making methodology used.

2 Decision Levels in Biomass to Energy Supply Chains

Decisions are made at different levels in supply chains and the difference between decisions at various levels lies in the scope of the design, operation and management choices made. There are three main decision levels in supply chain management; strategic, tactical and operational decisions.

2.1 Strategic Decisions

Strategic decisions are long term in their impact which may need revisions after a long time period usually 5 or more years. It decides what the chain's configuration will be, how resources will be allocated, and what processes each stage will perform. They are usually taken at the highest levels of management and carry higher levels of risk. However, effective strategic decisions lead to high levels of rewards for all stakeholders in the supply chain. Strategic decisions in biomass to energy supply chains include;

- Choosing the site, installed capacity and purpose of biomass based energy production facilities,
- Selection of energy conversion technologies and processes,
- Configuration of a network of suppliers, distributers and logistics handlers,
- Inventory and product management policies throughout the supply chain's life cycle,
- Long-term technological improvement and capacity planning analyses to meet variations in the end users' demand,
- Long term supply and demand contracts,
- Ensuring economic, environmental, technical and social sustainability.

2.2 Tactical Decisions

Tactical decisions are medium term decisions usually spanning between 6 months and 1 year. They are made within the constraints of the overarching strategic supply chain decisions. Tactical supply chain management decisions are made at a national or regional level to improve efficiencies, control cost and minimize risks. Common tactical decisions in biomass to energy supply chains are:

- Allocation of organizational resources to production and distribution activities
- Selection of transportation mode,
- Inventory solutions including selection of warehouse type (centralized or distributed),
- Development of the system products according to end users' preferences.

2.3 Operational Decisions

Operational decisions are short term decisions that are made weekly or daily and ensure that the products are efficiently produced and they move along the supply chain components with optimized performance. Given the constraints established by the configuration and planning policies, the goal during the operational level is to exploit the reduction of uncertainty and maximize benefit with short term activities. These activities include;

- Daily and weekly forecasting to estimate the end users' demand,
- Allocation of inventory or production to demands, setting delivery due dates and times and developing daily/weekly delivery schedules. Production and distribution schedules including production/distribution amounts and delivery routes,
- Allocation of a shipment to a particular transportation mode,
- Monitoring shortage of supply and backlogs, planning additional activities in daily basis to meet the shortages.

The design and management of efficient supply chains in today's competitive environment should focus on optimizing all the decisions involved. Such a holistic management approach ensures the design of robust and reliable supply chains.

3 Uncertainty and Sustainability Concepts in Biomass to Energy Supply Chains

3.1 Uncertainty Concept

Real-world supply chains operate in an uncertain environment due to incompletely defined objectives, incomplete and/or unavailable information about the supply chain or its environment. In an uncertain decision environment, the decision maker is unable to accurately predict the impact of possible actions on supply chain behavior. To identify the most effective strategies in a specific supply chain scenario, identification and management of the sources of uncertainties should be incorporated into the decision-making processes. In designing biomass to energy supply chains, consideration of inherent uncertainties should be of the primary concern. Some of the major uncertainties faced by bioenergy supply chains are about;

- Market demand and price
- Supply of biomass and other materials
- Production and processes
- Technique/technology
- Governmental and regulatory policies
- Transportation and logistics

Demand uncertainty refers to the unknown or unpredictable variations in the quantity and timing of demand (Awudu and Zhang 2012). Consumer demand can fluctuate due to some reasons such as seasonality changes to consumer preferences. Demand uncertainty directly affects inventory policy as it impacts the variation in sales during inventory replenishment. Probabilistic distributions are often used to model demand uncertainty through analyzing the market data. Price uncertainty can be defined as the chance or estimate that price of a product might change. As supply and product price usually changes within a small range affected by competitive market conditions, price uncertainty can be predicted more accurately than demand uncertainty. Price uncertainties in a biomass to energy supply chain include uncertainties about raw material cost (especially for crop biomass), gate fees, biofuel price, power/heat price and carbon price.

Supply uncertainties are characterized by natural variations in raw material yield, quality and transportation lead time. The supply of biomass and other necessary materials is sometimes hampered by bad weather conditions or congestion at biomass source. The reasons of uncertainty about quantity may be shortage of feedstock, seasonality of the supply, changes in harvesting technology, environmental and economic viability of alternative fuel options, unavailability of arable land and unreliable supply sources possibly due to the market establishment. Safety stock, multiple suppliers, lead time increase and future purchase of raw materials are some of the methods for overcoming the supply uncertainties.

Production and process uncertainties cause the inability to produce the planned quantity of products with the specified quality. Some of these uncertainties are engendered by delays in raw material procurement and they are closely related to logistics uncertainties. It is important to forecast natural conditions as they affect the yields of a renewable energy production system. Also, machine breakdowns and lead time constraints have important impacts on production efficiency. Setting maintenance and repair times in a timely manner by effective schedules can prevent the unexpected stops during the production process. Production and process uncertainties directly affect inventory decisions.

Technical/technological uncertainty exists when it is not clear that all system requirements can be implemented using the selected technologies at desired performance level. If the decision maker has a strong understanding of the requirements and implementation of the technology, negative impacts of these uncertainties can be eliminated. These types of uncertainties are encountered in renewable energy system implementations since renewable energy production technologies are relatively newer compared to fossil fuel based energy production technologies.

Uncertainties in governmental and regulatory policies occur when the future acts of government policy is uncertain. Policy uncertainty may refer to uncertainty about economic, environmental or social policy. For instance, changes in the tax rates and renewable energy incentives influence the operational decisions in biomass to energy supply chains.

Transportation and logistics uncertainties are caused by lack of information and coordination between shipper and customer as well as geographic and natural restrictions. They may result in unsuccessful deliveries of both raw materials and finished products in a timely and cost effective manner, delays in fleet scheduling, backlog and stockout, increased freight cost, volatile supply of raw materials and increased warehousing cost. Optimized distribution network design, optimized schedules for logistics operations and selection of the transportation mode according to the geographic conditions can mitigate the impacts of such uncertainties.

3.2 *Modelling Uncertainties in Biomass to Energy Supply Chains*

A challenge of supply chain management is to integrate operations across multiple stakeholders. Integration in operational activities of individual supply chain components is defined as supply chain synchronization. Supply chain synchronization seeks to coordinate the flow of materials, products, and information between supply chain partners to reduce duplication and redundancy (Bowersox et al. 2012).

For global logistics networks, traditional supply chain has some serious problems such as high transportation and logistics costs, long response time to changes of supply chain parameters, bullwhip effect, and difficulties in accurate inventory controlling. To avoid such problems, a flexible supply chain network has to be established. It is required to handle and treat the inherent uncertainties to cope with imprecise nature of flexible supply chains.

Conventional Linear Programming (LP) methodologies and solution algorithms do not have the ability to solve all imprecise flexible supply chain problems. Analytical methods are one of the methodologies that are used to solve problems with uncertainty. Some of these methodologies are Stochastic Mixed Integer Linear Programming (SMILP), Stochastic Mixed Integer Non-linear Programming (SMINLP), Markov Decision Process and Linear Programs with Scenario Generation. These methods might yield significant results if applied in biomass to energy supply chains. In addition, simulation methods are important tools for solving supply chain problems with uncertainties. Furthermore, fuzzy modeling approaches provides the appropriate framework to describe and treat uncertainty.

3.3 *Sustainability Concept*

In the design and management of a supply chain network, it is important to meet the environmental, social, economic and other requirements of present and future generations. Sustainability satisfies the conditions under which humans and nature can exist in a productive harmony. Sustainable systems simultaneously handle and integrate natural resource utilization, environmental management and pollution prevention, standard of living and job creation, social acceptability, economic growth, cost savings and profitability, research and development.

Besides economic behavior of the renewable energy supply chains, it is necessary to observe and analyze the environmental and social impacts of them to create and maintain a sustainable supply chain. The utilization of the natural resources without any regulatory policies for environment and society can damage the sustainability of biomass based supply chains.

4 Decision Making Approaches for Biomass to Energy Supply Chain Network Design

4.1 Review Studies

Since biomass to energy supply chain design problems include multiple stakeholders, different conflicting objectives and many sources of uncertainty, decision making in such problems requires selection of the suitable methodology as in many complex planning problems. In this section, pioneer review studies about decision making in biomass to energy supply chain design problems are overviewed.

Mckendry (2002a, b, c) proposed a three-series paper about biomass, biomass conversion technologies and biomass gasification technologies. His first paper presents an overview of biomass types, species, properties, harvesting and biomass to energy production issues. The second one gives a brief review of the main biomass to energy conversion processes and the third one refers only to the thermochemical conversion of biomass, specifically gasification technology. McCormick and Kaberger (2007) focused on bioenergy in European Union by identifying economic conditions, know-how and institutional capacity and supply chain coordination as the key barriers obstructing the expansion of bioenergy, and discussed issues pertinent to overcoming these barriers. Bekkering et al. (2010) conducted a review study on green gas supply chain optimization. Applications and developments of the biogas supply in The Netherlands are mentioned and then a literature research is conducted about dairy cattle manure production, digester types, biogas upgrading process, injection process and green gas supply chain. Iakovou et al. (2010) focused on waste biomass to energy supply chain management and presented a critical synthesis of the state of the art literature about design and management of waste biomass supply chains. After supply chain components, activities and characteristics are briefly mentioned, types of biomass to energy conversion technologies are presented in the paper. Gold and Seuring (2011) presented a literature review of papers from 2000 to 2009 that deal with bioenergy production, logistic issues and supply chain management. After giving brief definitions of sustainability, bio-energy, logistics management and supply chain management, main issues of bioenergy supply chain are presented. Calvert (2011) conducted a review study to deal with the ways in which geomatics has been used to provide geographic information about bioenergy feasibility in terms of resource and bioenergy assessments. Awudu and Zhang (2012) presented a review study on uncertainty and sustainability concepts in biofuel supply chain management. After explaining decision making process and decision levels in biofuel supply chain, the major uncertainties and methods for modeling these uncertainties are identified in the paper. Sustainability and approaches for modeling sustainability are explained. Sharma et al. (2013a, b) presented a comprehensive review study on basis, overview, modeling, challenges and future of biomass to energy supply chains. The study assesses energy trends and renewable energy targets and provides a brief description of biofuel feedstock and conversion technologies. Biomass supply chain design and modeling issues are also examined in the study. Mafakheri and Nasiri (2013) conducted a comprehensive review on modeling biomass to energy supply chain operations. After giving a literature review about main biomass supply chain operations, issues and challenges about design, planning and management of biomass to energy supply chains are presented in the paper. Yue et al. (2014) presented a study that describes key issues and challenges in modeling and optimization of biomass to bioenergy supply chains. Major energy conversion pathways from terrestrial and aquatic biomass to bioenergy/biofuel products are reviewed in the study. In addition, general structures and key activities in bioenergy supply chains from feedstock supply to end use are presented. De Meyer et al. (2014) presented an overview of optimization methods for design and management of bioenergy supply chains. Biomass to energy supply chains and decision making levels in biomass supply chain management are briefly described in the paper. The authors reviewed and compared 71 publications in terms of applied optimization method, decision level and main decision variables, and the objective to be optimized. Wolfsmayr and Rauch (2014) reviewed articles on the primary forest fuel supply chains that have been published from 1989 to 2011. They focused on the key issues of the transportation of primary forest fuel to heat/power plants, especially transportation modes, terminal types, primary forest fuel characteristics that effect transportation issues and energy wood types.

4.2 Biomass Potential Analyses

In locating and sizing the components of biomass to energy supply chains such as conversion plants, storages and distribution centers, it is important to accurately analyze the biomass potential of the handled region. Geographical information systems (GIS) have been introduced to biomass supply chain studies in order to compute more accurately the expected supply of biomass in a given region and the transportation distances and related costs as well as to assess the impacts of spatial feedstock subtraction for different chain designs. In this regard, Voivontas et al. (2001) proposed a GIS-based decision support system (DSS) to estimate the power production potential from agricultural residues. The procedure handles all possible restrictions to locate bioenergy units in candidate locations and establishes the needed cultivated area for biomass collection. The developed methodology introduces four levels of biomass potentials, namely theoretical, available, technological and economical potentials as sequential steps for the assessment of biomass potential. Shi et al. (2008) presented a case study to evaluate the feasibility of constructing biomass power plants and optimize the locations of plants in Guangdong Province, China. GIS is used to estimate theoretical and available biomass potentials considering factors such as vegetation type, ecological retaining, economical competition and harvest cost. After supply area of each candidate site and usable biomass within the supply area are determined, optimal sites for biomass power plants are identified.

4.3 *Studies with Uncertainty Concept*

This section presents a literature review on decision-making approaches that incorporate uncertainty for biomass to energy supply chain design and operation. Related studies are assessed in terms of their objective function structure. Firstly, studies with single objective are handled with a focus on the methods and modelling approaches used to optimize strategic, tactical and operational decisions. Then, studies that have multiple objectives are presented.

4.3.1 Studies with Single Objective

To facilitate the assessment of the methods and modelling approaches reviewed in this study, related publications are classified according to their main optimization methodology. In this concern, three groups of methodologies are distinguished: (1) LP approaches, (2) NLP approaches and (3) simulation and game theoretic methods.

Linear Programming Approach

LP has been widely used for modeling biomass to energy supply chains. Nagel (2000) developed a mixed integer linear programming (MILP) model to find the most economical and ecological biomass supply structure. The model determines optimal location, thermal and electrical capacity and technology of the energy supply units as well as combination of the consumers with supply units. Influences of different parameters on the objective function are analyzed by scenario and sensitivity analyses in the study. Tatsiopulos and Tolis (2003) developed a LP model that handles cotton-stalk biomass exploitation, collection, storage and transportation issues to optimize the overall biomass logistics network. They investigated the performances of four biomass to energy conversion plant scenarios to find the optimum combination of capacity-location scenario and transportation/storage methods in terms of logistics cost. To find the candidate storage locations, GIS is utilized in the paper. Freppaz et al. (2004) developed a DSS that integrates GIS with MILP to locate the biomass power plants, compute their optimal sizes in terms of energy production and feeding, determine the most appropriate kind of energy for a specific area, define the biomass collection plans by taking into account available biomass, technological aspects of plants and economical factors. Environmental analyses are also carried out to evaluate emissions. Dunnett et al. (2007) developed a modeling framework for strategic and operational decision making in a biomass to heat supply chain. The model is developed as MILP that optimizes design and operation scheduling of the supply chain simultaneously. The solutions are compared with those obtained through a simple heuristic methodology. Leduc et al. (2008) developed a MILP model to determine optimal locations and sizes of methanol plants and gas stations in Austria. The model covers all levels of the supply chain from biomass supply to biofuel production and sale. Leduc et al. (2010a, b) considered the integration of ethanol production with combined heat and power production. They used the MILP model proposed by Leduc et al. (2008) to determine optimal locations for ethanol plants to the aim of minimizing the total production cost and environmental impacts. Ekşioğlu et al. (2009) developed a mathematical model that integrates long-term biomass to biorefinery supply chain design and mid-term logistics decisions to determine the number, size and location of biorefineries, the amount of shipped, processed and stored biomass. The model also captures deterioration, supply seasonality and supply availability of the biomass. Leduc et al. (2010a, b) presented a dynamic MILP model to find the best location for a methanol production plant and minimize the specific biofuel production cost based on data about future biomass production potential. Results are analyzed both with and without district heat production cases. Lambert and Middleton (2010) proposed a MILP to identify all of the activities optimally from biomass production to energy conversion such as harvest, storage, transportation, pretreatment and refining in a field to biorefinery supply chain and evaluate the profitability of the supply chain. Parker et al. (2010) developed a MILP model for resource assessment and biorefinery siting optimization in a biofuel supply chain. The model assesses potential biofuel supply from different types of biomass, and determines the optimal locations, technology types and sizes of biorefineries. Huang et al. (2010) proposed a multi-period MILP model for strategic planning of bioethanol supply chains. The objective of the model is to minimize the costs of the entire supply chain from biomass fields to end users under demand, resource and technology constraints. The model determines locations and sizes of new refineries, expansion capacities of current biorefineries, ethanol production/distribution and feedstock distribution amounts for a given planning horizon. Čuček et al. (2010) developed a four-layer supply chain design model for the synthesis of regional renewable energy supply chains. The aim of the MILP model is to maximize the profit while evaluating environmental impact by the carbon footprint. The layers are; raw biomass generation, transferring the biomass to the collection centers, transferring the biomass from collection centers to processing plants and transferring the biomass from processing plants to end customers. Lam et al. (2011) deal with model size reduction techniques for the analyses of large-scale renewable production and supply networks. Techniques that bring computational time improvements with a little loss of accuracy such as reducing the connectivity in a biomass supply chain network, eliminating unnecessary variables and constraints and merging the collection centers are investigated. They proposed a model size reduction procedure with combining these three techniques and applied to Čuček et al. (2010)'s four-layer biomass supply chain design model. Leão et al. (2011) developed a mathematical model to optimize biomass to biodiesel supply chain sourced from family farms by taking into consideration agricultural, logistic, industrial and social aspects. The model is time dependent and determines quantity, location, technology and capacity for biomass conversion facilities, biomass production area and transportation routes of biomass. Aviso et al. (2011) developed a LP model with fuzzy input output analysis to optimize supply chains under water footprint constraints. They applied the model on a biofuel production case based on ethanol production from sugarcane and corn in the Republic of Philippines. Fuzzy tolerances of water availability and ethanol demand for three main geographic regions in the country are taken into consideration. An et al. (2011) proposed a model to design a lignocellulosic biomass to biofuel supply chain which has multiple commodity flows ranging from biomass suppliers to biofuel customers. The model deals with a production/distribution system, and determines facility locations and capacities, technologies and material flows while maximizing the profit. Kocoloski et al. (2011) developed a MILP model to optimize the cellulosic biomass to ethanol infrastructure investments for single and multi state regions. The model locates ethanol refineries, establish connections between biomass suppliers/ethanol demand points and these refineries. Kim et al. (2011a) formulated a MILP model to decide the optimal number, locations and sizes of various types of biofuel processing plants as well as fuel conversion technologies and logistics issues such as transportation amounts of biomass intermediate products and final products between supply areas to conversion plants and between conversion plants to final markets. The model is applied to design both a distributed and a more centralized conversion system. Kim et al. (2011b) improved Kim et al. (2011a)'s model by considering the uncertainties in the model parameters. They formulated a general MILP model for a simple biorefinery network structure for single and multiple design scenarios. The set of parameters that make dominant contributions to the overall variations of the profit are identified and multiple uncertainty scenarios from combinations of the dominant parameters are analyzed. Papapostolou et al. (2011a, b) developed a mathematical model to identify the best solutions for the optimal design and operation of biofuel supply chains that takes into account both technical and economic parameters affecting the performance of the supply chain. The authors state that their approach is capable to identify and solve a wide range of different scale and level problems such as facility location and design, raw materials selection and trading policy. Zhu and Yao (2011) proposed a multi-period, multi-echelon MILP model to design the entire supply chain and logistics system for the biomass to energy industry which incorporates multiple types biomass and system products. The model includes strategic decisions such as the distribution strategy, the locations and capacities of warehouses and tactical decisions like the types and amounts of biomass harvested/purchased, stored and processed in each month as well as transportation flows within the system in each month. Akgül et al. (2012a, b) developed a static MILP framework for strategic design of bioethanol supply chain network to determine locations and scales of biofuel production facilities, biomass cultivation and biofuel production rates, flow of biomass and biofuel between the components of supply chain, transportation modes of delivery for biomass and biofuel. The study also addresses sustainability issues such as the use of food crops as biofuel feedstock, land use requirements of energy crops and competition for biomass with other sectors. Marvin et al. (2012) developed a MILP model to determine the place and capacity of biorefineries, the amount of biomass to harvest and biorefineries that the harvested biomass transported to in a biomass to ethanol supply chain. They used their model to construct a biomass to ethanol supply chain in a region with nine states in Midwestern USA. Andersen et al. (2012) proposed a multiperiod MILP model for the optimal design and planning of biodiesel supply chain representing all components of the supply chain such as crop fields, storages, production plants and distribution centers. The model also considers crop land competition and alternative feedstocks. The model includes intermediate and final products such as seed, pellets, oil, biodiesel and glycerol. Lin et al. (2014) developed a MILP to design a biomass supply chain that includes a farm management module, a logistics planning module, a facility allocation module and an ethanol distribution module. Strategic decisions such as numbers, locations, capacities of plants and biomass/ethanol distribution patterns are made as well as tactical decisions like biomass production, storage, delivery and operating schedules. Mansoornejad et al. (2013) presented a scenario based strategic supply chain design approach for forest biorefineries using two kinds of scenarios, namely market scenarios and supply chain network scenarios. Once the supply chain network alternatives are identified and evaluated according to their profitability, flexibility and robustness, the alternatives are compared considering these metrics. Then a multi-product, multi-echelon MILP model is developed for the operational design of the supply chain network scenarios. Zhang and Hu (2013) developed a multi-period MILP methodology to determine facility number, location, capacity and biofuel production decisions at operational level such as biomass collection, fuel production, fuel distribution and biomass/biofuel inventory control and allocation for a biofuel supply chain design problem. The model includes biofuel supply shortage penalty and storage cost to accommodate different biomass supplies and biofuel demands by considering a number of biofuel demand scenarios. Ghaffariyan et al. (2013) developed a LP based approach to investigate the impact of five operational factors, namely energy demand, moisture mass fraction, interest rate, transportation distance and truck payload on forest residue based supply chain cost. The supply chain activities handled in the paper are; extraction of residues from the harvesting area to roadside, storage at roadside, materials chipping at the roadside and transportation of chips to energy plants. Awudu and Zhang (2013) proposed a single-period stochastic LP model for production planning in biofuel supply chains. Decisions such as amount of purchased and consumed biomass, biofuel production and sale amount as well as amounts of backlog and lost sales are made for the supply chain that includes biomass suppliers, biofuel refineries and distribution centers. The results are compared with those of a deterministic model using simulation technique. Osmani and Zhang (2013) developed a two-stage stochastic MILP for decision making in a multi-feedstock bioethanol supply chain. In the first stage, strategic decisions such as location and production capacity of biorefineries and allocation of farmlands for biomass cultivation are made. Tactical level decisions are made in the second stage. Results of the proposed stochastic model are compared with those of a deterministic model. Zhang et al. (2013) proposed a MILP to determine optimal bioethanol supply chain/logistics decisions. All of the stakeholders in bioethanol supply chain are considered in the study. Biomass cultivation sites are selected, biomass inventory level is determined, location and capacities of biorefineries and preprocessing plants are selected, and production/transportation volume of bioethanol is determined by the model. Ebadian et al. (2013) developed a methodology that combines MILP and simulation for determining network layout and operational design of agricultural biomass supply chain. Once the design of the supply chain is determined, operational decisions such as the number of required machines for each operation, the daily working schedule, utilization rates and storage capacities are made by the simulation model. Sharma et al. (2013a, b) developed a LP model to make strategic, tactical and operational decisions related to biomass supply chains considering weather uncertainty. A scenario optimization model is formulated to address the changes in the weather conditions over a one-year planning period. Foo et al. (2013) developed a LP model to minimize the environmental impact of a palm oil based regional supply chain. The direct source-link allocation based LP model is then extended to a MILP model to account for practical limitations in biomass logistics. Multiple biomass supply scenarios are considered to exhibit operational flexibility such as closure or expansion of mills. Roni et al. (2014) developed a MILP to model and analyze a hub and spoke supply chain network for biomass co-firing in coal-fired power plants. The study contributes to the existing literature in terms of evaluating the feasibility of using biomass for co-fire for coal based power generation and developing a hub and spoke supply chain network to optimize the biomass delivery costs. The model determines locations for hubs and the optimal flow of biomass to coal plants. Two delivery modes are taken into account, short distance delivery by trucks and long distance delivery by train. Xie et al. (2014) focused on integrating multimodal transport into the cellulosic biofuel supply chain design. To this aim, they proposed a multiperiod MILP considering three transport modes, truck, single railcar and unit train. Seasonal yields of feedstock and seasonal ethanol demands are used. Locations and capacities of transshipment hubs, refineries and terminals are determined by the model along with seasonal feedstock/biofuel storage and shipment amounts. Tong et al. (2014) developed a multi-period MILP model for optimal design and planning of a hydrocarbon biofuel supply chain integrated with existing petroleum refineries. The model simultaneously optimizes the supply chain design, insertion point selection for integrating the biofuel supply chain with a petroleum refinery and production planning. The model accounts for diverse conversion pathway selection, technology selection, deterioration, seasonality and geographical diversity of biomass, product distribution and government incentives.

Non Linear Programming Approach

Complex nature of the renewable energy problems can sometimes require to be modeled by NLP approaches. Heuristic methods are generally used for solving NLP models and look for not optimal but good enough solutions to solve complex problems in relatively reduced times.

In this regard, López et al. (2008) proposed a methodology based on NLP to determine the optimal location, biomass supply area, generated electricity power and net present value for biomass fueled electricity generation plants by taking into consideration three alternative conversion technologies. Particle Swarm Optimization (PSO) is used as the optimization method. The solutions obtained by the proposed algorithm compared to those obtained by Genetic Algorithms (GA).

Frombo et al. (2009) developed a GIS based environmental DSS for location, size and conversion technology choice of biomass-fueled power plants that takes into account different energy products, different harvesting definitions and pre-treatment operations. The DSS is formulated as a comprehensive long-term mixed integer non linear programming (MINLP) model and environmental constraints are considered. Rentizelas et al. (2009) developed a NLP-based DSS to design a system-wide biomass supply chain including biomass to bioenergy conversion facility and the district heating and cooling network. The model considers multiple types of biomass as feedstock and determines the location and size of bioenergy facility as well as types and amounts of biomass to be used in the facilities and maximum collection distance for each type. Gun and Smith (2011) developed a modeling framework to find the optimal bioenergy plant size and biomass supply area size based on minimization of total cost function. Impacts of various factors on the plant size, supply area and cost are analyzed. Bai et al. (2011) handled facility location and transportation planning problems in a modelling structure. They developed a MINLP model to determine optimal number and locations of biorefineries along with biomass supply and biofuel demand allocation and route selection for biomass and biofuel transportation. They use a Lagrangian relaxation based heuristic algorithm, and a branch and bound framework is developed. Corsano et al. (2011) proposed a MINLP model for sustainable design and behavior analyses of sugar to ethanol supply chain. The model integrates the supply chain design with ethanol plant design. It determines the values of supply chain related variables such as location and capacities of plants and warehouses, transportation amounts as well as plant design and operating variables such as processing times, material concentration and capacity of system units. Zhang et al. (2011) proposed a two-stage methodology to find the best location for biofuel production plants. In the first stage of the methodology, GIS is used to identify the candidate sites for the facilities, and mathematical modeling is utilized in the second stage to select the optimal location based on a total transportation cost model. Singh et al. (2011) presented a GIS and NLP based methodology to evaluate the feasibility of constructing biomass based power plants and optimize their location. GIS is used to assess the geographical distribution of crop residue and non-linear optimization techniques are employed to locate appropriate sites and sizes of plants. Sultana and Kumar (2011) investigate the delivery of multiple biomass types to a biorefinery in different combinations by emphasizing that total delivery cost of biomass includes harvest/collection, storage, pre-processing, transportation and in plant processing costs. They proposed an approach to determine the minimum cost delivery form at different truck capacities for a single biomass and the optimal delivery combination of different types of feedstock. Leboreiro and Hilaly (2011) developed a model to assess the economics of biomass collection, transportation and storage and to evaluate the optimum plant size for biorefineries considering the mathematical models for biomass collection radius, transportation distance, farm to plant and farm to storage transportation costs and production costs. Yagi and Nakata (2011) analyzed the economic feasibility of the small-scale forest biomass CHP system with gasification technology in Japan by finding optimal location, size and number of plants and production costs based on data about resource distribution and scale merit functions. In their two-series papers, Yu et al. dealt with the design and analysis of geographic distribution of biomass based power plants and satellite storages in China. In the first paper Yu et al. (2011) developed a new methodology to find the optimum geographic distribution of biomass power plant and its satellite storages with minimum collection costs considering China's specific delivery modes. A mathematical model optimizes the number of storages while GIS is used to determine the optimal locations among a number of candidate sites. The model can be only used for single type of biomass and straight-line delivery. In the second paper, Yu et al. (2012) improved their study so that it can be used for multi-types of biomass and road delivery which is a more realistic situation. Biomass in fields is delivered by tractors along a straight line to the nearest road junction, and then along the shortest road to the nearest storage. Sultana and Kumar (2012) developed a methodology to determine the suitable locations, optimal sizes and number of biomass based facilities to optimize the transportation costs by considering spatial and environmental factors. The methodology integrates GIS and AHP for spatial analysis to find suitable locations for biomass based facilities. A location-allocation model is used and to locate the facilities and allocate all collectable biomass to facilities with minimum transportation cost. Chen and Fan (2012) established a two-stage MISP model to support strategic planning of bioenergy supply chains and optimal feedstock allocation in considering potential future supply and demand uncertainties. Planning and operational decisions such as locations and sizes of biorefineries and terminals, feedstock procurement plans, ethanol production and distribution plans are made by the model. Avami (2012) developed a methodology that integrates resource, supply, biomass to bioenergy conversion, transportation, distribution and demand issues in an analytical framework. The model integrates agriculture sector (resource level) into the industrial technologies to produce biofuel. Optimal economic supply chain structure is determined with technical aspects and agricultural concerns. Avami (2013) adapted the model developed by Avami (2012) to bioethanol production and improved the model by introducing all bioethanol technologies to the model. Four scenarios are considered in the study. Different biomass resources are taken into account in the first and the second scenarios in which bioethanol is used directly as an automotive fuel. The third scenario investigates the centralized production systems in which biorefinery plants are located near the fuel demand zones. The fourth scenario considers the situation that bioethanol is converted to bio-ETBE as a fuel additive. Leboreiro and Hilaly (2013) developed a comprehensive non-linear biomass to energy supply chain model that integrates steps of biomass supply including collection, farm to plant and farm to storage transportation and storage with steps of biofuel production. Biomass collection and transportation costs are analyzed to search for the optimum plant capacity. A mathematical formulation to scale capital investment for fermentation-based biorefineries is proposed. Shabani and Sowlati (2013) presented a dynamic MINLP model to design the forest biomass to power supply chain. The model integrates tactical level activities of the supply chain such as biomass procurement, biomass storage, energy production and ash management. The model determines the amount of purchased, stored and consumed biomass and amount of electricity generation. Rentizelas et al. (2013) modeled and compared two different biomass to heat supply chain structures (a pellet production system and a centralized CHP unit) from the perspective of potential investor in the field of renewable energy generation. Both structures are optimized and location, capacity, biomass procurement, biomass inventory and final product inventory decisions are made for each structure. Finally, the supply chains are compared in terms of financial indicators.

Simulation and Game Theoretic Methods

Simulation methods are important tools for modeling and solving biomass to energy supply chain problems with uncertainties. In addition, game theoretic models are used to obtain optimal strategies for managing competitive biomass to energy supply chains. Competition and cooperation between many stakeholders can be modeled by game theoretic models.

Ebadian et al. (2011) proposed a stochastic simulation model based on IBSAL (see Sect. 4.3.2.2.) to provide optimal operational plan and schedule for logistics operations to supply a mixture of agricultural biomass to a proposed cellulosic ethanol plant considering variable data on weather conditions, harvest schedule and yields. The model is called IBSAL-MC and used to design all components in the biomass logistics system from field operations such as harvesting and collection to end user ethanol demand. Bai et al. (2012) developed a game-theoretic modeling approach that combines a non-cooperative bi-level Stackelberg leader-follower game model and a cooperative game model to design a biofuel supply chain. The model addresses possible business partnership scenarios between biomass suppliers and biofuel producers. The model determines number, locations and sizes of bi-orefineries as well as site-specific feedstock procurement prices incorporating Farmers' decisions on agricultural land use and biofuel manufacturers' market choice. Wang et al. (2013) incorporated the Renewable Identification Number (RIN) system into the biofuel supply chain design with non-cooperative farmers, manufacturers and blenders. RIN system is a tracking mechanism that was developed and executed by U.S. Environmental Protection Agency to monitor biofuel parties' compliance with the obligated mandates. The impacts of RIN system on supply chain decisions are analyzed using game-theoretic models. Sun et al. (2013) aimed to identify optimal management strategies for agri-biomass supply chains in a competitive environment. To this aim, a game model of a supply chain consisting of one supplier and two buyers is developed. The model enables to determine optimal strategies for the supplier and buyers. Windisch et al. (2013) applied a methodology that integrates business process mapping and simulation to monitor the business processes and stakeholders that constitute biomass supply chains. Two forest biomass supply chains in different operational environments (Finland and

Germany) are analyzed in terms of structure and functioning. Discrete event simulations are used to estimate process times for organizational and managerial activities.

Other Approaches

Decision making methods for biomass based supply chains except linear, non linear, simulation and game theoretic modeling approaches are proposed in this section. Gronalt and Rauch (2007) described an approach to configure a wood biomass supply network in a certain region in Austria. In the study, after specifying regionally available forest biomass and potential number of heating and energy plants, different woody biomass supply chains from forest to plant are evaluated according to their system costs for a number of alternative configurations. Amigun and Blottnitz (2007) emphasize that biogas technology is a promising energy system for low population and remotely situated communities in view of the fact that implementation of centralized energy generation and distribution systems are prohibitively costly and inefficient. The researchers investigated the significance of plant capacity and plant location on the capital investment cost of African biogas plants. Walla and Schneeberger (2008) investigated the relationships between the costs of biogas and electricity production from maize silage in relation to plant size. After a relationship between the capacity of a CHP unit and its electrical efficiency is derived, they developed a model to investigate unit costs of biogas and electricity production and transportation costs for feedstock and biogas slurry in relation to plant size. Rentizelas et al. (2009a, b) stated that larger fraction of cost in biomass to energy supply chains originates from the logistics operation and focused on biomass storage issue. After a literature review on biomass storages, typical layout and characteristics of biomass supply chains are mentioned briefly. Mansoornejad et al. (2010) developed a three-level hierarchical methodology that integrates market-based techno-economic analysis and supply chain based analysis. The levels of the procedure are; selection of product portfolio and conversion technology, targeting and designing the required manufacturing flexibility and design of strategic supply chain network. Zhang et al. (2011) developed a two-stage methodology to identify the optimal location for a biomass to biofuel conversion facility. In the first stage GIS is used to specify the candidate locations for biofuel facilities. In the second stage, the preferred location is selected among the specified locations in the first stage using a transportation cost model. Table 1 summarizes the reviewed studies with single objective that consider uncertainty.

4.3.2 Studies with Multiple Objectives

Supply chain design and management approaches that include multiple objectives satisfy the sustainability concerns with simultaneously optimizing economic,

Table 1 Studies with single objective that consider uncertainty

References	Objective	Uncertainty handling method	Design method	Case study location
Nagel (2000)	Economic—cost	Scenario analyses, sensitivity analyses	MILP	Brandenburg/Germany
Tatsiopulos and Tolis (2003)	Economic—cost	Scenario analyses	LP	Thessaly/Greece
Freppaz et al. (2004)	Economic—profit	Sensitivity analyses	MILP	Val Bormida/Savona/Italy
Dunnett et al. (2007)	Economic—cost	Sensitivity analyses	MILP	Hypothetical
Leduc et al. (2008)	Economic—cost	Scenario analyses	MILP	Austria
López et al. (2008)	Economic—profitability Index	Scenario analyses	NLP	Hypothetical
Rentizelas et al. (2009a, b)	Economic—NPV	Sensitivity analyses	NLP	Thessaly/Greece
Frombo et al. (2009)	Economic –cost	Scenario analyses	MINLP	Val Bormida/Savona/Italy
Ekşioğlu et al. (2009)	Economic—cost	Scenario analyses	MILP	Mississippi
Leduc et al. (2010a, b)	Economic—cost	Scenario analyses, sensitivity analyses	MILP	Sweden
Leduc et al. (2010a, b)	Economic—cost	Scenario analyses, sensitivity analyses	MILP	Norbotten/Northern Sweden
Lambert and Middleton (2010)	economic—profit	Sensitivity analyses	MILP	North Dakota/USA
Parker et al. (2010)	Economic—profit	Scenario analyses, sensitivity analyses	MILP	Western USA (Texas, Oklahoma, Kansas, Nebraska, South Dakota, North Dakota)
Huang et al. (2010)	Economic—cost	Sensitivity analyses	MILP	California/USA
Čuček et al. (2010)	Economic—profit	Sensitivity analyses, scenario analyses	MILP	Hypothetical (data based on central European conditions)
Leão et al. (2011)	Economic—cost	Scenario analyses, sensitivity analyses	LP	Brazil
Aviso et al. (2011)	Environmental—water footprint	Fuzzy input–output analyses	LP	Republic of Philippines
An et al. (2011)	Economic—profit	Scenario analyses, sensitivity analyses	MILP	Central Texas
Kocoloski et al. (2011)	Economic—cost	Sensitivity analyses, scenario analyses	MILP	Illinois/USA
Kim et al. (2011a, b)	Economic—profit	Sensitivity analyses, scenario analyses	MILP	Oklahoma, Arkansas, Louisiana, Mississippi, Alabama, Tennessee, Georgia, Florida, South Carolina, North Carolina

(continued)

Table 1 (continued)

References	Objective	Uncertainty handling method	Design method	Case study location
Kim et al. (2011a, b)	Economic—profit	Sensitivity analyses, scenario analyses	MILP	Oklahoma, Arkansas, Louisiana, Mississippi, Alabama, Tennessee, Georgia, Florida, South/North Carolina
Papapostolou et al. (2011a, b)	Economic—profit	Scenario analyses	MILP	Greece
Zhu and Yao (2011)	Economic—profit	Scenario analyses	MILP	Hypothetical

environmental and social objectives. Publications that include multiple types of objectives are presented in this section. Methodologies are classified into two groups: (1) LP approaches, (2) NLP and simulation approaches.

Linear Programming

Ayoub et al. (2009a, b) developed a multiobjective MILP model to design and evaluate biomass utilization networks in local areas that has three steps: classification, problem formulation and suggesting solution methods. The methodology is based on the biomass utilization superstructure which relates the biomass resources to system products, available processes and possible future processes as well as technology options. Dal Mas et al. (2010) developed a MILP model for the design and planning of multi-echelon biofuels supply chain under uncertainty of market conditions. The model is capable of representing the dynamic evolution of bioethanol supply chains based on financial criteria. The model determines supply chain design variables as well as biofuel market demand satisfaction rate, supply chain profit and financial risk. Giarola et al. (2011) presented a multi-period, multi-echelon MILP framework to optimize the environmental and financial performances of corn grain and stover based bioethanol supply chains simultaneously. Biomass type selection, supplier allocation, production technology determination, site selection, capacity assignment, production planning, logistic distribution and transportation mode selection issues are taken into account simultaneously. Dal-Mas et al. (2011) focused on biomass based ethanol supply chain design under uncertainty conditions. A multi-period, multi-echelon MILP is developed to model entire corn to ethanol supply chain determining biomass cultivation site locations, capacities and locations of ethanol production facilities and logistic issues. Akgül et al. (2012a, b) presented a multiobjective static modeling framework for the optimization of biofuel supply chains considering economic and environmental objectives. The model determines GHG emissions for each stage in the life cycle besides the decision variables about biomass cultivation, biofuel production, location, scale and material flows. The impact of technological learning on the performance of the supply chain is analyzed through a multiperiod model. Pérez-Fortes et al. (2012) developed a multiobjective

MILP model to support designing and planning of biomass based supply chains. The model determines the most suitable pre-processing technologies with their capacities and locations, connections between the units of the supply chain, biomass storage periods, matter transportation flows and electricity network among the different communities. Giarola et al. (2013) proposed a multi-period, multi-echelon life cycle analysis (LCA) integrated MILP modeling framework to support strategic design and planning decisions for bioethanol supply chains. The model is a two-stage stochastic decision making tool that includes economic and environmental models. Risk mitigation, technology learning and carbon trading issues are also considered in the study. Kanzian et al. (2013) developed a multiobjective optimization model for design of forest biomass supply chains. The handled supply chain consists of roadside stocks, terminals for fuel storages, shipping stations and district heating or CHP plants. Network layout decisions such as locations of plants are made as well as decisions about transportation mode, production volumes and usage of terminals. Osmani and Zhang (2014) proposed a two-stage stochastic MILP to maximize the expected profit while minimizing environmental impact of a lignocellulosic bioethanol supply chain under biomass supply, bioethanol demand and biomass/bioethanol price uncertainties. Environmental impact is defined in terms of carbon credits and directly incorporated into the objective function. In the first stage, strategic decisions are made while tactical/operational decisions are given in the second stage.

Nonlinear Programming and Simulation

Sokhansanj et al. (2006) developed a dynamic integrated biomass supply analysis and logistics model (IBSAL) to simulate the flow of biomass from field to biorefinery. IBSAL is a modeling approach to plan resources and estimate costs, energy use and GHG emissions for different collection and transportation systems. Climatic and operational constraints that have influence on the availability of biomass to a biorefinery are modeled using the proposed approach. Also resources such as labour or equipment are allocated for biomass supply and transportation. Kumar and Sokhansanj (2007) used IBSAL to evaluate delivery systems for switchgrass supply options according to cost, energy input and carbon emissions criteria. The cost of collecting switchgrass, storing it and delivering it to a biorefinery using several collection and transportation options are evaluated. Mobini et al. (2011) developed a comprehensive simulation model based on IBSAL to investigate the logistics of supplying forest biomass to a biomass power plant. The model evaluates the cost of delivered biomass, estimates the equilibrium moisture content of biomass in the gate of power plant and carbon emissions resulting from the logistic operations. The supply chain included in the study is an extension of the supply chain investigated by Mahmoudi et al. (2009). Three harvesting systems are considered in the model. An and Searcy (2012) utilized IBSAL to simulate the conceptual logistics system including machines as a part of the system and analyze the overall system performance. To do so, new IBSAL models to model the machine functions are developed for the new machine concepts. Ayoub et al. (2007) developed a multi-level decision making strategy and DSS named gBEDS (general Bioenergy Decision System) for biomass based power generation considering different stakeholders along the biomass supply chains which divides the bioenergy generation decisions into national and local regional levels. The gBEDS integrates data visualization, simulation and optimization methods for locating biomass storages and biomass to energy conversion plants for two planning levels. Ayoub et al. (2009a, b) developed an optimization model and solution approach to design and evaluate bioenergy production supply chains in local level. After briefly describing gBEDS, they proposed a multiobjective NLP model which minimizes cost, total gas emissions, energy used and number of labors subject to biomass availability and investment cost constraints. GA is used to solve the model for a case study Iida city in Japan. Mahmoudi et al. (2009) proposed a simulation model for forest biomass supply logistics which is developed that has the potential to provide estimates of the quantity, unit cost, supply time, moisture content and equipment carbon emissions associated with biomass type. Gómez et al. (2010) assessed the potential of electricity generation in Spain from agro-industrial residues. After analyzing the energy potentials of various biomass feedstocks, they proposed a methodology based on non linear formulations for the economic analysis of biomass to electricity production considering the characteristics and geographical distribution of biomass, the conversion technology, the economy of scale and income from electricity feed-in tariffs. Zhang et al. (2012) developed a simulation model for design of biofuel supply chains considering key supply chain activities such as harvesting, processing, transportation and storage. The model determines optimal biofuel facility location, logistics design and inventory management decisions for different facility locations and plant sizes. Perpiña et al. (2013) presented a multicriteria assessment approach integrated with GIS for selection of suitable sites for biomass plants. To this aim, principal criteria are defined, evaluated and weighted based on AHP divided into three groups, namely environmental, economic and social Mobini et al. (2013) used discrete-event system simulation to facilitate the design and analysis of the wood pellet supply chains. The main activities that modeled are transportation of the raw material and pellets, raw material storage, drying, size reduction, pelletizing and cooling. The model can also be used to determine the best locations of the pellet plants. Table 2 summarizes the reviewed studies with multiple objectives that consider uncertainty.

4.4 Studies Without Uncertainty Concept

This section presents a literature review of articles that do not consider uncertainty of the system parameters. The publications are classified into two categories, namely single objective and multiple objective studies, as in the Sect. 4.3.

Table 2 Studies with multiple objectives that consider uncertainty

References	Objective	Uncertainty handling method	Design method	Case study location
Sokhansanj et al. (2006)	Economic, technical, environmental—cost, energy consumption, carbon emissions	Sensitivity analyses	Simulation	Hypothetical
Kumar and Sokhansanj (2007)	Economic, technical, environmental—cost, energy consumption, carbon emissions	Scenario analyses	Simulation	Midwestern USA
Ayoub et al. (2007)	Economic, environmental, social—cost, emissions, number of labors	Fuzzy clustering	Simulation	Japan
Mahmoudi et al. (2009)	Economic, environmental—cost, carbon emissions	Sensitivity analyses	Simulation	Quesnel/British Columbia/Canada
Ayoub et al. (2009a, b)	Economic, technical, environmental—cost, emissions, energy consumption	Scenario analyses	MILP	Aomori Prefecture/Japan
Ayoub et al. (2009a, b)	Economic, environmental, social—cost, emissions, energy used, number of labors	Scenario analyses, sensitivity analyses	NLP	Iida/Japan
Gómez et al. (2010)	Economic—NPV, profitability index	Scenario analyses, sensitivity analyses	NLP	Spain
Dal Mas et al. (2010)	Economic—expected profit, the risk on investment	Scenario analyses, sensitivity analyses	MILP	Northern Italy
Giarola et al. (2011)	Economic, environmental—NPV, GHG emissions	Scenario analyses	MILP	Northern Italy
Mobini et al. (2011)	Economic, technical, environmental—cost, equilibrium moisture content, carbon emissions	Sensitivity analyses	Simulation	Quesnel/British Columbia/Canada
Dal-Mas et al. (2011)	economic—profit, financial risk	Scenario analyses, sensitivity analyses	MILP	Northern Italy
Zhang et al. (2012)	Economic, technical, environmental—cost, energy consumption, GHG Emissions	Scenario analyses	Simulation	Lower peninsula of Michigan

(continued)

Table 2 (continued)

References	Objective	Uncertainty handling method	Design method	Case study location
An and Searcy (2012)	Economic, technical, environmental—cost, energy consumption, carbon emissions	Sensitivity analyses, scenario analyses	Simulation	Texas
Akgül et al. (2012a, b)	Economic, environmental—cost, GHG emissions	Scenario analyses, sensitivity analyses	MILP	United Kingdom
Pérez-Fortes et al. (2012)	Economic, environmental, social—NPV, environmental impacts, social impact (max. number of technologies constructed)	Sensitivity analyses, scenario analyses	LCA, MILP	Atebubu District/Ghana
Giarola et al. (2013)	Economic, environmental—NPV, GHG emissions	Stochastic modeling, scenario analyses	MILP	Northern Italy

4.4.1 Studies with Single Objective

Methodologies are classified into two categories in this section: (1) LP approaches, (2) NLP and simulation approaches.

Linear Programming

Dyken et al. (2010) developed a MILP approach which consists of two models, namely an operational model and an investment model. The operational model optimizes energy carriers and conversion components of the bioenergy plants where the investment model finds the optimal set of investments during the period of analysis with handling both economical and environmental aspects. Elia et al. (2011) developed a MILP model to analyze the United States energy supply chain network for the hybrid coal, biomass, and natural gas to liquids (CBGTL) facilities. The model selects the optimal locations of CBGTL facilities, the feedstock combination, and size of each facility that gives the minimum overall production cost. Lam et al. (2013) proposed a two-stage green strategy for design of waste to energy supply chain network which includes mainly three layers, supply layer, processing layer and demand layer. The strategy includes micro and macro stages. The processing layer is included in micro stage where waste type biomass is allocated to processing technologies and converted to final products. Macro stage deals with synthesis and optimization of waste to energy supply network via supply and demand layers. The stage determines transportation mode and location of collection and processing facilities. Geraili et al. (2014) developed an iterative decision support strategy for optimal design of renewable energy production systems which consists of three steps, strategic optimization, process modeling/simulation and operational optimization. Strategic design stage includes feedstock selection, product portfolio design, technological design, supply chain design and capacity planning by LP. The optimized design is used to simulate the renewable process. The operating conditions are then optimized using metaheuristic algorithms.

Non Linear Programming, Simulation and Heuristic Methods

Krukanont and Prasertsan (2004) proposed a study to determine the location and capacity of the biomass power plants that uses rubber wood as feedstock. GIS, mathematical modeling and simulation methods are utilized respectively to map the rubber wood area and to pinpoint the suitable locations for biomass power plants, to derive the biomass costs and to determine the appropriate sites and sizes of the power plants. Bruglieri and Liberti (2008) developed mathematical programming models for solving problems arising from planning and running biomass based energy production processes. Whereas the "running model" is a MILP model, they proposed two "planning models", both of which are MINLP problems. Reche-López et al. (2009) applied and compared several metaheuristic techniques to optimize the location and biomass supply area of biomass based power plants. For this purpose two trajectory (Simulated Annealing and Tabu Search) and two population-based (GA and PSO) methods are applied. After explaining the four considered metaheuristic methods, a new PSO algorithm is proposed, all above mentioned methods are applied and comparative results are given. Rentizelas and Tatsiopoulos (2010) proposed a study to find the best location of a bioenergy generation facility that optimizes the system-wide operational and investment costs. Since the optimization problem is nonlinear the above mentioned hybrid solution procedure is employed. The solutions of the hybrid methodology are compared to those of GA and Sequential Quadratic Programming. Velazquez-Marti and Fernandez-Gonzalez (2010) proposed a mathematical calculation methodology to determine the optimal points for bioenergy production facilities to supply energy to a selected set of cities. The cities are grouped in sets due to two criteria: all the energy produced by the plant must be consumed and the transportation cost of energy must be minimized. Vera et al. (2010) developed a methodology to determine the best location, biomass supply area and power plant size that offer the best profitability to the investor. The methodology includes Binary Honey Bee Foraging (BHBF) Approach which based on particles swarm. Results are compared with those of Binary PSO and GA. Gómez-González et al. (2013) developed a hybrid heuristic approach to find the best location and size of biomass fuelled electricity generation facilities. To this aim, PSO is used to search a range of location combinations in the distribution network and Optimal Power Flow (OPF) is utilized to define available capacity for each combination. Local distribution of biomass sources, biomass transportation and extraction costs, operation and maintenance

Table 3 Studies with single objective that do not consider uncertainty

References	Objective	Design method	Case study location
Dyken et al. (2010)	Economic—cost	MILP, dynamic programming	–
Elia et al. (2011)	Economic—cost	MILP	United States
Lam et al. (2013)	Economic for each model- NPV/cost	MILP	Malaysia
Geraili et al. (2014)	Economic—NPV	LP, simulation	Hypothetical
Krukanont and Prasertsan (2004)	Economic—cost	NLP, simulation	Southern Thailand
Bruglieri and Liberti (2008)	Economic—cost	MINLP	Italy
Reche-López et al. (2009)	Economic—profitability index	Simulated annealing, tabu search, PSO, GA	Hypothetical
Rentizelas and Tatsiopoulos (2010)	Economic—NPV	GA, sequential quadratic programming	Thessaly/Greece
Velazquez-Marti and Fernandez-Gonzalez (2010)	Economic—cost	NLP	La Hoya de Bunyol/Comunidad Valenciana/Spain
			La Plana de Utiel/Comunidad Valenciana/Spain
Vera et al. (2010)	Economic—profitability index	NLP, BHBF, binary PSO, GA	Úbeda/Jaén/Andalusia/Spain
Gómez-González et al. (2013)	Economic—cost	JFPSO, OPF, GA, PSO	–

costs, power losses costs, network operation costs and technical constraints are taken into account. In the study a novel discrete PSO is proposed, namely Jumping Frog PSO (JFPSO) and integrated this method with OPF. Table 3 summarizes the reviewed studies with single objective that do not consider uncertainty.

4.4.2 Studies with Multiple Objectives

Biomass based supply chain design studies without uncertainty concept that consider multiple objectives are presented in this section. Erkut et al. (2008) developed a multiobjective MILP model to solve the location allocation problem of municipal solid waste (MSW) facilities in regional and prefectural levels with respect to economic and environmental criteria. Firstly, they assumed that each prefecture is self-sufficient and located MSW facilities in each prefecture. Then, they unite all prefectures and investigated the optimal solution to cover the needs of all prefectures in the region. Ayoub et al. (2008) developed a methodology to design and analyze local biomass utilization network. In the study, various networks in local areas are analyzed, the bottleneck processes are defined, the existing problem is optimized and suitable solutions are suggested to overcome the bottlenecks. GA is used as solution approach considering three optimization criterions, namely cost minimization, emissions minimization and energy consumption minimization. Zamboni et al. (2009) proposed a modeling framework for the strategic design of entire biomass based supply chains using MILP considering both economic and environmental objectives in a multiobjective structure. Decisions such as biomass cultivation site locations, biofuel production capacities and locations of facilities, distribution processes of biomass and biofuel are specified. Mele et al. (2009) developed a MILP model to design supply chains for sugar and ethanol production considering economic and environmental factors with simultaneously minimizing total cost and environmental impacts of the supply chain. Environmental impacts are analyzed through Life Cycle Assessment. Papapostolou et al. (2011) developed a multiperiod mathematical model to support the planning and operational issues in biomass based supply chains such as investment decision making, the quantities of cultivated biomass and the extent of cultivation area, the quantities of biofuels, heat or power to be produced. Čuček et al. (2012) proposed a two stage multiobjective MINLP model to optimize regional biomass to energy supply chains. They extended Čuček et al. (2010)'s four layer methodology for assessment of the supply chain performance based on LCA. Santibañez-Aguilar et al. (2014) developed a deterministic, multiperiod MILP model to design and plan biorefinery supply chains considering economic, environmental and social performance characteristics. The model considers availability and seasonality of biomass sources, multiple products that are produced using diverse production technologies, demands and prices in different market conditions and various transportation modes. Table 4 summarizes the reviewed studies with multiple objectives that do not consider uncertainty.

5 Discussion

5.1 Conclusions

This study describes the key issues in decision making for biomass to energy supply chains such as decision levels, uncertainty and sustainability concepts. It also provides a comprehensive review and systematic classification of the current literature on decision making approaches for design, management and operation of biomass to energy supply chains.

The research on biomass to energy supply chain network design indicates that the implementation of such supply chains are technically feasible by means of basic principles although some challenges can be encountered related to economical, environmental and operational consideration. Such practical challenges have to be overcome by adequate evaluation methods to select the promising options for the locations and capacity levels of the facilities and logistics activities, and to obtain the optimum configuration of the supply chain. A well-designed supply chain can

Table 4 Studies with multiple objectives that do not consider uncertainty

References	Objective	Design method	Case study location
Erkut et al. (2008)	Environmental, technical, economic—greenhouse effect, final disposal amount, energy recovery, material recovery, cost	MILP	Macedonia
Ayoub et al. (2008)	Environmental, technical, economic—cost, emissions, energy consumption	GA	–
Zamboni et al. (2009)	Environmental, economic—cost, greenhouse gas emissions	MILP, LCA	Northern Italy
Mele et al. (2009)	Environmental, economic—cost, environmental impact	MILP, LCA	Argentina
Papapostolou et al. (2011a, b)	Economic—revenue	LP	Thessaloniki/Greece
Čuček et al. (2012)	Environmental, social, economic—profit, environmental footprint, social footprint	MINLP, LCA	Hypothetical (data based on central European conditions)
Santibañez-Aguilar et al. (2014)	Environmental, social, economic—supply Chain profit, environmental impact, number of generated jobs	MILP, LCA	Mexico

be highly reliable and cost effective, meets economic constraints, uses appropriate technologies and improves the quality of life.

As in almost all complex real life systems, there exist uncertainties and risks in renewable energy systems that have to be tackled in the investment-planning phase. We observed that 88 of 104 papers overviewed in this study incorporate uncertainties in decision making for biomass to energy supply chain design, management and operation. Scenario and sensitivity analyses are the most frequently used techniques while stochastic modelling and fuzzy programming are among rarely utilized methods to tackle inherent uncertainties.

Figures 1 and 2 illustrate the distribution of the methods used in the studies surveyed in this paper. As demonstrated in the figures, various modeling and optimization methodologies have been employed for handling the problems related to the design and operation of biomass to energy supply chains. Figure 1 depicts that LP is the most widely used modelling technique for both single and multi objective studies with uncertainty concept. MILP models have the largest area of utilization amongst the LP models. The figure also reveals that simulation and simulation-based design methodologies are frequently preferred techniques for multi objective studies due to their flexibility in adapting fluctuant system conditions and ability to reflect the impacts of different criteria simultaneously. Furthermore, NLP technique is also employed especially for the studies with single objective as it enables to represent the real energy system behavior more realistically. In a few studies, game theoretic models are used to model the competition and cooperation between the stakeholders.

Figure 2 exhibits that heuristic methods such as PSO, GA, simulated annealing and tabu search are the most preferred techniques for single objective studies that do not consider uncertainties where simulation, LP and NLP are used less frequently. In

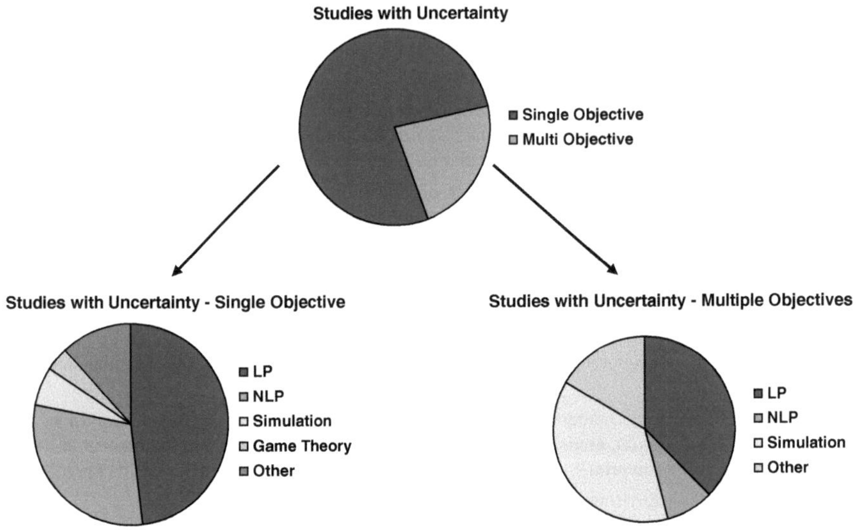

Fig. 1 Distribution of the methods used in the studies that consider uncertainty

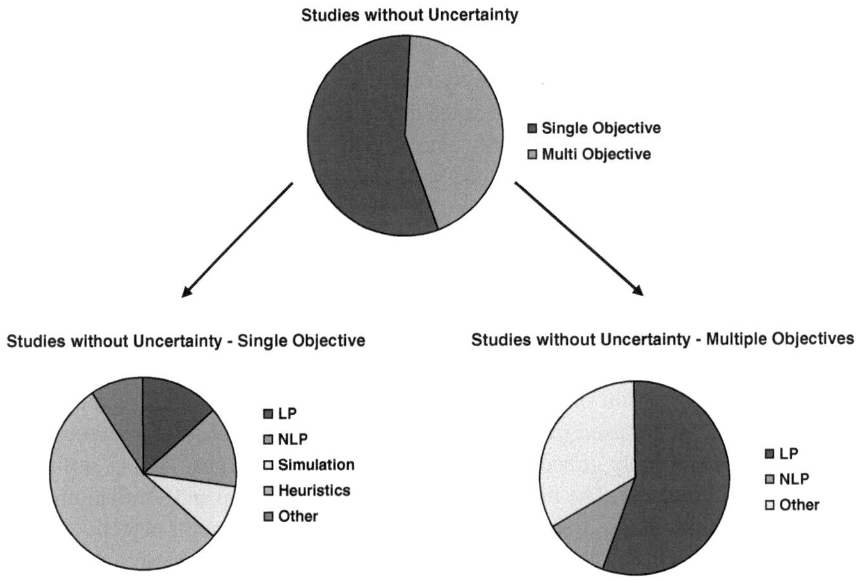

Fig. 2 Distribution of the methods used in the studies that do not consider uncertainty

addition, MILP is widely utilized to model the multi objective problems that do not incorporate uncertainties. Moreover, life cycle analyses and GIS-based methods are employed as subsidiary methods to model and analyze the environmental impacts of the supply chains and geographic distribution of the biomass sources.

5.2 *Future Research Directions*

Most of the real world problems dealing with the renewable energy supply chain design consider different objectives such as satisfying minimum system costs and minimum level of harmful gas emissions, and a set of technological, economic, environmental, social and energetic constraints. Despite this fact, this study demonstrates that the number of multi objective studies is less than single objective studies in the current literature about biomass to energy supply chain design and operation. Multi dimensionality of the sustainability goal and complexity of socioeconomic and biological systems make multi criteria and multi objective methodologies increasingly required in decision making for sustainable energy systems. Although goal programming is one of the most powerful approaches in practical decision making for multi objective problems which aims to minimize unwanted deviations from target values for objectives, this study reveals that it is not used in biomass to energy conversion systems design.

Recent technological developments enforce the decision makers to deal with more complex energy planning and optimization problems. Dynamic programming, which is a method for giving complex planning decisions by breaking the problems down into simpler subproblems, is used when a sequence of interrelated decisions are required. This study reveals that dynamic programming is rarely used in biomass to energy supply chain design and planning problems. It is also observed in this paper that most of studies that consider uncertainty in system parameters utilize scenario and sensitivity analyses. Fuzzy modeling is a method to describe and treat uncertainties in design and planning problems. We think that fuzzy modeling and dynamic programming approaches can be more widely utilized to provide the appropriate framework to complex biomass to energy supply chain network design problems.

References

Akgül O, Shah N, Papageorgiou LG (2012a) An optimisation framework for a hybrid first/second generation bioethanol supply chain. Comput Chem Eng 42:101–114

Akgül O, Shah N, Papageorgiou LG (2012b) Economic optimisaiton of a UK advanced biofuel supply chain. Biomass Bioenergy 41:57–72

Amigun B, von Blottnitz H (2007) Capacity-cost and location-cost analyses for biogas plants in Africa. Resour Conserv Recycl 55:63–73

An H, Searcy SW (2012) Economic and energy evaluation of a logistics system based on biomass modules. Biomass Bioenergy 46:190–202

An H, Wilhelm WE, Searcy SW (2011) A mathematical model to design a lignocellulosic biofuel supply chain system with a case study based on a region in Central Texas. Bioresour Technol 102:7860–7870

Andersen F, Iturmendi F, Espinosa S, Diaz MS (2012) Optimal design and planning of biodiesel supply chain with land competition. Comput Chem Eng 47:170–182

Avami A (2012) A model for biodiesel supply chain: a case study in Iran. Renew Sustain Energy Rev 16:4196–4203

Avami A (2013) Assessment of optimal biofuel supply chain planning in Iran: technical, economic, and agricultural perspectives. Renew Sustain Energy Rev 26:761–768
Aviso KB, Tan RR, Culaba AB, Cruz JB Jr (2011) Fuzzy input-output model for optimizing eco-industrial supply chains under water footprint constraints. J Clean Prod 19:187–196
Awudu I, Zhang J (2012) Uncertainties and sustainability concepts in biofuel supply chain management: a review. Renew Sustain Energy Rev 16:1359–1368
Awudu I, Zhang J (2013) Stochastic production planning for a biofuel supply chain under demand and price uncertainties. Appl Energy 103:189–196
Ayoub N, Martins R, Wang K, Seki H, Naka Y (2007) Two levels decision system for efficient planning and implementation of bioenergy production. Energy Convers Manag 48:709–723
Ayoub N, Seki H, Naka Y (2008) A methodology for designing and evaluating biomass utilization networks. In: Braunschweig B, Joulia X (eds) ESCAPE'18: 18th European Symposium on computer aided process engineering, Lyon, June 2008. Computer aided chemical engineering, vol 25. Elsevier, p 1053
Ayoub N, Seki H, Naka Y (2009a) Superstructure-based design and operation for biomass utilization networks. Comput Chem Eng 33:1770–1780
Ayoub N, Elmoshi E, Seki H, Yaka Y (2009b) Evolutionary algorithms approach for integrated bioenergy supply chains optimization. Energy Convers Manag 50:2944–2955
Bai Y, Hwang T, Kang S, Ouyang Y (2011) Biofuel refinery location and supply chain planning under traffic congestion. Transp Res Part B 45:162–175
Bai Y, Ouyang Y, Pang JS (2012) Biofuel supply chain design under competitive agricultural land use and feedstock market equilibrium. Energy Econ 34:1623–1633
Bekkering J, Broekhuis AA, van Gemert WJT (2010) Optimisation of a green gas supply chain—a review. Bioresour Technol 101:450–456
Bowersox DJ, Closs DJ, Cooper MB, Bowersox JC (2012) Supply chain logistics management. McGraw-Hill, Irwin
Bruglieri M, Liberti L (2008) Optimal running and planning of a biomass-based energy production process. Energy Policy 36:2430–2438
Calvert K (2011) Geomatics and bioenergy feasibility assessments: taking stock and looking forward. Renew Sustain Energy Rev 15:1117–1124
Chen CW, Fan Y (2012) Bioethanol supply chain system planning under supply and demand uncertainties. Transp Res Part E 48:150–164
Corsano G, Vecchietti AR, Montagna JM (2011) Optimal design for sustainable bioethanol supply chain considering detailed plant performance model. Comput Chem Eng 35:1384–1398
Čuček L, Lam HL, Klemeš JJ, Varbanov PS, Kravanja Z (2010) Synthesis of regional networks for the supply of energy and bioproducts. Clean Technol Environ Policy 12:635–645
Čuček L, Varbanov PS, Klemeš JJ, Kravanja Z (2012) Total footprints-based multi-criteria optimisation of regional biomass energy supply chains. Energy 44:135–145
Dal Mas M, Giarola S, Zamboni A, Bezzo F (2010) Capacity planning and financial optimization of the bioethanol supply chain under price uncertainty. In: Pierucci S, Buzzi Ferraris G (eds) ESCAPE'20: 20th European symposium on computer aided process engineering, Ischia, June 2010. Computer aided chemical engineering, vol 28. Elsevier, p 97
Dal-Mas M, Giarola S, Zamboni A, Bezzo F (2011) Staregic design and investment capacity planning of the ethanol supply chain under price uncertainty. Biomass Bioenergy 35:2059–2071
De Meyer A, Cattrysse D, Rasinmäki J, Van Orshoven J (2014) Methods to optimise the design and management of biomass-for-bioenergy supply chains: a review. Renew Sustain Energy Rev 31:657–670
Dunnett A, Adjiman C, Shah N (2007) Biomass to heat supply chains-applications of process optimization. Chem Eng Res Des Part B 85:419–429
Dyken S, Bakken BH, Skjelbred HI (2010) Linear mixed-integer models for biomass supply chains with transport, storage and processing. Energy 35:1338–1350

Ebadian M, Sowlati T, Sokhansanj S, Stumborg M, Townley-Smith L (2011) A new simulation model for multi-agricultural biomass logistics system in bioenergy production. Biosyst Eng 110:280–290

Ebadian M, Sowlati T, Sokhansanj S, Townley-Smith L, Stumborg M (2013) Modeling and analysing storage systems in agricultural biomass supply chain for cellulosic ethanol production. Appl Energy 102:840–849

Ekşioğlu SD, Acharya A, Leightley LE, Arora S (2009) Analyzing the design and management of biomass-to-biorefinery supply chain. Comput Ind Eng 57:1342–1352

Elia JA, Baliban RC, Xiao X, Floudas CA (2011) Optimal energy supply network determination and life cycle analysis for hybrid coal, biomass, and natural gas to liquid (CBGTL) plants using carbon-based hydrogen production. Comput Chem Eng 35:1399–1430

Erkut E, Karagiannidis A, Perkoulidis G, Tjandra SA (2008) A multicriteria facility location model for municipal solid waste management in North Greece. Eur J Oper Res 187:1402–1421

Foo DCY, Tan RR, Lam HL, Aziz MKA, Klemeš JJ (2013) Robust models for the synthesis of flexible palm oil-based regional bioenergy supply chain. Energy 55:68–73

Freppaz D, Minciardi R, Robba M et al (2004) Optimizing forest biomass exploitation for energy supply at a regional level. Biomass Bioenergy 26:15–25

Frombo F, Minciardi R, Robba M, Sacile R (2009) A decision support system for planning biomass-based energy production. Energy 34:362–369

Geraili A, Sharma P, Romagnoli JA (2014) A modeling framework for design of nonlinear renewable energy systems through integrated simulation modeling and metaheuristic optimization: applications to biorefineries. Comput Chem Eng 61:102–117

Ghaffariyan MR, Acuna M, Brown M (2013) Analyzing the effect of five operational factors on forest residue supply chain costs: a case study in Western Australia. Biomass Bioenergy 59:486–493

Giarola S, Zamboni A, Bezzo F (2011) Spatially explicit multi-objective optimisation for design and planning of hybrid first and second generation biorefineries. Comput Chem Eng 35:1782–1797

Giarola S, Bezzo F, Shah N (2013) A risk management approach to the economic and environmental strategic design of ethanol supply chains. Biomass Bioenergy 58:31–51

Gold S, Seuring S (2011) Supply chain and logistics issues of bio-energy production. J Clean Prod 19:32–42

Gómez A, Zubizarreta J, Rodrigues M, Dopazo C, Fueyo N (2010) An estimation of the energy potential of agro-industrial residues in Spain. Resour Conserv Recycl 54:972–984

Gómez-González M, López A, Jurado F (2013) Hybrid discrete PSO and OPF approach for optimization of biomass fueled micro-scale energy system. Energy Convers Manag 65:539–545

Gronalt M, Rauch P (2007) Designing a regional forest fuel supply network. Biomass Bioenergy 31:393–402

Gun J, Smith CT (2011) Optimal plant size and feedstock supply radius: A modelling approach to minimize bioenergy production costs. Biomass Bioenergy 35:3350–3359

Huang Y, Chen CW, Fan Y (2010) Multistage optimization of the supply chains of biofuels. Transp Res Part E 46:820–830

Iakovou E, Karagiannidis A, Vlachos D, Toka A, Malamakis A (2010) Waste biomass-to-energy supply chain management: a critical synthesis. Waste Manag 30:1860–1870

Kanzian C, Kühmaier M, Zazgornik J, Stampfer K (2013) Design of forest energy supply networks using multi-objective optimization. Biomass Bioenergy 58:294–302

Kim J, Realff MJ, Lee JH, Whittaker C, Furtner L (2011a) Design of biomass processing network for biofuel production using an MILP model. Biomass Bioenergy 35:853–871

Kim J, Realff MJ, Lee JH (2011b) Optimal design and global sensitivity analysis of biomass supply chain networks for biofuels under uncertainty. Comput Chem Eng 35:1738–1751

Kocoloski M, Griffin WM, Matthews HS (2011) Impacts of facility size and location decisions on ethanol production cost. Energy Policy 39:47–56

Krukanont P, Prasertsan S (2004) Geographical distribution of biomass and potential sites of rubber wood fired power plants in Southern Thailand. Biomass Bioenergy 26:47–59
Kumar A, Sokhansanj S (2007) Switchgrass (Panicum vigratum, L.) delivery to a biorefinery using integrated biomass supply analysis and logistics (IBSAL) model. Bioresour Technol 98:1033–1044
Lam HL, Klemeš JJ, Kravanja Z (2011) Model-size reduction techniques for large-scale biomass production and supply networks. Energy 36:4599–4608
Lam HL, Ng WPQ, Ng RTL et al (2013) Green strategy for sustainable waste-to-energy supply chain. Energy 57:4–16
Lambert DK, Middleton J (2010) Logistical design of a regional herbaceous crop residue-based ethanol production complex. Biomass Bioenergy 34:91–100
Leão RRCC, Hamacher S, Oliveira F (2011) Optimization of biodiesel supply chains based on small farmers: A case study in Brazil. Bioresour Technol 102:8958–8963
Leboreiro J, Hilaly AK (2011) Biomass transportation model and optimum plant size for the production of ethanol. Bioresour Technol 102:2712–2723
Leboreiro J, Hilaly AK (2013) Analysis of supply chain, scale factor, and optimum plant capacity for the production of ethanol from corn stover. Biomass Bioenergy 54:158–169
Leduc S, Schwab D, Dotzauer E, Schmid E, Obersteiner M (2008) Optimal location of wood gasification plants for methanol production with heat recovery. Int J Energy Res 32:1080–1091
Leduc S, Starfelt F, Dotzauer E et al (2010a) Optimal location of lignocellulosic ethanol refineries with polygeneration in Sweden. Energy 35:2709–2716
Leduc S, Lundgren J, Franklin O, Dotzauer E (2010b) Locaiton of a biomass based methanol production plant: a dynamic problem in northern Sweden. Appl Energy 87:68–75
Lin T, Rodríguez LF, Shastri YN, Hansen AC, King KC (2014) Integrated strategic and tactical biomass-biofuel supply chain optimization. Bioresour Technol. doi: 10.1016/j.biortech.2013.12.121
López PR, Jurado F, Reyes NR, Galán SG, Gómez M (2008) Particle swarm optimization for biomass-fuelled systems with technical constraints. Eng Appl Artif Intell 21:1389–1396
Mafakheri F, Nasiri F (2013) Modelling of biomass-to-energy supply chain operations: applications, challenges and research directions. Energy Policy. doi:10.1016/j.en.pol.2013.11.071
Mahmoudi M, Sowlati T, Sokhansanj S (2009) Logistics of supplying biomass from a mountain pine beetle-infested forest to a power plant in British Columbia. Scand J For Res 24:76–86
Mansoornejad B, Chambost V, Stuart P (2010) Integrating product portfolio design and supply chain design for the forest biorefinery. Comput Chem Eng 34:1497–1506
Mansoornejad B, Pistikopoulos EN, Stuart PR (2013) Scenario-based strategic supply chain design and analysis for the forest biorefinery using an operational supply chain model. Int J Prod Econ 144:618–634
Marvin WA, Schmidt LD, Benjaafar S, Tiffany DG, Daoutidis P (2012) Economic optimization of a lignocellulosic biomass-to-ethanol supply chain. Chem Eng Sci 67:68–79
McCormick K, Kåberger T (2007) Key barriers for bioenergy in Europe: economic conditions, know-how and institutional capacity, and supply chain co-ordination. Biomass Bioenergy 31:443–452
Mckendry P (2002a) Energy production from biomass (part 1): overview of biomass. Bioresour Technol 83:37–46
Mckendry P (2002b) Energy production from biomass (part 2): conversion technologies. Bioresour Technol 83:47–54
Mckendry P (2002c) Energy production from biomass (Part 3): gasification technologies. Bioresour Technol 83:55–63
Mele FD, Guillén-Gosálbez G, Jimenéz L (2009) Optimal planning of supply chains for bioethanol and sugar production with economic and environmental concerns. In: Jeżowski J, Thullie J (eds) ESCAPE'19: 19th European symposium on computer aided process engineering, Cracow, June 2009. Computer aided chemical engineering, vol 26. Elsevier, p 997

Mobini M, Sowlati T, Sokhansanj S (2011) Forest biomass supply logistics for a power plant using the discrete-event simulation approach. Appl Energy 88:1241–1250
Mobini M, Sowlati T, Sokhansanj S (2013) A simulation model for the design and analysis of wood pellet supply chains. Appl Energy 111:1239–1249
Nagel J (2000) Determination of an economic energy supply structure based on biomass using a mixed-integer linear optimization model. Ecol Eng 16:91–102
Osmani A, Zhang J (2013) Stochastic optimization of a multi-feedstock lignocellulosic-based bioethanol supply chain under multiple uncertainties. Energy 59:157–172
Osmani A, Zhang J (2014) Economic and environmental optimization of a large scale sustainable dual feedstock lignocellulosic-based bioethanol supply chain in a stochastic environment. Appl Energy 114:572–587
Papapostolou C, Kondili E, Kaldellis JK (2011a) Modelling biomass and biofuels supply chains. In: Pistikopoulos EN, Georgiadis MC, Kokossis AC (eds) ESCAPE'21: 21st European symposium on computer aided process engineering, Chalkidiki, May 2011. Computer aided chemical engineering, vol 29. Elsevier, p 1773
Papapostolou C, Kondili E, Kaldellis JK (2011b) Development and implementation of an optimisation model for biofuels supply chain. Energy 36:6019–6026
Parker N, Tittmann P, Hart Q et al (2010) Developmet of a biorefinery optimized biofuel supply curve for the Western United States. Biomass Bioenergy 34:1597–1607
Pérez-Fortes M, Lainez-Aguirre JM, Arranz-Piera P, Velo E, Puigjaner L (2012) Design of regional and sustainable bio-based networks for electricity generation using a multi-objective MILP approach. Energy 44:79–95
Perpiña C, Martínez-Llario JC, Pérez-Navarro A (2013) Multicriteria assessment in GIS environments for siting biomass plants. Land Use Policy 31:326–335
Reche-López P, Ruiz-Reyes N, Galán SG, Jurado F (2009) Comparison of metaheuristic techniques to determine optimal placement of biomass power plants. Energy Convers Manag 50:2020–2028
Rentizelas AA, Tatsiopoulos IP (2010) Locating a bioenergy facility using a hybrid optimization method. Int J Prod Econ 123:196–209
Rentizelas AA, Tatsiopoulos IP, Tolis A (2009a) An optimization model for multi-biomass tri-generation energy supply. Biomass Bioenergy 33:223–233
Rentizelas AA, Tolis AJ, Tatsiopoulos IP (2009b) Logistics issues of biomass: The storage problem and the multi-biomass supply chain. Renew Sustain Energy Rev 13:887–894
Rentizelas AA, Tolis AI, Tatsiopoulos IP (2013) Optimisation and investment analysis of two biomass-to-heat supply chain structures. Biosyst Eng. doi: 10.1016/j.biosystemseng.2013.07.012
Roni MS, Eksioglu SD, Searcy E, Jha K (2014) A supply chain network design model for biomass co-firing in coal-fired power plants. Transp Res Part E 61:115–134
Santibañez-Aguilar JE, González-Campos JB, Ponce-Ortega JM, Serna-González M, El-Halwagi MM (2014) Optimal planning and site selection for distributed multiproduct biorefineries involving economic, environmental and social objectives. J Clean Prod 65:270–294
Shabani N, Sowlati T (2013) A mixed integer non-linear programming model for tactical value chain optimization of a wood biomass power plant. Appl Energy 104:353–361
Sharma B, Ingalls RG, Jones CL, Khanchi A (2013a) Biomass supply chain design and analysis: basis, overview, modeling, challenges, and future. Renew Sustain Energy Rev 24:608–627
Sharma B, Ingalls RG, Jones CL, Huhnke RL, Khanchi A (2013b) Scenario optimization modeling approach for design and management of biomass-to-biorefinery supply chain system. Bioresour Technol 150:163–171
Shi X, Elmore A, Li X et al (2008) Using spatial information technologies to select sites for biomass power plants: a case study in Guangdong Province, China. Biomass Bioenergy 32:35–43
Singh J, Panesar BS, Sharma SK (2011) Geographical distribution of agricultural residues and optimum sites of biomass based power plant in Bathinda, Punjab. Biomass Bioenergy 35:4455–4460

Sokhansanj S, Kumar A, Turhollow AF (2006) Development and implementation of integrated biomass supply analysis and logistics model (IBSAL). Biomass Bioenergy 30:838–847

Sultana A, Kumar A (2011) Optimal configuration and combination of multiple lignocellulosic biomass feedstocks delivery to a biorefinery. Bioresour Technol 102:9947–9956

Sultana A, Kumar A (2012) Optimal siting and size of biorefinery facilities using geographic information system. Appl Energy 94:192–201

Sun J, Lin J, Qian Y (2013) Game-theoretic analysis of competitive agri-biomass supply chain. J Clean Prod 43:174–181

Tatsiopoulos IP, Tolis AJ (2003) Economic aspects of the cotton-stalk biomass logistics and comparison of supply chain methods. Biomass Bioenergy 24:199–214

Tong K, Gleeson MJ, Rong G, You F (2014) Optimal design of advanced drop-in hydrocarbon biofuel supply chain integrating with existing petroleum refineries under uncertainty. Biomass Bioenergy 60:108–120

Velazquez-Marti B, Fernandez-Gonzalez E (2010) Mathematical algorithms to locate factories to transform biomass in bioenergy focused on logistic network construction. Renew Energy 35:2136–2142

Vera D, Carabias J, Jurado F, Ruiz-Reyes N (2010) A Honey Bee Foraging approach for optimal location of a biomass power plant. Appl Energy 87:2119–2127

Voivontas D, Assimacopoulos D, Koukios EG (2001) Assessment of biomass potential for power production: a GIS based method. Biomass Bioenergy 20:101–112

Walla C, Schneeberger W (2008) The optimal size for biogas plants. Biomass Bioenergy 32:551–557

Wang X, Ouyang Y, Yang H, Bai Y (2013) Optimal biofuel supply chain design under consumption mandates with renewable identification numbers. Transp Res Part B 57:158–171

Windisch J, Röser D, Mola-Yudego B, Sikanen L, Asikainen A (2013) Business process mapping and discrete-event simulation of two forest biomass supply chains. Biomass Bioenergy 56:370–381

Wolfsmayr UJ, Rauch P (2014) The primary forest fuel supply chain: a literature review. Biomass Bioenergy 60:203–221

Xie F, Huang Y, Eksioglu S (2014) Integrating multimodal transport into cellulosic biofuel supply chain design under feedstock seasonality with a case study based on California. Bioresour Technol 152:15–23

Yagi K, Nakata T (2011) Economic analysis on small-scale forest biomass gasification considering geographical resources distribution and technical characteristics. Biomass Bioenergy 35:2883–2892

Yu H, Wang Q, Ileleji KE et al (2011) Design and analysis of geographic distribution of biomass power plant and satellite storages in China. Part 1: Straight-line delivery. Biomass Bioenergy. doi: 10.1016/j.biombioe.2011.10.022

Yu H, Wang Q, Ileleji KE et al (2012) Design and analysis of geographic distribution of biomass power plant and satellite storages in China. Part 2: Road delivery. Biomass Bioenergy. doi: 10.1016/j.biombioe.2012.06.028

Yue D, You F, Snyder SW (2014) Biomass-to-bioenergy and biofuel supply chain optimization: overview, key issues and challenges. Comput Chem Eng. doi: 10.1016/j.compchemeng.2013.11.016

Zamboni A, Bezzo F, Shah N (2009) Supply chain optimization for bioethanol production system in Northern Italy: environmentally conscious strategic design. In: Alves RMB, Nascimento CAO, Biscaia EC Jr (eds) PSE2009: 10th International symposium on process systems engineering, Chalkidiki, August 2009. Computer aided chemical engineering, vol 27. Elsevier, p 2037

Zhang L, Hu G (2013) Supply chain design and operational planning models for biomass to drop-in fuel production. Biomass Bioenergy 58:238–250

Zhang F, Johnson DM, Sutherland JW (2011) A GIS-based method for identifying the optimal location for a facility to convert forest biomass to biofuel. Biomass Bioenergy 35:3951–3961

Zhang F, Johnson DM, Johnson MA (2012) Development of a simulation model of biomass supply chain for biofuel production. Renew Energy 44:380–391

Zhang J, Osmani A, Awudu I, Gonela V (2013) An integrated optimization model for switchgrass-based bioethanol supply chain. Appl Energy 102:1205–1217

Zhu X, Yao Q (2011) Logistics system design for biomass-to-bioenergy industry with multiple types of feedstocks. Bioresour Technol 102:10936–10945

BLOMST—An Optimization Model for the Bioenergy Supply Chain

Michal Kaut, Ruud Egging, Truls Flatberg and Kristin Tolstad Uggen

Abstract In this chapter, we present a new model for optimal strategic and tactical planning of the bioenergy supply chain under uncertainty. We discuss specific challenges, characteristics and issues related to this type of model. The technological details, variability in supply and demand, and uncertainty in virtually all aspects of the supply chain require advanced modeling techniques. Our model provides a broad modeling approach that addresses the entire supply chain using an integrated perspective. The broad applicability of the approach is illustrated by the two cases discussed at the end of the chapter. The first case presents a forest to bioenergy supply chain in a region of the Norwegian west coast. The second case presents the miscanthus supply chain to a transformation plant in Burgundy, France and takes into consideration uncertain final demand.

1 Introduction

Renewable energy is the fastest-growing source of energy generation, according to the IEO2013 Reference case (U.S. Energy Information Agency 2013). Total power generation from renewables is projected to increase by 2.8 % annually until 2040. Although about 80 % of the total increase is in hydroelectric and wind power, bioenergy generation is expected to grow at about the same pace. The growth of non-hydro renewable energy sources in OECD Europe is stimulated by renewable energy policies. The EU mandates that 20 % of total energy production must come

M. Kaut (✉) · R. Egging · T. Flatberg · K.T. Uggen
SINTEF Technology and Society, Trondheim, Norway
e-mail: michael.kaut@sintef.no

R. Egging
Department of Industrial Economics and Technology Management, NTNU, Trondheim, Norway
e-mail: Ruud.Egging@sintef.no

S.D. Eksioglu et al. (eds.), *Handbook of Bioenergy*,
Energy Systems, DOI 10.1007/978-3-319-20092-7_2

from renewables by 2020 (according to the 20-20-20 target); this is up from about 13 % in 2010. In addition to the EU targets, several countries provide incentives that promote the expansion of renewable energy generation. For example, Germany, Spain, Denmark, and the United Kingdom have enacted feed-in tariffs that guarantee minimum prices for energy generated from renewable sources.

The market for bioenergy plants is expected to grow quickly in the coming years (ecoprog GmbH 2013). In 2020, 3500 bioenergy plants are expected to be operational worldwide, implying a growth of installed capacity by about 46 % in an eight-year period.

A beneficial characteristic of biomass is that it can be stored, in contrast with solar or wind-based generation, which means that bioenergy generation can more easily be matched with varying demand. Unfortunately, biomass often requires large production and collection areas, has low energy density, is expensive to harvest and transport, and has high maintenance and logistics costs. This makes it challenging for the bioenergy industry to compete with the highly developed fossil-fuel supply chains (De Meyer et al. 2014). Also, biomass quality can be affected by transport and storage. Although *passive* drying reduces the moisture content, which is a favorable outcome, storage can also induce fiber deterioration, which reduces the energy content (Wolfsmayr and Rauch 2014). Furthermore, bioenergy supply chains have to deal with the geographical spread of supply sources and weather and season induced supply variations. In addition, the supply chain is challenged by complex logistics and inventory management aspects, as well as variety of uncertain factors.

Almost every step in the supply chain may have uncertainty factors. Some uncertainties are inherited from the biomass supply part, others from the energy generation and demand parts. Weather conditions and technical disruptions affect harvesting time windows as well as biomass yield and quality. Transportation and logistic uncertainties involve fleet availability, storage and road conditions, all of which induce unpredictable supply. Technological innovation and government policies and incentives greatly affect the competitiveness of investments and operations.

Researchers have looked at biomass-bioenergy markets in settings varying from a single feedstock and a single consumer to integrated settings in the local economy or other energy markets. To a large extent the modeling approaches reported in the literature are inspired by a specific case, resulting in many specialized optimization models for certain parts of the value chain or for specific supply chains.

Van Tilburg et al. (2006) present the BIOTRANS model, a general model for the biomass-to-biofuel supply chain. This model is deterministic, has a cost-minimization focus, and does not consider seasonality, storage, or biomass quality variations, although yields are location specific. Ekşioğlu et al. (2009) develop a mixed-integer design and management model for the supply chain of bioethanol produced from lignocellulosic biomass. It is a deterministic cost-minimization model and considers seasonal variation and storage.

Several papers consider uncertainties in specific biomass-to-biofuel supply chains. Marufuzzaman et al. (2014a, b) develop stochastic biofuel supply chain models considering various types of risks and apply enhanced variants of Benders Decomposition to solve the models. Marufuzzaman et al. (2014a) consider technological progress and biomass supply uncertainties, whereas Marufuzzaman et al. (2014b) consider supply disruption risks at intermodal hubs due to various natural disasters with varying impact levels. There results indicate significant expected cost savings due to the explicit consideration of uncertainty in the modeling approaches.

Uncertain events will materialize frequently over a multi-year planning horizon, and there are multiple decision types, and associated costs and profits, in bioenergy supply chain design and management that can (should) be affected by this. To adequately represent the random events and multiple decision moments (recourse options) we believe a multi-stage modeling approach is necessary. We are aware of only two multi-stage stochastic optimization models which aim to cover the entire supply chain. Cundiff et al. (1997) develop a multi-stage linear program that considered the impact of weather conditions during the growth season and harvesting period. They include four scenarios, which take into consideration good and bad weather in each period. They considered one type of biomass and allowed storage capacity expansions with fixed locations, but no discrete investment decisions about new capacities or locations were allowed. Walther et al. (2012) present a strategic stochastic optimization model for investments related to production networks for synthetic bio-diesel in North-west Germany. Neither of these models include active drying, or the terminal concept which we include in the model presented in this chapter.

Many papers discuss challenges and recommended future research directions. Sharma et al. (2013) indicate that the combined complexity and uncertainty plus non-financial objectives require advanced multi-objective approaches. Yue et al. (2014) note that models covering uncertainty in biomass quality as well as energy prices and correlation between uncertain parameters are not handled by any known optimization model. For instance Shabani et al. (2013), De Meyer et al. (2014), Mafakheri and Nasiri (2013), and Wolfsmayr and Rauch (2014) argue for a holistic, integrated approaches taking into account interrelationships, interdependences, and coordination needs between all stakeholders in the supply chain, rather than a single agent. In what follows we propose a stochastic biofuel supply chain model for a general network structure of the various supply chain components.

Considering the list of challenges and lack of generalized stochastic models in the literature, we have developed a general framework for optimizing the supply chain for bioenergy production under uncertainty. This generic stochastic bioenergy optimization model can be used in many types of analysis, independent of technologies considered, types of biomass used, the user operating in a specific part of the supply chain, or the geographical region. Our model is flexible in taking a strategic and/or tactical planning perspective, and considers uncertainty in virtually all aspects relevant to the biomass bioenergy supply chain (although in this chapter we do not cover the model extensions needed to handle mandatory crop rotation and perennial crops). Through an adequate composition of the stochastic scenario tree, even correlations between uncertain parameters could be captured. We illustrate the broad applicability of our approach with two case studies that are entirely different in nature; these studies are presented at the end of the chapter.

The rest of the chapter is organized as follows: in the next section (Sect. 2), we present the core model, i.e. the minimal set of constraints needed for a functional deterministic model. Since uncertainty is such an important characteristic, Sect. 3 deals with reformulation of the model into a stochastic one. In Sect. 4, we extend the core model with additional features. Finally, in Sect. 5, we present two test cases, one dealing with forest biomass in Norway and the other with miscanthus in France.

2 The Core Model

The model is based on a network representation of the supply chain from production to consumption. The nodes of the network represent activities and processes that the products can undergo. Currently, the model has nodes for production, transformation, storage, and consumption. Arcs between the nodes are used to model flow of commodities or equipment between nodes.

This structure means that the model naturally decomposes into several parts: one for each node type and one for the flows between them. The actual network structure is then provided by data, while the model itself is case-independent. The result is a flexible model, capable of handling complex supply chains, such as the one shown in Fig. 1. There, we can see that 'chipping' and 'pelletizing' are represented by the same node type, 'transformation'. This means that both the input and output products have to be specified as data. We could save the amount of required data by having separate node types for the two processes, but that would increase the model complexity dramatically; not only would we have to model more node types, but we would also have to model more possible links between them.

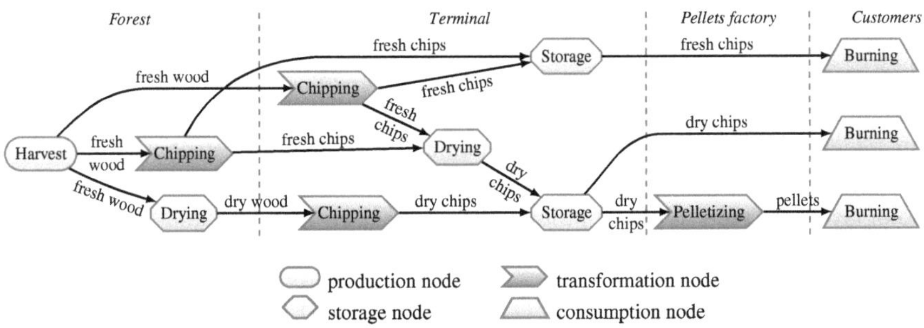

Fig. 1 Structure of a stylized forest-biomass supply chain

2.1 *Notation*

2.1.1 Indices, Sets, and Subsets

The model supports the flow of multiple products and their transformation to other products. Each actual product is specified by its product type (referred to simply as 'product') and by the crop that it is made from. Hence, birch chips are a product 'chips' made of crop 'birch'.

Name	Description	Information
$a \in \mathscr{A}$	Transportation arcs	
$c \in \mathscr{C}$	All crops	
$d \in \mathscr{D}$	Dimensions	
$i \in \mathscr{I}$	Production limits	
$j \in \mathscr{J}$	Alternatives	
$n \in \mathscr{N}$	Nodes in production network	
$p \in \mathscr{P}$	All products: input, intermediate and final	
$t \in \mathscr{T}$	Time periods	$\mathscr{T} = 1, \ldots, T$
$n \in \mathscr{N}^{\mathrm{P}}$	Production nodes	$\mathscr{N}^{\mathrm{P}} \subset \mathscr{P}$
$n \in \mathscr{N}^{\mathrm{T}}$	Transformation nodes	$\mathscr{N}^{\mathrm{T}} \subset \mathscr{P}$
$n \in \mathscr{N}^{\mathrm{S}}$	Storage nodes	$\mathscr{N}^{\mathrm{S}} \subset \mathscr{P}$
$n \in \mathscr{N}^{\mathrm{C}}$	Consumption nodes	$\mathscr{N}^{\mathrm{C}} \subset \mathscr{P}$
$p \in \mathscr{P}^{\mathrm{B}}$	All basic biomass (untreated, possibly harvested) crops	$\mathscr{P}^{\mathrm{B}} \subset \mathscr{P}$
$p \in \mathscr{P}^{\mathrm{I}}$	All intermediate products (treated, possibly ready for use)	$\mathscr{P}^{\mathrm{I}} \subset \mathscr{P}$
$p \in \mathscr{P}^{\mathrm{F}}$	All final products (ready for use)	$\mathscr{P}^{\mathrm{F}} \subset \mathscr{P}$

Note that $\mathscr{P}^{\mathrm{B}}$, $\mathscr{P}^{\mathrm{I}}$ and $\mathscr{P}^{\mathrm{F}}$ may overlap partially, while sets $\mathscr{N}^{\mathrm{P}}$, $\mathscr{N}^{\mathrm{T}}$, $\mathscr{N}^{\mathrm{S}}$, and $\mathscr{N}^{\mathrm{C}}$ must form a partition of $\mathscr{N}$.

2.1.2 Constants

The flow of products can be measured in a number of dimensions; in bioenergy models, a typical unit is the energy content [GJ] or dry mass content [t^{dm}], while in other applications, a volume measure is more appropriate, e.g., [lm^3] (loose m^3, e.g., for wood chips). In the model, subscript d in q_d denotes the unit of measurement. Furthermore, each product p has its default dimension d_p^* along which it is measured; it follows that the default units used for the product are $\mathrm{q}_{\mathrm{d}_p^*}$, which we denote by q_p^*. This can then be converted into other units using the unit-conversion parameters $U_{c,p,d}$.

Name	Description	Unit
$C^{A}_{n_1,n_2,p,d}$	Transport costs	€/q_d
C^{N}_{n}	Operating costs of node n, per period	€
$C^{P}_{n,c}$	Production costs, per area	€/ha
$C^{Q}_{n,c}$	Production costs, per volume	€/$q^*_{p^b_c}$
$C^{R-}_{n,d}$	Transformation costs (per input unit)	€/q_d
$D^{min}_{n,t,d}$	Demand minimum	q_d
$D^{max}_{n,t,d}$	Demand maximum	q_d
$E^{D}_{n,p}$	Efficiency of product p in satisfying demand at node n	
$P_{n,t,d}$	Market price	€/q_d
$\bar{Q}^{A}_{n1,n2,p,d}$	Transportation capacity	q_d
$\bar{Q}^{R-}_{n,d}$	Input capacity of a transformation	q_d
$\bar{Q}^{R+}_{n,d}$	Output capacity of a transformation	q_d
$Q^{P}_{n,c,p,t}$	Actual produced volumes at each period	q^*_p
$\bar{Q}^{P}_{n,c}$	Production capacity (area)	ha
Q^{min}_{i}	Minimum production of production limit i	d_i
Q^{max}_{i}	Maximum production of production limit i	d_i
$S_{n,d}$	Storage capacity	q_d
$U_{c,p,d}$	Unit conversion ('density')	q_d/q^*_p
$Y^{R}_{n,p}$	Transformation efficiency (output per input)	
$Y^{P}_{n,c}$	Production yield	$q^*_{p^b_c}$/ha

2.1.3 Variables

The model has one set of binary decision variables for the node usage, used for calculating operating costs, plus several continuous, non-negative variables for modeling the flows and volumes.

Name	Description	Unit
$f_{n_1,n_2,c,p,t}$	Transportation flow	q^*_p
$q^{P}_{n,c,p,t}$	Production quantity	q^*_p
$q^{D}_{n,c,p,t}$	Consumption quantity	q^*_p
$r^{in}_{n,c,p,t}$	Transformation input quantity	q^*_p
$r^{out}_{n,c,p,t}$	Transformation output quantity	q^*_p
$s_{n,c,p,t}$	Storage levels at the end of a period	q^*_p
$z_{n,t}$	Whether a node is used in a period	0/1

Note that we model flows using continuous variables, ignoring the fact that they are carried by (discrete) vehicles. This is a natural simplification for a tactical/strategic model, with periods spanning weeks or even months.

2.2 Production Nodes

Production nodes are the sources of the biomass, i.e. the fields or forests. Each node can produce multiple crops and products, because one field or forest can be planted with more than one crop and some crops can be harvested in several ways. Note that the latter effect could also be achieved by letting the production node only 'produce' the plant, and by modeling the harvesting process in separate transformation nodes. This would give us more control over the harvesting techniques and allow for associating different efficiencies and costs to the harvesting methods—assuming that we (the optimizing agent) actually can control these processes.

It follows that the constraints describing the production nodes are necessarily case-dependent. If we, for example, do not have any control over the production, the produced volumes are input data to the model:

$$q^{\mathrm{P}}_{n,c,\mathrm{p}^{\mathrm{b}}_{c},t} = Q^{\mathrm{P}}_{n,c,\mathrm{p}^{\mathrm{b}}_{c},t},$$

where $\mathrm{p}^{\mathrm{b}}_{c} \in \mathscr{P}^{\mathrm{B}}$ denotes the product of harvesting crop c.

If we, on the other hand, control the production, then we have to distinguish between cases where there is a harvest time when we harvest all of the crop (as with grains) and cases where the crops grow continuously and we harvest only a part of the crop (as with forests). In the former situation, production is limited by the planted area $\bar{Q}^{\mathrm{P}}_{n,c}$, and the yields $Y^{\mathrm{P}}_{n,c}$:

$$\sum_{t \in \mathscr{T}} q^{\mathrm{P}}_{n,c,\mathrm{p}^{\mathrm{b}}_{c},t} \leq \bar{Q}^{\mathrm{P}}_{n,c} Y^{\mathrm{P}}_{n,c}, \quad n \in \mathscr{N}^{\mathrm{P}}, \tag{1}$$

where the unspecified indices run over their default sets, i.e. $c \in \mathscr{C}$. We use this simplification throughout the paper.

In the latter case with partial harvesting, we specify the production limits for each product in some given time period, to ensure sustainability. Each production limit $i \in \mathscr{I}$ is specified for every production node (n_i), crop (c_i), and dimension (d_i) and time interval $\left[\mathrm{t}^{S}_{i}, \mathrm{t}^{\mathrm{E}}_{i}\right]$ at which it was measured:

$$\mathrm{Q}^{\min}_{i} \leq \sum_{\mathrm{t}^{\mathrm{S}}_{i} \leq t \leq \mathrm{t}^{\mathrm{E}}_{i}} U_{c,p,d} q^{\mathrm{P}}_{\mathrm{n}_i,\mathrm{c}_i,\mathrm{p}^{\mathrm{b}}_{\mathrm{c}_i},t} \leq \mathrm{Q}^{\max}_{i} \quad i \in \mathscr{I}. \tag{2}$$

If a harvested crop results in several products, as it does for trees, we can model this through a transformation node, attached to the production node.

Note that the planted area is taken as input data, i.e., the model does not control what gets planted where in the first model version. This is because in all of our test cases, we optimized other parts of the supply chain and, therefore, could not make these decisions. However, the model is easily extensible to situations where the planting/sowing is a part of the decision process.

2.3 Transformation Nodes

Transformation nodes convert products; they represent processes like chipping of trees, production of pellets from chips, and gasification of biomass.

There are many possible types of transformation nodes, depending on the number of input and output products, and by the way they combine. In our model, we use two different types. First, we have a one-to-many transformation node n, in which an input product $\mathrm{P}_n^{\mathrm{in}}$ gets transformed into one or more output products. This is used to model, for example, tree harvesting, where we separate output products logs and branches.

The produced volumes of each output product p are controlled using the transformation yield per unit of input product. In addition, each transformation node has a conversion capacity in terms of the input volume, possibly along several dimensions (weight, volume).

$$r^{\mathrm{out}}_{n,c,p,t} = Y^{\mathrm{R}}_{n,\mathrm{p}_n^{\mathrm{in}}} \cdot r^{\mathrm{in}}_{n,c,\mathrm{p}_n^{\mathrm{in}},t} \qquad n \in \mathcal{N}^{\mathrm{T}} \tag{3}$$

$$\sum_{c\in\mathscr{C}} U_{c,\mathrm{p}_n^{\mathrm{in}},d} r^{\mathrm{in}}_{n,c,\mathrm{p}_n^{\mathrm{in}},t} \leq z_{n,t} \cdot \bar{Q}^{\mathrm{R}^-}_{n,d} \qquad n \in \mathcal{N}^{\mathrm{T}} \tag{4}$$

The other type of transformation nodes in our model represents many-to-one transformations, where the input products are alternatives, i.e., the output product can be made using any of the inputs or their mixture. For example, in the miscanthus supply chain that we present in Sect. 5.2, pellets can be made from both bales and baled chips.[1]

The constraints are similar to the ones listed above, except that we have one output product $\mathrm{P}_n^{\mathrm{out}}$ and the conversion capacity is given in terms of the output volumes.

[1]Note that this transformation is different from another common many-to-one process, namely the assembly of parts into one product. There, all inputs are needed to make the output, typically with fixed proportions—something we have not encountered in any biomass chain that we have studied.

$$r^{\text{out}}_{n,c,\text{P}^{\text{out}}_{\text{n}},\text{t}} = \sum_{p\in\mathcal{P}} Y^{\text{R}}_{n,p} \cdot r^{\text{in}}_{n,c,p,t} \quad n \in \mathcal{N}^{\text{T}} \tag{3'}$$

$$\sum_{c\in\mathcal{C}} U_{c,\text{P}^{\text{out}}_{\text{n}},\text{d}} r^{\text{out}}_{n,c,\text{P}^{\text{out}}_{\text{n}},\text{t}} \leq z_{n,t} \cdot \bar{Q}^{\text{R}+}_{n,d} \quad n \in \mathcal{N}^{\text{T}} \tag{4'}$$

Another example of this type of transformation is gasification of biomass, though this would require removing the crop-subscript c from the output product in (3'):

$$r^{\text{out}}_{n,\text{P}^{\text{out}}_{\text{n}},\text{t}} = \sum_{p\in\mathcal{P},c\in\mathcal{C}} Y^{\text{R}}_{n,p} \cdot r^{\text{in}}_{n,c,p,t} \quad n \in \mathcal{N}^{\text{T}}$$

Note that having one many-to-one transformation node is very similar to having one transformation node for each input product p, with a one-to-one transformation from p to P^{out}_n. The difference is that this would imply separate capacity for each input product, and to represent a shared input capacity additional transformation nodes would have to be introduced.

For both presented transformation node types, it is natural to attach costs to the incoming products. In the former, there is only one product that gets split into several different ones, while in the latter, the output can be produced from several inputs, where each one can have different transformation costs.

2.4 *Storage Nodes*

A storage node models the storage of one or several products between time periods. Each such node is specified by its capacity limits, which can be given along several dimensions (volume, weight). Storage levels for each product are tracked throughout all time periods:

$$\sum_{c\in\mathcal{C}} \sum_{\substack{p\in\mathcal{P}:\\ \exists s_{n,c,p,t}}} U_{c,p,d} s_{n,c,p,t} \leq z_{n,t} \cdot S_{n,d} \quad n \in \mathcal{N}^{\text{S}}, d : \exists S_{n,d} \tag{5}$$

$$s_{n,c,p,t} = \gamma_n \cdot s_{n,c,p,t-1} - \sum_{\substack{n'\in\mathcal{N}:\\ (n,n')\in\mathcal{A}}} f_{n',n,c,p,t} + \sum_{\substack{n'\in\mathcal{N}:\\ (n',n)\in\mathcal{A}}} f_{n',n,c,p,t} \quad n \in \mathcal{N}^{\text{S}}, c, p, t : \exists S_{n,c,p,t}, \tag{6}$$

where $1 - \gamma_n$ represents the mass loss due to storage for one period at node n; γ_n is typically a number very close to one. Note that this model assumes that if the storage can handle several products, then they all share the whole capacity. If each product has its own dedicated part of the storage place, each should be modeled by their own storage node, all placed in the same location.

Initial storage levels can be provided or the model can use cyclic behavior where the storage level in the last period is used as the initial level. The first is modeled as $s_{n,c,p,0} = S_{n,c,p}$ and the second one is modeled as $s_{n,c,p,0} = \gamma_n \cdot S_{n,c,p,T}$. In the second case, the initial storage level is a decision variable whose optimal value becomes an important part of the solution.

2.5 Consumption Nodes

Consumption nodes are nodes consuming some of the products. They are specified by a minimum and maximum demand for each period:

$$z_{n,t} \cdot D^{\min}_{n,t,d} \leq \sum_{c \in \mathscr{C}} \sum_{p \in \mathscr{P}^F} E^{D}_{n,p} U_{c,p,d}\, q^{D}_{n,c,p,t} \leq z_{n,t} \cdot D^{\max}_{n,t,d} \quad n \in \mathscr{N}^{C}, \tag{7}$$

where each inequality is created only if the associated parameter exists. Note that if $D^{\min}_{n,t,d} > 0$, then we can fix $z_{n,t} = 1$. Parameters $E^{D}_{n,p}$ can be used to limit which products can satisfy the demand at each given node.

Furthermore, the customers may have limits on different crop shares in the total amount that they buy. This is modeled similarly to the production limits, i.e., we have a set $\mathscr{K}$ of limits, each specified by its node n^{K}_{k} and a minimum and/or maximum proportion of crop c^{K}_{k} in the mix, denoted by $k^{\min}_{k}$ and $k^{\max}_{k}$. The constraints are then

$$k^{\min}_{k} \sum_{c \in \mathscr{C}} \sum_{p \in \mathscr{P}^F} q^{D}_{n^K_k,c,p,t} \leq \sum_{p \in \mathscr{P}} q^{D}_{n^K_k,c^K_k,p,t} \leq k^{\max}_{k} \sum_{c \in \mathscr{C}} \sum_{p \in \mathscr{P}F} q^{D}_{n^K_k,c,p,t} \quad k \in \mathscr{K}, \tag{8}$$

where each inequality is created only if the corresponding limit $k^{\min}_{k}$ or $k^{\max}_{k}$ exists for given k.

2.6 Transportation and Flows

Transport capacity can be provided for any measurement unit:

$$\sum_{c \in \mathscr{C}} U_{c,p,d} f_{n_1,n_2,c,p,t} \leq \bar{Q}^{A}_{n_1,n_2,p,d} \quad (n_1, n_2) \in \mathscr{A} \tag{9}$$

Moreover, in all nodes except storage nodes, we need conservation-of-flow constraint, that is, the total consumption, outward transportation flows, and transformation inputs must equal production, inward transportation flows, and transformation output, in every time period:

$$q^{\mathrm{P}}_{n,c,p,t} + \sum_{n' \in \mathcal{N}} f_{n',n,c,p,t} + r^{\mathrm{out}}_{n,c,p,t} = q^{\mathrm{D}}_{n,c,p,t} + \sum_{n' \in \mathcal{N}} f_{n,n',c,p,t} + r^{\mathrm{in}}_{n,c,p,t}, \tag{10}$$

for all $n \in \mathcal{N} \setminus \mathcal{N}^{\mathrm{S}}$. In the model implementation, only the required variables will actually be created at the different types of nodes: production nodes will only have production and outward transportation flows, etc.

2.7 *Objective Function*

The exact form of the objective function is case dependent: if there is a given demand that has to be satisfied, then it is natural to minimize the costs of doing so. If, on the other hand, we can freely choose how much to deliver to each customer, then one also has to take into account income and maximize the overall profit instead.

Therefore, instead of stating the complete objective function, we list its components: the income and different types of costs.

$$\text{income from sale} \quad \sum_{n,c,p,t,d} U_{c,p,d} P_{n,t,d} q^{\mathrm{D}}_{n,c,p,t} \tag{11a}$$

$$\begin{array}{l}\text{production/harvesting costs,}\\ \text{using constraint (1)}\end{array} \quad \sum_{n,c,p,t} \left(\frac{C^{\mathrm{P}}_{n,c}}{Y^{\mathrm{P}}_{n,c}} + C^{\mathrm{Q}}_{n,c} \right) q^{\mathrm{P}}_{n,c,p,t} \tag{11b}$$

$$\text{transportation costs} \quad \sum_{n_1,n_2,c,p,t,d} C^{\mathrm{A}}_{n_1,n_2,p,d} U_{c,p,d} f_{n_1,n_2,c,p,t} \tag{11c}$$

$$\text{transformation costs} \quad \sum_{n,c,p,t,d} C^{\mathrm{R-}}_{n,d} U_{c,p,d} r^{\mathrm{in}}_{n,c,p,t} \tag{11d}$$

$$\text{node-usage costs} \quad \sum_{n,t} C^{\mathrm{N}}_{n} z_{n,t} \tag{11e}$$

Note that the node-usage variables $z_{n,t}$ are really needed only for nodes with non-zero costs C^{N}_{n}; without these costs, we could omit the variables, though this would require slight reformulation of some of the constraints.

3 Stochastic Formulation

In most bioenergy applications, the problems are subject to multiple uncertainties, as future demand, supply (yield), and prices are seldom known with precision ahead of time. It is therefore natural to use a model that can handle at least some of the uncertainties. There are several approaches to optimization under uncertainty, of which we have chosen stochastic programming, in particular its formulation with stochastic variables represented by discrete scenarios.

It is relatively easy to convert a deterministic model to a two-stage stochastic model since all we need to add is a scenario index to all the stochastic entities. However, we want a general multi-stage formulation, because this gives us the freedom to change the complexity of the stochastic representation solely using data, i.e., without changing the model.

There are two basic approaches to converting a deterministic model into a general stochastic one, illustrated graphically in Table 1.

In a scenario-based formulation, we simply add a scenario index to all time-dependent model entities and then enforce the scenario-tree structure using so-called non-anticipativity constraints (Birge and Louveaux 1997). While adding the extra index is easy, adding the constraints is not a trivial task. In addition, it will make the model bigger. Even in the very simple scenario tree in Table 1, we have five groups of constraints, one with four nodes and four with two nodes. To enforce the equality, one typically sets all nodes equal to the first one, resulting in $k - 1$

Table 1 Comparison of the two types of stochastic model representations, on a five-period tree with binary branching in periods 2 and 4

Model type	Graphical representation	Example constraint
Deterministic	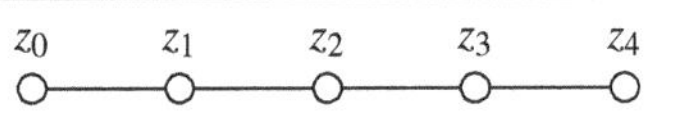	$\sum_{t=0}^{4} z_t = 1$
Stochastic Scenario-based	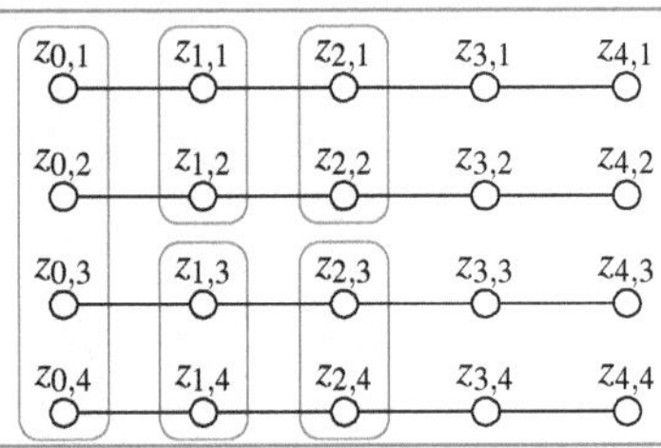	$\sum_{t=0}^{4} z_{t,s} = 1,$ $\forall s \in \{1,\ldots,4\}$
Stochastic Node-based	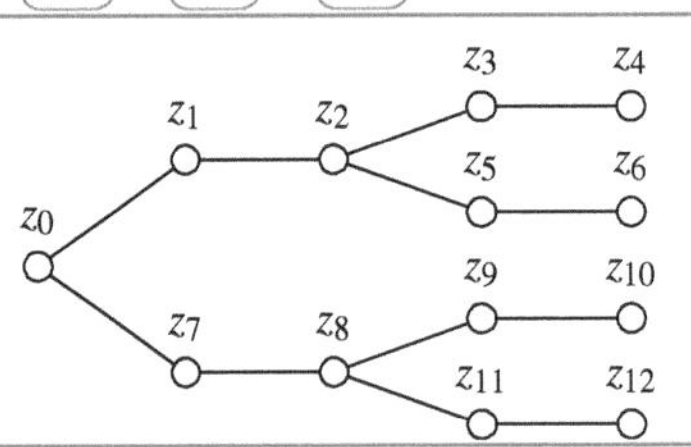	$z_v + z_{\mathrm{Pa}(v)} + z_{\mathrm{Pa}(\mathrm{Pa}(v))} + z_{\mathrm{Pa}(\mathrm{Pa}(\mathrm{Pa}(v)))} + z_{\mathrm{Pa}(\mathrm{Pa}(\mathrm{Pa}(\mathrm{Pa}(v))))} = 1,$ $\forall v \in \{4, 6, 10, 12\}$

The blue boxes in the scenario-based formulation denote groups of nodes that have to be set equal using non-anticipativity constraints

constraints for a group of k nodes. In our case, this means $3 + 4 \times 1 = 7$ constraints for every stochastic variable.

Alternatively, in a node-based approach, we replace the time index by a scenario-tree-node index for all time-dependent model entities. The structure of the scenario tree is then described by providing a parent (or *predecessor*) node Pa(v) to each node $v \in 2\mathscr{V}$: Pa(1) = 0, Pa(2) = 1,..., Pa(11) = 8, Pa(12) = 11. While this avoids extra constraints, it makes the model more difficult to read, especially if we have constraints that cover more than two periods (as we can see in Table 1).

Hence, we use a 'hybrid' approach where we use the node-based approach for all model entities, but provide extra data structures that allow us to use scenario numbers in the model formulation where needed. In particular, we define SP(t, s) to specify the node in period $t \in \mathscr{T}$ of scenario $s \in \mathscr{S}$; taking an example from the last figure of Table 1, the scenario-tree node corresponding to $t = 2$ in scenario $s = 3$ is z_8, so we have $\mathrm{SP}(2,3) = 8$. Using this parameter, we can write the constraint from Table 1 as

$$\sum_{t=0}^{4} z_{\mathrm{SP}(t,s)} = 1, \quad s \in \{1,\ldots,4\}$$

This way, we have eliminated the major disadvantage of the node-based formulation, without the extra constraints needed for the scenario-based approach.

With this notation in place, rewriting the deterministic model from Sect. 2 into a stochastic one becomes relatively straightforward. For all of the variables, we replace the time index t with the new stochastic-node index $v \in \mathscr{V}$. For links between periods, $t - 1$ becomes Pa(v), while links further back in time are easiest written using the SP(t,s) table.

As we have discussed in Sect. 2, one can construct several different objective functions using the elements presented in (11a)–(11s), depending on the optimizing agent. Here, we present a stochastic version of one of the variants, namely an agent maximizing the overall profit in the supply chain. For simplicity, we assume a risk-neutral agent and therefore maximize the expected profit, i.e., the probability-weighted profit over all scenario nodes.

The whole model from Sect. 2, with equation numbers referring to the original deterministic ones, is then as follows:

$$\begin{aligned} \text{maximize} \sum_{v \in \mathscr{V}} \Pr(v) \Bigg(& \sum_{n,c,p,d} U_{c,p,d} P_{n,\mathrm{Per}(,)v} d q^{\mathrm{D}}_{n,c,p,v} - \sum_{n,c,p} \left(\frac{C^{\mathrm{P}}_{n,c}}{Y^{\mathrm{P}}_{n,c}} + C^{\mathrm{Q}}_{n,c} \right) q^{\mathrm{P}}_{n,c,p,v} \\ & - \sum_{n_1,n_2,c,p,d} C^{\mathrm{A}}_{n_1,n_2,p,d} U_{c,p,d} f_{n_1,n_2,c,p,v} \\ & - \sum_{n,c,p,d} C^{\mathrm{R}-}_{n,d} U_{c,p,d} r^{\mathrm{in}}_{n,c,p,v} - \sum C^{\mathrm{N}}_{n} z_{n,v} \Bigg) \end{aligned} \qquad (11s)$$

subject to

$$\sum_{t\in\mathcal{T}} q^{\mathrm{P}}_{n,c,p,\mathrm{SP}(t,s)} \leq \bar{Q}^{\mathrm{P}}_{n,c} Y^{\mathrm{P}}_{n,c}, \quad n \in \mathcal{N}^{\mathrm{P}}, p \in \mathcal{P}^{\mathrm{B}} \tag{1s}$$

$$Q^{\min}_i \leq \sum_{\mathrm{t}^{\mathrm{S}}_i \leq t \leq \mathrm{t}^{\mathrm{E}}_i} U_{c,p,d} q^{\mathrm{P}}_{n_i,c_i,\mathrm{p}^{\mathrm{b}}_{c_i},\mathrm{SP}(t,s)} \leq Q^{\max}_i \quad i \in \mathcal{I} \tag{2s}$$

$$r^{\mathrm{out}}_{n,c,p,v} = Y^{\mathrm{R}}_{n,\mathrm{p}^{\mathrm{in}}_n} \cdot r^{\mathrm{in}}_{n,c,\mathrm{p}^{\mathrm{in}}_n,v} \quad n \in \mathcal{N}^{\mathrm{T}} \tag{3s}$$

$$\sum_{c\in\mathcal{C}} U_{c,\mathrm{p}^{\mathrm{in}}_n,d} r^{\mathrm{in}}_{n,c,\mathrm{p}^{\mathrm{in}}_n,v} \leq z_{n,v} \cdot \bar{Q}^{\mathrm{R}-}_{n,d}, \quad n \in \mathcal{N}^{\mathrm{T}} \tag{4s}$$

$$\sum_{c\in\mathcal{C}} \sum_{\substack{p \in \mathcal{P}: \\ \exists s_{n,c,p,v}}} U_{c,p,d} s_{n,c,p,v} \leq z_{n,v} \cdot S_{n,d} \quad n \in \mathcal{N}^{\mathrm{S}} \tag{5s}$$

$$\begin{aligned} s_{n,c,p,v} = {} & \gamma_n \cdot s_{n,c,p,\mathrm{Pa}(v)} \\ & - \sum_{\substack{n' \in \mathcal{N}: \\ (n,n') \in \mathcal{A}}} f_{n,n',c,p,t} + \sum_{\substack{n' \in \mathcal{N}: \\ (n',n) \in \mathcal{A}}} f_{n',n,c,p,t} \quad n \in \mathcal{N}^{\mathrm{S}} \end{aligned} \tag{6s}$$

$$z_{n,v} \cdot D^{\min}_{n,t,d} \leq \sum_{c\in\mathcal{C}} \sum_{p\in\mathcal{P}^{\mathrm{F}}} E^{\mathrm{D}}_{n,p} U_{n,p,d} q^{\mathrm{D}}_{n,c,p,t} \leq z_{n,v} \cdot D^{\max}_{n,t,d} \quad n \in \mathcal{N}^{\mathrm{C}} \tag{7s}$$

$$k^{\min}_k \sum_{c\in\mathcal{C}} \sum_{p\in\mathcal{P}} q^{\mathrm{D}}_{n^{\mathrm{K}}_k,c,p,v} \leq \sum_{p\in\mathcal{P}} q^{\mathrm{D}}_{n^{\mathrm{K}}_k,c^{\mathrm{K}}_k,p,v} \leq k^{\max}_k \sum_{c\in\mathcal{C}} \sum_{p\in\mathcal{P}^{\mathrm{F}}} q^{\mathrm{D}}_{n^{\mathrm{K}}_k,c,p,v} \quad k \in \mathcal{K} \tag{8s}$$

$$\sum_{c\in\mathcal{C}} U_{c,p,d} f_{n_1,n_2,c,p,v} \leq \bar{Q}^{\mathrm{A}}_{n_1,n_2,p,d} \quad (n_1, n_2) \in \mathcal{A} \tag{9s}$$

$$q^{\mathrm{P}}_{n,c,p,v} + \sum_{n'\in\mathcal{N}} f_{n',n,c,p,v} + r^{\mathrm{out}}_{n,c,p,v} = q^{\mathrm{D}}_{n,c,p,v} + \sum_{n'\in\mathcal{N}} f_{n,n',c,p,v} + r^{\mathrm{in}}_{n,c,p,v} \quad n \in \mathcal{N} \backslash \mathcal{N}^{\mathrm{S}} \tag{10s}$$

Just like in the deterministic model, we have to decide how to deal with the initial storage levels in (6s). If we used the initial-storage constraint ($s_{n,c,p,0} = s_{n,c,p}$) in the deterministic model, then we can use it without change. If, on the other hand, we used the cyclical storage approach ($s_{n,c,p,0} = \gamma_n \cdot s_{n,c,p,T}$.), then we have to decide how to interpret this in the stochastic settings. There are several possible approaches to this, each with merits and problems:

$$s_{n,c,p,0} = \sum_{v\in\mathcal{V}_T} \Pr(v) \cdot s_{n,c,p,v} \tag{12a}$$

$$s_{n,c,p,0} = s_{n,c,p,v} \quad \forall v \in \mathcal{V}_T \tag{12b}$$

$$s_{n,c,p,0} \leq s_{n,c,p,v} \quad \forall v \in \mathcal{V}_T, \tag{12c}$$

where $v = 0$ is the root node of the scenario tree and $\mathcal{V}_T$ is the set of last-period nodes, $\mathcal{V}_T = \{v \in \mathcal{V} : \text{Per}(v) = T\}$.

The first formulation, (12a), ensures that *on average*, we will finish with the same amount in storage at the end as at the start. This is probably the most natural extension of the deterministic version. It can, however, give undesired effects; for example, assume that we include demand as the only stochastic parameter in the model and that the optimal storage level in the first period is $s_0^* > 0$. Constraint (12a) will then cause the end-of-horizon storage level to be higher than s_0^* in low-demand scenarios and lower—possibly even zero—in scenario(s) with the highest demand. The end-of-horizon storage level could even be zero in the 'average' scenario, which would imply that the results found by the model would not provide proper guidance for two successive average years (since the average year needs positive initial storage levels, but finishes with empty storage).

This problem is avoided in version (12b), where the final storage has to be equal in all scenarios. Unfortunately, this creates new problems. If we, for example, have one scenario with very low demand, it may be optimal to end up with a high level of storage at the end. With (12b), however, the final storage at the end of the low-demand scenario has to be the same as in high-demand scenarios. The optimal solution might then be to produce lower amounts of biomass, which may imply a huge loss of profit potential. The last issue is resolved in (12c), where we allow the low-demand scenarios to end up with more storage. Another way of resolving the issues in (12b) is to add extra time at the end of the horizon, to allow the models to settle on common storage levels (Thapalia et al. 2009).

In contrast to (12a) that ensures that *on average*, we will finish with enough storage to continue another *average* year, (12c) ensures that this happens in every scenario. In this sense, the latter represents a more conservative risk attitude. For this reason, this is the formulation we use in our implementation.

4 Extensions to the Model

In this section, we present several extensions to the model. We formulate them for the stochastic version of the model, but note that they can be used for the deterministic version as well. Each of these adds specific functionality to the model. They can be combined, depending on what is needed for each case.

4.1 Terminals

By *terminal*, we mean an area with multiple facilities (e.g., transformation and storage). In the context of the model, a *terminal* is a grouping of one or more nodes. There are at least two uses for terminals in the model: they allow us to model costs associated with running the terminal in addition to costs for each facility/node. For this, we just need to associate a binary variable y_g to each terminal $g \in \mathcal{G}$ and then require that the nodes belonging to the terminal, $n \in \mathcal{N}_g^G$, can be used only if y_g is equal to one:

$$z_{n,v} \leq y_g \quad n \in \mathcal{N}_g^G, v, g : \exists z_{n,v} \text{ and } \exists y_g.$$

More importantly, in some applications the goal of the model is to establish a new terminal, by choosing one from a list of candidate locations. We model this in a more general way, using a set $\mathcal{J}$ of 'alternatives'. For each alternative $j \in \mathcal{J}$, we then specify a group of terminals, $\mathcal{G}_j$, with associated lower and upper bounds on the number of terminals to be opened, $\underline{G}_j$ and $\bar{G}_j$:

$$\underline{G}_j \leq \sum_{g \in \mathcal{G}_j} y_g \leq \bar{G}_j \quad j \in \mathcal{J}$$

If we are to choose exactly one from all of the potential terminals, we do this using $\mathcal{J} = \{1\}, \mathcal{G}_1 = \mathcal{G}$, and $\underline{G}_1 = \bar{G}1 = 1$. This formulation assumes that the decision regarding terminals is made at the start of the first period and is valid for the entire duration of the model. An alternative would be to add a time index to these decisions, so that they can be postponed. Note that this would increase the number of binary variables, and hence, the solution time.

In addition to the above constraints, we would also need to add the costs of opening the terminals, $\sum_g C_g^{\mathrm{G}} y_g$, to the objective function. Note that we could easily add terminal-usage costs (per period) as well, though it would require extra binary variables.

4.2 Drying

In forestry applications, it is important to model drying of the wood. There are at least two approaches to modeling the process in an optimization model, where the difference is whether we treat the moisture content continuously or discretize it. In the former case, we attach a variable for moisture content to each product that needs drying and then track how this value decreases over time. The difficulty with this approach is that one needs to express the energy content of the product as a function of the moisture content—and since this relationship is non-linear, this would mean employing some kind of piecewise-linear approximation (Van Dyken et al. 2010).

In the latter approach, we expand the set of products to contain additional information about moisture content. For example, we replace the product type ‘log’ by ‘log 20 %,’ ‘log 40 %,’ ‘log 60 %,’ and ‘log 80 %,’ each with the appropriate energy content. With this approach, drying is modeled as a change in the product type, after given time in storage.

In our implementations, we have used the latter approach, where the transformation does not depend only on the time spent in the storage, but also on the actual period—the model was developed for a customer in Norway, where wood dries only during the summer.

In the model, the possibility that a product p of crop c can transition into another product in storage node n is signaled by existence of $r = \mathrm{R}_{n,c,p}$. The transition is then specified by its output product $\mathrm{p}_r^{\mathrm{out}}$, the last period in which the product must arrive to the storage in order to undergo the transition $\mathrm{t}_r^{\mathrm{in}}$, the first time period when the transition is finished and the output product can be taken out of the storage $\mathrm{t}_r^{\mathrm{out}}$, and finally, the mass loss during the transition κ_r. In our particular case, we say that all fresh wood that is in storage by the end of April is dry at the start of September, so we have $\mathrm{t}_r^{\mathrm{in}} = 3$ and $\mathrm{t}_r^{\mathrm{out}} = 9$.

With this notation, (6) is modified in the following way:

$$\begin{aligned} s_{n,c,p,v} = \gamma_n \cdot s_{n,c,p,\mathrm{Pa}(v)} &- \sum_{\substack{n' \in \mathcal{N}: \\ (n,n') \in \mathcal{A}}} f_{n,n',c,p,v} + \sum_{\substack{n' \in \mathcal{N}: \\ (n',n) \in \mathcal{A}}} f_{n',n,c,p,v} \\ &- \gamma_n^{\mathrm{t}_r^{\mathrm{out}} - \mathrm{t}_r^{\mathrm{in}}} \cdot s_{n,c,p,\mathrm{t}_r^{\mathrm{in}}} \quad \text{if exists } r = \mathrm{R}_{n,c,p} \text{ and } \mathrm{Per}(v) = \mathrm{t}^{\mathrm{out}}_{\mathrm{R}_{n,c,p}} \\ &+ \sum_{\substack{p' \in \mathcal{P}: \\ \exists r = \mathrm{R}_{n,c,p'} \\ p = \mathrm{p}_{\mathrm{r}}^{\mathrm{out}},\, \mathrm{Per}(\mathrm{v}) = \mathrm{t}_{\mathrm{r}}^{\mathrm{out}}}} \gamma_n^{\mathrm{t}_r^{\mathrm{out}} - \mathrm{t}_r^{\mathrm{in}}} \cdot (1 - \kappa_r) \cdot s_{n,c,p',\mathrm{t}_r^{\mathrm{in}}} \end{aligned} \tag{6'}$$

With the values stated above, the second line ensures that in September, the fresh wood will be removed from storage, while the third line replaces it with a corresponding amount of dry wood (minus the losses κ_r). Note that this assumes that no fresh trees are removed from storage during the drying period.[2]

An alternative, and simpler, way of modeling the drying process would be to assume a constant drying rate throughout the year, that is, to assume that crop c stored at node n needs $\Delta t_{c,n}^{\mathrm{d}}$ time periods to transition from product p^{w} to product p^{d}. If we assume that only fresh crops enter storage and only dry crops leave storage, then the storage Eq. (6) would become

[2]If this was a problem, one could add additional product types for ‘drying wood' that would be forbidden to be taken out of the storage.

$$
\begin{aligned}
s_{n,c,p^{\mathrm{w}},v} &= \gamma_n \cdot s_{n,c,p^{\mathrm{w}},\mathrm{Pa}(v)} + \sum_{\substack{n' \in \mathcal{N}: \\ (n',n) \in \mathcal{A}}} f_{n',n,c,p^{\mathrm{w}},v} - \sum_{\substack{n' \in \mathcal{N}: \\ (n',n) \in \mathcal{A}}} f_{n',n,c,pw,\mathrm{Per}(v)-\Delta t^{\mathrm{d}}_{c,n}} \\
s_{n,c,p^{\mathrm{d}},v} &= \gamma_n \cdot s_{n,c,p^{\mathrm{d}},\mathrm{Pa}(v)} - \sum_{\substack{n' \in \mathcal{N}: \\ (n,n') \in \mathcal{A}}} f_{n,n',c,p^{\mathrm{d}},v} + \sum_{\substack{n' \in \mathcal{N}: \\ (n',n) \in \mathcal{A}}} f_{n',n,c,pw,\mathrm{Per}(v)-\Delta t^{\mathrm{d}}_{c,n}}
\end{aligned}
\tag{6''}
$$

Note that this approach can be used to model *active drying*, we just need to set the storage-using costs accordingly and perhaps introduce a volume-dependent cost as well.

4.3 Tracking Equipment

The transformation Eq. (4) assumes that the capacity is given and constant for each transformation node. This, however, ignores the fact that some transformations need extra equipment—which we refer to as a 'transformation device'—to be present in the transformation node. An example of such a requirement is a mobile chipper needed for off-terminal chipping of wood. Since we normally have only a limited amount of these devices, we need to ensure that each is used only in one place at a time.

There are several ways of handling this in the model. If the devices are so mobile that they can be used in several nodes during one period, we can simply add constraints that limit the overall transformation capacity (summed over all transformation nodes) to the capacity of the available devices. Note that this means that we do not control the amount of nodes a device is used at during one period and it also ignores relocation costs. On the other hand, the approach does not add any new variables, and therefore, should not increase the solution time of the model. This approach is used, for example, in Gunnarsson et al. (2004).

If, on the other hand, the transformation devices are costly and/or time-consuming to move, we would need a more detailed representation in our model. There, each transformation device $h \in \mathcal{H}$ is specified by its output product $P^{\mathrm{H}}_h \in \mathcal{P}$, set of nodes where it may be needed $\mathcal{N}^{\mathrm{H}}_h$, and transport costs $C^{\mathrm{H}}_{h,n_1,n_2}$ between nodes n_1 and n_2. We then introduce binary variables $w^h_{n,v}$ denoting whether a transformation device h is present in node n in scenario-tree node v, plus additional variables for tracking their movements where: $w^h_{n,n',v}$ is equal to one if the transformation device h is moved from node n to node n' at the end of period of the scenario-tree node v.

$$w^h_{n,v} = w^h_{n,\mathrm{Pa}(v)} - \sum_{n' \in \mathcal{N}^{\mathrm{H}}_h \setminus \{n\}} w^h_{n,n',\mathrm{Pa}(v)} + \sum_{n' \in \mathcal{N}^{\mathrm{H}}_h \setminus \{n\}} w^h_{n',n,\mathrm{Pa}(v)} \quad h \in \mathcal{H}, n \in \mathcal{N}^{\mathrm{H}}_h \tag{13}$$

$$\sum_{n \in \mathcal{N}^{\mathrm{H}}_h} w^h_{n,1} = 1 \quad h \in \mathcal{H} \tag{14}$$

$$z_{n,v} \leq \sum_{h \in \mathcal{H} : n \in \mathcal{N}^{\mathrm{H}}_h} w^h_{n,v} \quad n \in \mathcal{N}^{\mathrm{T}} \tag{15}$$

Note that it is enough to have (14) only for the first period, since the flow-conservation constraints (13) guarantee that they will hold in all periods. Finally, constraints (15) ensure that transformation nodes cannot be used without the required device in place.

Furthermore, the movement-tracking variables $w^h_{n,n',v}$ do not need to be declared as binary in the model, since they will be automatically integer because of constraints (13). Nevertheless, these variables might increase the solution time significantly, so they should be included only if relocation costs of the devices are high enough compared to other costs. If not, we can simply remove these variables and constraints (13) from the model.

5 Illustrative Examples

In this section we present two different cases. In one case, we used a deterministic version of the model, while in the other we included uncertainty. The first case is based on the harvesting of trees, while the raw material in the second one is cultivated. The examples are meant to illustrate the flexibility of the model and show benefits from using an optimization based decision support tool. For this reason, to emphasize the illustrative insights, only three future scenarios are considered in the case with uncertainty.

Both examples have been implemented in the Mosel modelling language and solved using FICO™ Xpress Optimizer, on an dual-core 2.4 GHz machine with 8 GB of RAM.

5.1 Wood-Based Bioenergy Supply Chain

In this example we consider a supply chain where wood chips are used in heating plants in western Norway. Wood is available in large quantities in the area, but difficult to harvest due to the steep terrain, which translates to high production costs. The wood chips can be produced from different species, each with different energy contents: pine, spruce, and hardwood.

Production of wood chips can be performed with a mobile wood chipper at the felling areas or at the terminals. The difference between these two alternatives is that transportation of wood chips is cheaper than transportation of logs, but the operational costs of a mobile chipper are larger than those of a stationary chipper at a terminal.

The wood can be used directly (as fresh chips) in some heating plants, but, more commonly, drying is needed before use. Drying can take place at felling areas or at the terminal, and before or after chipping, which means that there are a lot of choices to be made in the upstream part of the supply chain. Trees cut in winter/spring can be used the fall, while trees cut in summer/fall cannot be used until the next fall. This means that timing of felling and storage capacity are important factors to consider in order to meet demand (Fig. 2).

Terminals are usually a large area where logs and chips can be dried and stored for protection against rain and snow. Storage and drying at the felling sites is more uncertain and the losses are higher due to less controlled conditions. The heating plants have limited capacity for storage of chips and their demand for chips is

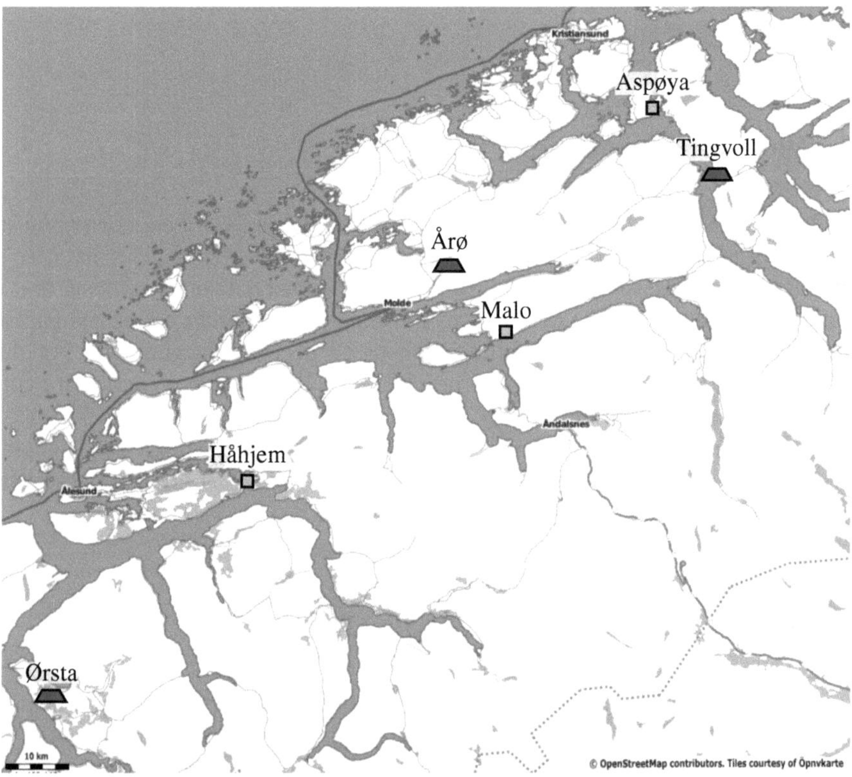

Fig. 2 The area used in the test case in Sect. 5.1. The heat plants are denoted by '⏢', while '◘' denotes the possible locations of terminals

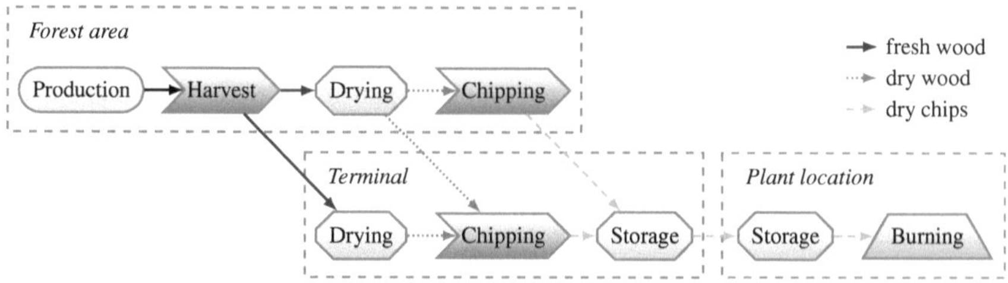

Fig. 3 Structure of the forest supply chain in Sect. 5.1. Note that the case includes three heat plants and three possible terminal locations, each with three forest areas attached. The forest areas are connected only to the nearest terminal, while each terminal can supply all three heat plants

largely correlated with weather conditions, which results in seasonal variation. Some of the plants are even closed during summer. An illustration of the supply chain in this example is given in Fig. 3.

In the example case from the Møre region in Norway, we have the perspective of the society of forest owners. A map of locations in the example case can be seen in Fig. 2. Note that ferries are needed for transportation between Håhjem and Ørsta, Håhjem and Årø, and Aspøya and Årø, but Malo and Årø are connected by tunnel.

We want to decide timing and area for felling, when and where to chip, and where to deliver. We also want to make decisions regarding if and where a terminal should be opened. In the example, we have three existing heating plants and three candidate locations for a terminal. Another important input parameter is the available wood from different species; these data are generated by using a GIS tool based on data from the Norwegian government. The costs incurred are production costs for felling, transportation costs, chipping costs, and costs related to storage. Sales amounts to heating plants are measured in energy output of delivered chips (not volume). We do not include rental costs for the terminal areas. In negotiations, profit can be used as a guideline for the forest owners to decide how much they are willing to pay in rent for a terminal. The objective function is to maximize total profit.

We want to decide timing and area for felling, when and where to chip, and where to deliver. We also want to make a decision on if and where a terminal should be opened. In the example we have three existing heating plants and three candidate locations for a terminal. Another important input parameter is available wood of different species, this data is generated by the use of a GIS tool based on data from the Norwegian government. The costs incurred are production costs for felling, transportation costs, chipping costs, costs related to terminal and costs related to storage. Sales amounts to heating plants are measured in energy output of delivered chips (not volume). The objective function is to maximize total profit.

In this example case, we use all of the extensions presented in Sect. 4: we have multiple terminals to choose from (4.1), the wood or chips need drying before they can be sent to the heat plants (4.2), and we have one mobile chipper whose location and movement have to be tracked (4.3). For initial storage levels, we use the 'steady-state' formulation with initial levels equal to the end-of-horizon levels, minus some losses.

In addition, we added extra constraints to prohibit harvesting, drying and chipping around a closed terminal. In the core model, the storage constraints (6) allow flow of products through a storage node, even if the node is closed. In our particular case, this was not realistic, so we added the following constraints for the storage nodes at terminals:

$$\sum_{\substack{n' \in \mathcal{N}: \\ (n',n) \in \mathcal{A}}} f_{n',n,c,p,t} \leq \mathrm{M}_{n,c,p,t} \cdot z_{n,t},$$

where the constants $\mathrm{M}_{n,c,p,t}$ are upper bounds on the left-hand side sums, ensuring that the constraints are inactive when $z_{n,t} = 1$. Note that this type of 'big-M constraints' is known to lead to bad LP relaxation, especially with large values of the constants. It is therefore advisable to find as tight upper bounds of the left-hand sides as possible. Note that in our case, it is actually possible to avoid the 'big-M formulation' of these constraints altogether, though at the cost of more changes to the model.

The time perspective of the example case is one year, with each time period being a month in length. The heat plants' demand varies throughout the year. We do not require the demand to be fully satisfied, as there are other sources from which the heat plants can buy their fuel.

5.1.1 Numerical Results

The example case has been tested imposing how many of the three potential terminal locations should be opened. The differences between opening one, two, and three terminals were then been analyzed and compared. Solution times for the three cases were, respectively, 7, 8, and 31 s. Most of the complexity comes from the binary variables used for tracking the mobile chippers—without them, the model solves much quicker.

In the case where we allow only one terminal to be opened, the model uses the one at Aspøya. Looking at the map in Fig. 2, this may be surprising, as one could expect the middle terminal (Malo) to be preferred. This, however, can be explained by the topology of the Norwegian west coast. Because of the fjords, the shortest way from Malo to Tingvoll passes by both Årø and Aspøya, and to Ørsta by Årø and Håhjem.

In all the cases, the terminals are used only for storing chips. In other words, all drying and chipping is done in the forest area adjacent to the terminals. This is due to the fact that the disadvantages (higher costs and losses) of forest storage and chipping are more than compensated by lower transportation costs of chips versus whole trees—a finding that is in concordance with Kanzian et al. (2009).

The numerical results are summarized in Table 2. There, we can see that having two terminals increases the expected profit by 274 thousand NOK, due to decreases in transportation costs. This value should then be compared to the cost of opening

Table 2 Summary of results of the case from Sect. 5.1. All financial values are in thousand NOK

Number of terminals	1	2	3
Revenues	11,020	11,020	11,211
Transp cost	1,572	1,296	1,271
Profit	4,276	4,550	4,642
Total consumption [MWh]	40	40	41
Avg. demand satisfaction	80.6 %	80.6 %	93.2 %
Cost increase for 100 % sat.	0.77 %	0.19 %	0.06 %
Value of extra terminal	–	274	92

the extra terminal. For three terminals, the advantage of the extra terminal decreases to 92 thousand NOK.

While the optimal strategy with one and two open terminals is to satisfy only 80.6 % of the energy demand at the three heat plants (on average), the extra costs of satisfying all off the demand are very low. It can therefore be expected that, in practice, one would aim to satisfy all demand. This would decrease the plant operators' incentives to look for alternative fuel sources giving competitors access to the region. For this reason, we assume 100 % demand satisfaction in the following figures.

Figure 4 presents the storage levels of the final product (chips) at the terminal and heat plant storages in the case with full demand satisfaction. The graph for the results with one open terminal is on the left, two in the middle, and three to the right. We can see that there is a big difference between one and two open terminals. Having the extra terminal allows for more storage to build up, which then leads to a decrease in transportation costs, and therefore higher profit. Common to all of the three cases is the buildup of storage during the fall, caused by the fact that the wood has to first dry during the summer.

Finally, Fig. 5 presents the flow of chips through storage at the heat plants, for the case of three open terminals and full demand satisfaction. In addition, this figure shows chipping activity at and around the terminals.

We can see that even with all three terminals open, only the heat plant at Tingvoll gets its supply solely from the closest terminal; the Ørsta plant uses two terminals and the Årø plant, being in the middle, gets supplied from all three. This

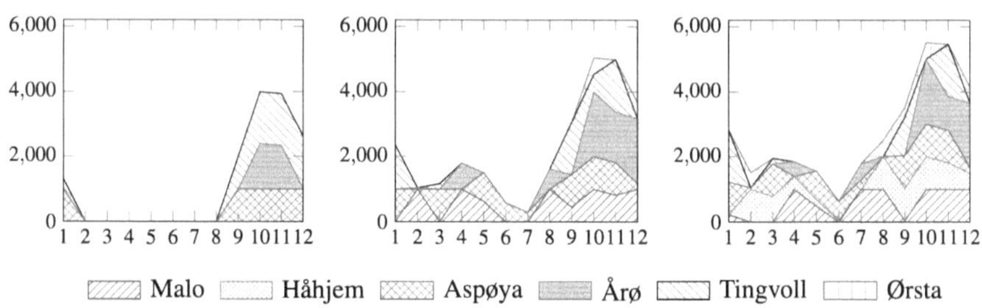

Fig. 4 Storage levels of chips at the heat plants and terminals

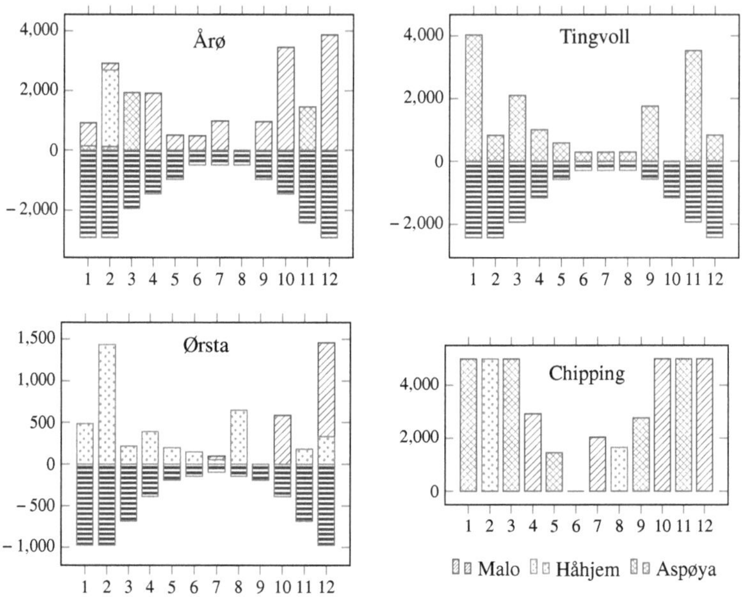

Fig. 5 Flow of chips through the storage at the heat plants, plus the chip production, per month. The *negative values* in the first three charts denote the flow of chips to the plant (i.e. consumption)

is caused by the fact that we only have one mobile chipper, combined with limited storage volume at both terminals and heat plants.

5.2 *Miscanthus Transformation Plant*

Miscanthus is a perennial grass that is increasingly being used for bioenergy purposes. We consider a miscanthus transformation plant located in the Burgundy region of France. The plant previously produced pellets from sugar beets; it now produces pellets from perennial grasses and wood.

Miscanthus is harvested in March and April with three possible options: harvesting and chipping to small chips (2–3 cm) transported directly to the plant, or harvesting and baling with 8–10 cm strands or with 20–30 cm strands. The small chips are compressed, packed, and sold in bags for use in gardening (mulching). The bales can be stored locally at the farmer's location and transported to the plant as needed. All bales should have been picked up by the end of July to make room for autumn crops. The bales with short strands, also called *baled chips*, can be used for pellet production and animal bedding, as well as for energy purposes. Bales with long strands are only used for pellet production. Due to the smaller strands,

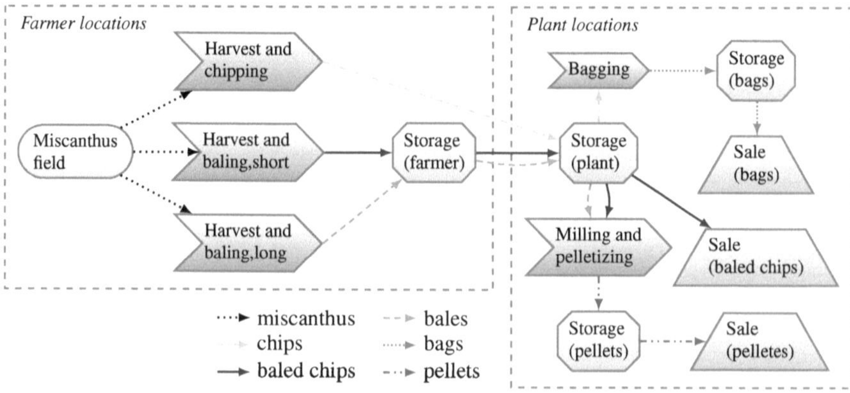

Fig. 6 Structure of the flow of miscanthus

producing baled chips will incur greater losses both during baling and later handling, but provides flexibility due to multiple sales options. Each harvesting option is modeled as a separate transformation node, see Fig. 6. We do not consider drying explicitly in the model as miscanthus is only harvested when moisture levels are acceptably low.

We consider a case based on current fields planted with miscanthus using estimated yields for mature crops (Rizzo et al. 2014) and increased storage capacity compared to the existing state. To illustrate the value of a stochastic model, we consider a situation where the demand of pellets is uncertain. This uncertainty is revealed after harvesting decisions have been made, with demand at the expected level, or 20 % higher or lower than that level, with probabilities 0.4, 0.3, and 0.3, respectively. This is an example of a two-stage stochastic programming problem, where the first-stage decisions (how to harvest) are based only on the data available at that time. Table 3 gives a summary of the test case characteristics; note that pellets sell at a slightly higher price than bagged and baled chips. We use a profit-maximizing objective similar to Eq. (11a)–(11s); we have twelve periods, corresponding to a monthly granularity and we use a cyclic storage behavior.

We start by comparing solutions of the stochastic and deterministic versions of the model, where the latter uses expected values for the stochastic parameter, as presented in the left chart of Fig. 7. There, we can see that the results are quite intuitive: the deterministic model produces more bales (long strands) due to smaller losses compared to baled chips (short strands). The stochastic model, on the other hand, produces more baled chips, because these can be used to produce pellets (albeit more expensive than using bales with long strands), and thus compensates for the uncertainty of the pellets' demand.

We proceed by comparing how the two solutions fare in the stochastic environment. To do this, we solve the stochastic model with the first-stage decisions (production variables) fixed to the solution of the deterministic model, and compare

Table 3 Test case characteristics

Production:	Producers	60 farmer locations
	Yield	10–18 t^{dm}/ha
	Total area	320 ha
Harvesting:	Chipping	Cost €6.7 /t^{dm}, loss 5 %
	Baling, long	Cost €29.2 /t^{dm}, loss 5 %
	Baling, short	Cost €29.2 /t^{dm}, loss 10 %
Transport:	Distances	Road distances based on OpenStreetMap data
	Cost	€0.4–1.2 /t^{dm} km
Storage:	Capacity	Plant 6720 m^3
	Cost	No storage cost or loss is considered
Sales:	Chips	Max 100 t^{dm} /month, price €75 /t^{dm}
	Baled chips	Max 200 t^{dm} /month, price €75 /t^{dm}
	Pellets	Max 300 t^{dm} /month, price €85 /t^{dm}

the results to the optimal solution of the stochastic model.[3] The results of this comparison are in the right chart of Fig. 7. We can see that the deterministic solution fares better in the mean-value and high-demand scenario, but suffers in the low-demand scenario, so it is worse on average. Again, this is due to its higher use of the cheaper, yet inflexible, bales. This can be seen explicitly in Fig. 8, which shows negligible differences in the sold amounts in the first two scenarios, but higher sales of baled chips for the stochastic solution in the last, low-demand, scenario. Again, this shows that the stochastic solution covers part of the pellets' demand in the first two scenarios using baled chips (causing smaller profits in these two scenarios), which gives it the opportunity to sell more when the demand for pellets is low. In other words, the baled chips are used as a buffer against uncertainty in pellets demand.

Finally, we compute the *value of stochastic solution* (VSS, see Birge and Louveaux 1997), given as the difference between the expected profit of using the optimal stochastic and deterministic solution. This is the difference between the last two columns shown in the right chart of Fig. 7:

$$\mathrm{VSS} = €\,116\,902 - €\,114\,666 = €\,2236\,. \tag{16}$$

This means that in this case, the stochastic solution adds only about 2 % to the deterministic one. This is mainly because we have only one stochastic parameter in the model (the demand of pellets). If we let more of the parameters (other demands and prices) be stochastic, the stochasticity would have a higher impact and the value

[3]In our case, the model becomes infeasible because in the low-demand scenarios, we are left with more unsold products than we have storage for. For this reason, we have added additional variables that allow 'throwing away' products (with neither cost nor income). Obviously, these variables are all zero in the optimal stochastic solution.

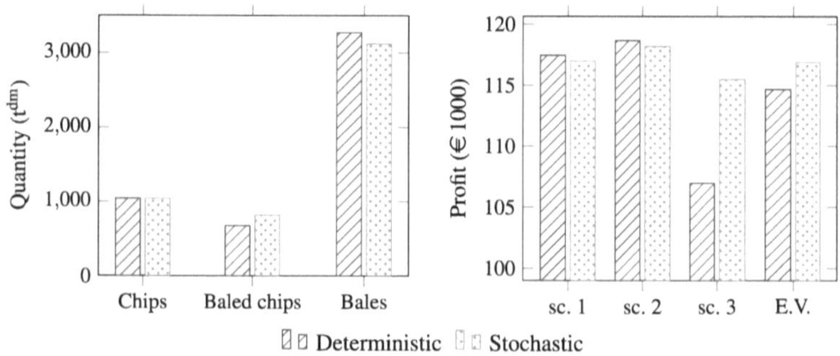

Fig. 7 Results of the miscanthus case using the deterministic and stochastic solutions are presented. The *left chart* displays the transformed quantities of harvested miscanthus. The *right chart* displays the expected profit of the two solutions when evaluated on the scenario tree

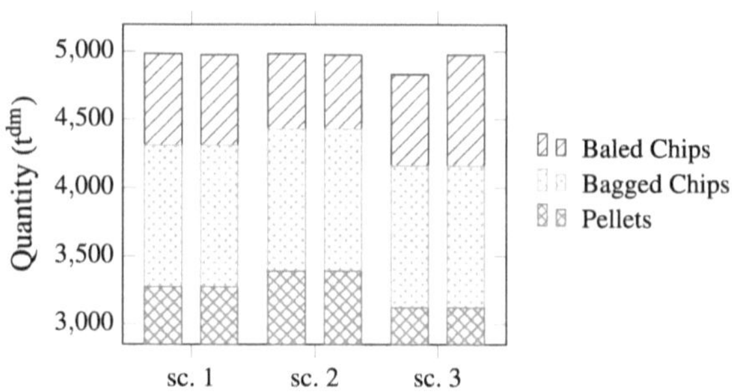

Fig. 8 Sold amounts are presented per scenario. The *left column* in each set represents the deterministic solution; the *right column* represents the stochastic solution

of stochastic solution would increase. On the other hand, it would make interpretation of the results significantly more complicated.

Let us go back to the second value in (16), that is, the expected value of using the expected-value solution (EEV) and compare it to the objective value of the deterministic model, which is €117467. This shows that if we ignore the uncertainty and solve the deterministic model, the reported profit is an over-estimation of the actual profit in the uncertain world (this is a known, and general, observation). It is also interesting to note that the expected profit of the stochastic solution is very close to this figure, showing that its flexibility can almost compensate for the uncertainty.

5.3 Adding More Scenarios

So far, we have only tested the stochastic formulations with three scenarios. While this allowed us to examine the difference between scenario solutions, one needs

Table 4 Results of the case from Sect. 5.2 for different number of scenarios

#sc.	obj. value	VSS	#variables	sol. time
1	117.5	–	6437	0.3
3	116.9	2.2	12773	0.4
10	116.4	2.0	34949	0.8
30	116.1	1.8	98309	2.2
100	116.0	1.8	320069	7.7
300	116.0	1.8	953669	28.0
1000	115.9	1.8	3171269	206.7

Objective values and the VSS are in thousand Euro, while the solution time is in seconds and includes also the time to build the model

more scenarios to obtain reliable results. To test how many, we have solved the same problem with varying number of scenarios. To keep the model consistent with the three-scenario case, we have used normal distribution with the mean and variance computed from the three scenarios.

The results of the test are presented in Table 4. There, we can see that both the objective value and VSS stabilize at about hundred scenarios. Furthermore, we can see that the case with one thousand scenarios takes three and a half minutes to solve; the time is equally split between building the model and solving it.

6 Concluding Remarks

We present a new, generic optimization model for strategic and tactical planning of the biomass to bioenergy supply chain under uncertainty. The model structure is flexible and capable of representing relevant characteristics and issues related to the biomass-bioenergy supply chain, including technological process details, capacity limitations in multiple units of measurement, time variability in supply and demand, and uncertainty in virtually all aspects. Two cases of different supply chains illustrate how the model can be parameterized for different types of analysis, and give insight in the effects of uncertainty on optimal decisions.

The model presented can be used by actors in all parts of the supply chain considered. It can improve the decision making processes by giving results and enabling analysis of different possibilities much faster than traditional planning. Integrated in the companies' software this can be a powerful tool for planners and decision makers in an industry with high competition and tight margins.

The flexibility of the model opens for easy expansion and improvement of the model. More details about different stages in the model or tailor-made setups for different companies are examples of likely requests from the industry that can be included. Further development can make the model even more powerful and the value of using the model can be increased, making the bioenergy industry more competitive.

Acknowledgments This work was partly funded under the EU seventh Framework Programme by the LogistEC project No. 550 311858: Logistics for Energy Crops' Biomass. The views expressed in this work are the sole responsibility of the authors and do not necessary reflect the views of the European Commission. This work was partly funded by Regionalt forskningsfond Midt-Norge through the project 'Fra skog til energi' (ES 217558).

We are grateful to Philippe Béjot (Bourgogne Pellets Cooperative) who kindly provided details on the miscanthus case.

Maps and distance matrices were created using data from OpenStreetMap, © OpenStreetMap contributors.

References

Birge J, Louveaux F (1997) Introduction to stochastic programming. Springer, New York

Cundiff JS, Dias N, Sherali HD (1997) A linear programming approach for designing a herbaceous biomass delivery system. Bioresour Technol 59(1):47–55

De Meyer A, Dirk C, Rasinmäki J, Van Orshoven J (2014) Methods to optimise the design and management of biomass-for-bioenergy supply chains: a review. Renewable and sustainable energy reviews, 31:657–670. ISSN:1364-0321. doi:10.1016/j.rser.2013.12.036

van Dyken S, Bakken BH, Skjelbred HI (2010) Linear mixed-integer models for biomass supply chains with transport, storage and processing. Energy 35(3):1338–1350. doi:10.1016/j.energy.2009.11.017

ecoprog GmbH (2013) Biomass to energy 2013/2014—The world market for biomass power plants. Technical report. http://www.ecoprog.com/en/publications/energy-industry/biomass-to-energy.htm

Ekşioğlu SD, Acharya A, Leightley LE, Arora S (2009) Analyzing the design and management of biomass-to-biorefinery supply chain. Comput Ind Eng 57(4):1342–1352

Gunnarsson H, Rönnqvist M, Lundgren JT (2004) Supply chain modelling of forest fuel. Eur J Oper Res 158(1):103–123. ISSN:0377-2217. doi:http://dx.doi.org/10.1016/S0377-2217(03)00354-0

Kanzian C, Holzleitner F, Stampfer K, Ashton S (2009) Regional energy wood logistics—optimizing local fuel supply. Silva Fennica 43(1):113–128. http://www.metla.fi/silvafennica/full/sf43/sf431113.pdf

Mafakheri F, Nasiri F (2013) Modeling of biomass-to-energy supply chain operations: applications, challenges and research directions. Energy Policy 67:116–126

Marufuzzaman M, Eksioglu SD, Huang YE (2014a) Two-stage stochastic programming supply chain model for biodiesel production via wastewater treatment. Comput Oper Res 49(0):1–17. ISSN:0305-0548. doi:http://dx.doi.org/10.1016/j.cor.2014.03.010. http://www.sciencedirect.com/science/article/pii/S0305054814000653

Marufuzzaman M, Li X, Eksioglu SD, Wang J (2014b) Designing a reliable intermodal hub and spoke system for biofuel supply chain network. In: Transportation research board 93rd annual meeting compendium of papers

Rizzo D, Martin L, Wohlfahrt J (2014) Miscanthus spatial location as seen by farmers: a machine learning approach to model real criteria. Biomass Bioenergy 66:348–363. doi:10.1016/j.biombioe.2014.02.035

Shabani N, Akhtari S, Sowlati T (2013) Value chain optimization of forest biomass for bioenergy production: a review. Renew Sustain Energy Rev 23:299–311. doi:10.1016/j.rser.2013.03.005

Sharma B, Ingalls R, Jones C, Khanchi A (2013) Biomass supply chain design and analysis: basis, overview, modeling, challenges, and future. Renew Sustain Energy Rev 24:608–627

Thapalia BK, Wallace SW, Kaut M (2009) Using inventory to handle risks in the supply of oil to Nepal. Int J Bus Perform Supply Chain Model 1(1):41–60. ISSN:1758-941X (online), 1758-9401 (print). doi:10.1504/IJBPSCM.2009.026265

van Tilburg X, Egging R, Londo H (2006) Biotrans functional and technical description. ECN Policy Studies ECN-RX–06-013, Energy research Centre of the Netherlands. http://www.ecn.nl/publications/ECN-RX–06-013

U.S. Energy Information Agency (2013) International energy outlook 2013. Technical Report DOE/EIA-0484(2013). http://www.eia.gov/forecasts/ieo/

Walther G, Schatka A, Spengler TS (2012) Design of regional production networks for second generation synthetic bio-fuel–a case study in northern germany. Eur J Oper Res 218(1):280–292

Wolfsmayr UJ, Rauch P (2014) The primary forest fuel supply chain: a literature review. Biomass Bioenergy 60:203–221

Yue D, You F, Snyder SW (2014) Biomass-to-bioenergy and biofuel supply chain optimization: Overview, key issues and challenges. Comput Chem Eng (In press). doi:10.1016/j.compchemeng.2013.11.016

Optimal Allocation of Lignocellulosic Biomass Feedstocks for Biofuel Production: A Case Study of California

Chien-Wei Chen and Yongxi Huang

Abstract The economic potentials of producing cellulosic ethanol from biowaste resources as an alternative to corn ethanol is explored in this paper. A portfolio of eight types of biowastes, including crop residues, municipal wastes, and forest residues, is considered and a multilayer biofuel supply chain system is developed based on a systems optimization technique. A case study of converting lignocellulosic biomass to biofuel in California is presented. The biowaste resources can produce up to 900 million gallons of ethanol per year. Through smart modeling of the biofuel supply chain in an integrative manner, a low delivered ethanol cost can be achieved at $1.85 for the use of near-term (2015) conversion technology and $1-1.1 per gallon for the use of mid-term (2015–2025) conversion technology.

Keywords Bioethanol · Lignocellulosic biomass · Energy supply chain · Deterministic model · Sustainability

1 Introduction

In an effort to address climate change, biofuel as one kind of renewable energy has been highly promoted from federal to state policies. For instance, the federal Energy Independence and Security Act (EISA) of 2007 sets a goal of supplying renewable fuels of 36 billion gallon per year by 2022, in which 16 billion gallons are from cellulosic biofuel (Congress 2007). In California, the Bioenergy Action Plan (California Bioenergy Interagency Working Group 2006) has targeted specific goals of increasing the share of state consumption of biofuels produced in-state

C.-W. Chen
Taiwan Research Institute, 29 F., no. 27, Sec. 2, Zhongzheng E. Rd., Danshui Dist., 251
New Taipei City, Taiwan, Rebublic of China

Y. Huang (✉)
Glenn Department of Civil Engineering, Clemson University, Clemson 29634, USA
e-mail: yxhuang@clemson.edu

S.D. Eksioglu et al. (eds.), *Handbook of Bioenergy*,
Energy Systems, DOI 10.1007/978-3-319-20092-7_3

from 20 % in 2010 to 40 % by 2020 and to 75 % by 2050. In addition, pressing energy security issues such as reducing oil dependence and enhancing energy diversity have also urged the government to develop more reliable and sustainable future energy systems.

In California, ethanol has been widely used in blending with gasoline at a rate of 5.7 % (i.e., E5.7), which requests nearly 900 million gallons of ethanol per year (Fuel Delivery Temperature Study 2008). However, the current in-state production capacity from the four existing corn-based ethanol facilities is only 81 million gallons per year (MGY) (Jenkins et al. 2007), accounting for less than 10 %. Achieving the in-state biofuel production goals as articulated in the Bioenergy Action Plan requires the use of a portfolio of feedstock sources and advanced cellulosic biofuel conversion technologies.

In this study, lignocellusloic biomass is considered as vital alternative renewable sources to the corn grain for biofuel production, which features a number of traits (Jenkins et al. 2007; Hill et al. 2006; Farrell et al. 2006; Perlack et al. 2005): (1) higher per-acre ethanol yields, (2) lower agronomic inputs compared to purpose-grown energy crops, (3) better efficiency in terms of life-cycle environmental performance, greenhouse gas (GHG) balances, and net-energy ratio, (4) less restriction in land use, which could reduce competition with food crops, and (5) plenty of resources including agricultural residues, forest thinning, and urban wastes.

In improving the cost effectiveness of biofuel production and delivery, the entire biofuel pathway should be considered in an integrative manner (Hill et al. 2006; Farrell et al. 2006), which includes all involved facilities and operations in the supply chain from the raw feedstock supply to the end users distribution. For example, a large-size biorefinery can take the advantage of improved economy of scale with reduced production cost, but may impose higher transportation cost between feedstock resources and demand centers. In recognition of the importance, systems approach has been employed in developing forest biomass allocation strategy (Freppaz et al. 2004), optimizing investments of converting biomasses into a tri-generation applications (Rentizelas et al. 2009), and optimizing the process of converting agricultural residues to hydrogen in California (Parker 2007). Recent studies on biofuel supply chain design primarily focused on developing a cost-effective supply chain by solving an optimization based on social-surplus maximization or cost-minimization principles in the line with the primary goal of general supply chain design problems (Melo et al. 2009). The related studies in biofuel supply chain design can be broadly classified into three general categories: deterministic biofuel supply chain optimization (Bai et al. 2011; Ekşioğlu et al. 2009; Ekşioğlu et al. 2010; Parker et al. 2007), biofuel supply chain design under uncertainty (Chen and Fan 2012; Cundiff et al. 1997; Huang et al. 2014), and multistage biofuel supply chain expansions (Acharya et al. 2008; Huang et al. 2010). For a comprehensive reviews on biofuel supply chain design, interested readers can be referred to a recent review paper (Awudu and Zhang 2012).

This study on biofuel supply chain is the integration of physical design and operational management as a whole to achieve the best system economics based on an optimization model aided multilayer biofuel supply chain network, including feedstock procurements, biofuel productions, biofuel blending and storage, and end-user distributions. Modeling this supply chain is not trivial, due to the complexity of the multilayer supply chain structure and geographic distributions of facilities over the layers. For example, a centralized facility takes the advantage of economies of scale, but may result in higher transport cost. Note that a choice of modeling methods between stochastic and deterministic methods for this study can be debatable. On one hand, stochastic model advances the modeling practice in the sense of incorporating uncertainties of demand or supply into the integrated system modeling. On the other hand, the increased model complexity results in escalated efforts in data collection and simulation and consequently the high cost of computational implementation. Very often, solution methods based on decomposition are used (Huang et al. 2010; 2014) in order to attain a "good" solution within a predefined optimality gap (e.g., 5 %) and an affordable time. In addition, results from a prior study (Huang et al. 2010) indicate that the benefit of the stochastic solution can be small (e.g., even lower than 1 %), which left a question whether the high cost of implementing the stochastic model and the solution algorithm is justifiable.

We formulated a mixed integer deterministic programming model for the biofuel supply chain optimization, which will also serves as the cornerstone in this handbook for the biofuel supply chain design and planning models. The study is focused on the analysis of *economic impacts of the supply curves*, *biomass-to-biofuel conversion technology performance*, and *demand fluctuations* on the biofuel system layout and the delivered fuel cost. Through the case study of California, we offer the managerial insights on the system planning and management according to conversion technology advancements and demand growth.

The remaining of the chapter is organized as follows. The background of California study is presented in Sect. 2. The deterministic optimization model is presented in Sect. 3. Section 4 includes study results and findings from sensitivity analyses. The conclusion and future research will be outlined in Sect. 5.

2 Background of California Study

2.1 Conversion Technology Assessment

The LignoCellulosics Ethanol (LCE) via hydrolysis and fermentation conversion technology with specific Dilute Acid pretreatment process is explicitly considered in this study. This biofuel conversion technology is expected to be ready for commercialization between 2015 (near-term) and 2025 (mid-term), which in general is more efficient but less costly in biofuel production than its near-term performance around 2015 (Parker et al. 2007). The differences in terms of conversion efficiency and associated costs will be discussed in the Sect. 2.3.

It is assumed that multiple types of feedstock could be converted to ethanol with single conversion technology through advanced uniform-feedstock preprocessing prior to the conversion process at refineries, based on the report published by Idaho National Laboratory (Program 2008). However, the preprocessing costs associated with different feedstock types are unknown yet, therefore are not included in this study.

2.2 Feedstock Resources

Eight types of waste-based feedstocks are considered in the study, which are corn stover, rice straw, wheat straw, forest residues, MSW (Municipal Solid Waste) wood, MSW paper, MSW yard, and cotton residual.

The existing feedstock annual yields and location data are adopted from the Western Government Association (WGA) report (Ekşioğlu et al. 2009) and have been aggregated at county or city centroids in GIS. These feedstocks are widely dispersed across the state, and different feedstocks tend to cluster in different regions. In general, the agricultural residues are concentrated in central valley area, MSW is available in metropolitan areas such as the San Francisco bay area and the Los Angeles areas, and forest residues are mainly located in Northern part of the state.

Associated parameters including total yields, moisture content and conversion rate are summarized in Table 1. The conversion rates are measured by the gallons of ethanol converted from one dry ton of the feedstock, which are varied with the status of conversion technology. In general, the conversion rates increase with improvement of the conversion technology.

The feedstock procurement cost is the expense of transporting feedstocks from fields to the roadside in a transportable form. The feedstock supply amount is

Table 1 Feedstock parameters

Feedstock types	Total yields (thousand dry ton)	# of nodes	Near-term LEC conversion rate (gallons/dry ton biomass)	Mid-term LEC conversion rate (gallons/dry ton biomass)	Moisture content (% weight)
Cornstover	563	27	72.6	80.6	15
Wheat straw	368	32	69.3	76.8	15
Forest	4268	47	80.9	90.2	50
Rice straw	867	14	69.3	76.8	15
Cotton residual	439	10	64.3	71.4	50
MSW-paper	1743	57	76.8	86.0	10
MSW-wood	899	57	71.1	78.9	50
MSW-yard	2124	57	63.9	70.0	50

Resource Parker et al. (2007)

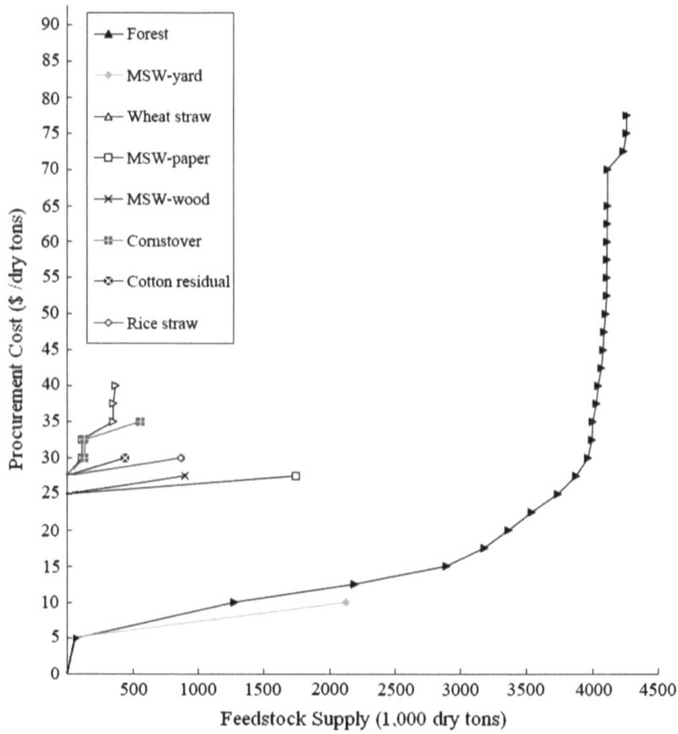

Fig. 1 Feedstock supply by procurement cost

piecewise linear to the procurement cost. The feedstock supply by procurement cost is adapted from the California data set in WGA report (Parker et al. 2007) and plotted in Fig. 1. Using forest as an example, there is about 12.5 million dry tons of forest feedstock available at a price of $10/dry ton; and an additional 17.5 million dry tons become available when the price increases to $15 per dry ton. The figure also indicates the maximum availability of each feedstock type (where the vertical lines correspond to and they are invariant of procurement price); for instance, forest residues can maximally provide 4.3 million dry tons at very high cost of $72 per dry ton.

2.3 *Refinery, Terminal, and Demand Center*

In this study, there are 28 potential sites for refineries based on the siting criteria used in the WGA report (Parker et al. 2007). In addition, there are 29 terminals for fuel blending and storage purposes and 143 cities chosen as demand centers. Their geographic distributions are presented in Fig. 2.

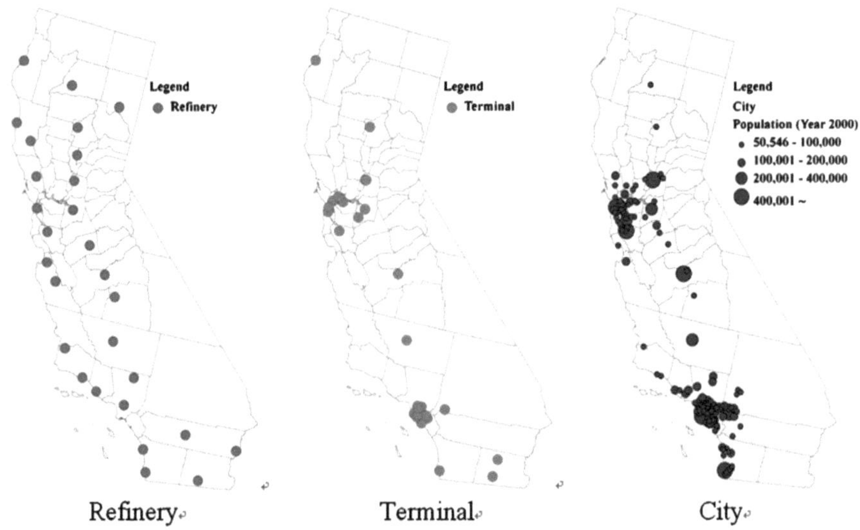

Fig. 2 Potential sites for refineries and terminals, and demand clusters

Table 2 Refinery parameters and production cost

	Near-term (present—2015)	Mid-term (2015—2025)
Fixed capital cost ($ million)	8.2	6.2
variable capital cost ($/gallon)	0.821	0.314
Ethanol production cost ($/gallon)	0.64	0.26
Maximum capacity (MGY)	60	100

Resource Ekşioğlu et al. (2009)

Refinery capital cost consists of fixed capital cost (facility setup cost) and variable capital cost (facility size-dependent cost). The fixed capital cost was annualized assuming a real discount rate of 10 % and a lifetime of 20 years. The refinery associated parameters and biofuel production cost are summarized in Table 2, which vary with the performances of near- and mid- term conversion technologies.

The single-size tank with the capacity of 100 thousand barrels (equivalent to 4.2 million gallons) at *terminals* is considered. The total capital cost is $1.57 million, which consists of tank cost of $1.26 million, blending system cost of $0.3 million, and a cost of $10,000 for product receipt by truck (Reynolds 2000).

Cities with a population of 50,000 are selected, mainly clustered in the four metropolitan areas—the Sacramento area, the San Francisco Bay Area, the Los Angeles area, and the San Diego area. The city fuel demand is assumed to be proportional to its population.

2.4 *Transportation Network*

California road network was used to transport feedstocks and fuels, which contains local, rural, urban arterial road, and interstate and state highways. The shortest traveling distance and time between facilities are calculated. The transportation cost is different in terms of transporting liquids and bulk solids and measured by three major cost components—loading/unloading costs, time- and distance- dependent costs. The transportation parameters are presented in Table 3.

3 Mathematical Model

A deterministic model is developed to achieve the best economics for the entire biofuel supply chain system as shown in Fig. 3. Note that the supply chain ends at city gate and further fuel distribution is not included in this study. Based on the model, we will be able to answer the following specifics:

Table 3 Transportation parameters

	Liquids	Bulk solids
Loading/unloading	\$0.02/gallon	\$5/wet ton
Time dependent	\$32/hr/truckload	\$29/hr/truckload
Distance dependent	\$1.30/mile/truckload	\$1.20/mile/truckload
Truck Capacity	8,000 gallons	25 wet tons
Travel Speed	55 miles/hr	35 miles/hr

Source Parker et al. (2007)

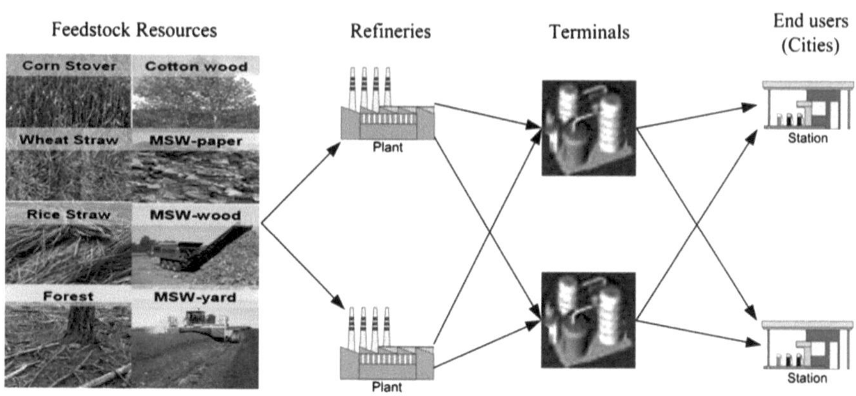

Fig. 3 Cellulosic ethanol supply chain

- The best locations and design capacities for ethanol refineries and terminals;
- The procurement amount of each type of feedstock and locations; and
- Feedstock and fuel flows in the supply chain.

Formulation notation is summarized in Table 4.

3.1 Objective function

The object is to minimize the cost of the entire ethanol supply chain including ethanol refinery capital cost (A_1), production cost (A_2), feedstock procurement cost (A_3), terminal cost (A_4), feedstock transportation cost (A_5), and ethanol delivery costs (A_6):

$$\min_{z,y,x,w,s,t} \{A_1 + A_2 + A_3 + A_4 + A_5 + A_6\} \tag{1}$$

A detailed explanation of the six cost components follows.

Refinery cost A_1 includes annual refinery fix capital cost and operating and maintenance (O&M) cost:

$$A_1 = \sum_{j \in J} RC_j^f z_j + \sum_{j \in J} RC_j^v x_j \tag{2}$$

where RC_j^f is fixed capital cost of ethanol refinery at location j; and RC_j^v is variable O&M cost of ethanol refinery at location j. In the study, the value of RC_j^f and RC_j^v equal to m$6.157 and $0.314 per gallon, respectively.

Annual ethanol production cost A_2 is:

$$A_2 = \sum_{j \in J} RP \cdot x_j \tag{3}$$

where RP is the estimated cellulose enzyme cost in ethanol production processes. In the study, RP was $0.26 per ethanol gallon.

Total feedstock procurement cost A_3 is:

$$A_3 = \sum_{f \in F} \sum_{i_f \in I_f} \sum_{j \in J} \sum_{p \in P} PR_{fp} w_{i_f j}^p \tag{4}$$

where PR_{fp} is procurement cost of feedstock type f at price level p. Note that the feedstock availability will change with the procurement costs.

Annual terminal capital cost A_4 is:

$$A_4 = \sum_{k \in K} TC_k y_k \tag{5}$$

Table 4 Notation

Parameters
F: index f, set of feedstock types
I_f: index i_f, set of feedstock fields for feedstock type f
J: index j, set of candidate refinery locations
K: index k, set of candidate terminal locations
L: index l, set of demand cities
P: index p, set of price level
$dist_{ij}$: distance between node i and j
PR_{fp}: procurement amount of feedstock type f at price level p (\$/dry ton)
RC_j^f: fixed refinery capital cost at location j
RC_j^v: variable refinery capital cost at location j (\$/gallon capacity)
RP: ethanol production cost (\$/gallon)
$CapR_j$: maximum refinery capacity at location j (gallon/year)
TC_k: terminal capital cost at location k
$CapT_k$: the maximum terminal service capacity at location k (gallon)
$TDBS$: distance dependent transportation cost (\$/mile/truckload) of bulk solids (including fuel, insurance, maintenance, and permitting cost)
$TTBS$: travel time dependent transportation cost (\$/hr/truckload) of bulk solids (including labor and capital cost)
$LUBS$: truck loading and unloading cost of bulk solids (\$/wet ton)
$CapBS$: truck capacity of bulk solids (wet ton)
$SPBS$: average truck travel speed for transporting bulk solids (mile/hr)
$TDLQ$: distance dependent transportation cost (\$/mile/truckload) of liquids
$TTLQ$: travel time dependent transportation cost (\$/hr/truckload) of liquids
$LULQ$: truck loading and unloading cost of liquids (\$/gallon)
$CapLQ$: truck capacity of liquids (gallon)
$SPLQ$: average truck travel speed for transporting liquids (mile/hr)
MC_f: moisture content of feedstock type f, used to convert the feedstock dry ton to wet ton
C_f: conversion rate of feedstock type f (gallon/dry ton)
$Yield_{i_f}^p$: the maximum annual yields of feedstock f in its field i_f at price level p
$Supply_{i_f}$: aggregated annual feedstock yields of type f in field i_f over all price levels
$Demand_l$: demand amount at city l
Decision Variables
z_j: 1 if a refinery is located at j; 0 otherwise
y_k: 1 if a terminal will be located at k; 0 otherwise
x_j: the total amount of ethanol produced at refinery j (gallon)
$w_{i_f j}^p$: the total amount of feedstock of type f transported from field i_f to refinery j at price level p (dry ton);
s_{jk}: amount of ethanol transported from refinery j to terminal k (gallon)
t_{jl}: amount of ethanol transported from terminal k to city l

where TC_k is terminal capital cost at location k. In this study, terminal is where ethanol is blended with gasoline.

Feedstock transportation cost A_5 is:

$$A_5 = \sum_{f \in F} \sum_{i_f \in I_f} \sum_{j \in J} \left(\frac{(TDBS + \frac{TTBS}{SPBS}) \times dist_{i_f j}}{CapBS} + LUBS \right) \times \frac{\sum_{p \in P} w_{i_f j}^p}{(1 - MC_f)} \tag{6}$$

Feedstock transportation cost consists of loading/unloading cost, time dependent travel cost, and distance dependent travel cost between feedstock fields and refineries.

Ethanol delivery cost A_6:

$$A_6 = \begin{array}{l} \sum_{j \in J} \sum_{k \in K} \left(\frac{(TDLQ + \frac{TTLQ}{SPLQ}) \times dist_{jk}}{CapLQ} + LULQ \right) \times s_{jk} \\ + \sum_{k \in K} \sum_{l \in L} \left(\frac{(TDLQ + \frac{TTLQ}{SPLQ}) \times dist_{kl}}{CapLQ} + LULQ \right) \times t_{kl} \end{array} \tag{7}$$

Similarly, the ethanol delivery cost contains the same three cost transportation components between refineries and terminals and between terminals and demand cities.

3.2 Constraints

3.2.1 Constraints on feedstock sites

At every price level p, the total procured feedstock of type f cannot exceed its availability at this price level.

$$\sum_{j \in J} w_{i_f j}^p \leq Yield_{i_f}^p, \ \forall i_f \in I_f, p \in P \tag{8}$$

where $Yield_{i_f}^p$ is the availability of feedstock f in its field i_f at price level p.

3.2.2 Constraints on ethanol refineries

The total amount of biofuel produced at refinery j equals the amount of feedstock transported to that refinery multiplied by their associated energy conversion rates C_f.

$$\sum_{f \in F} \sum_{i_f \in I_f} \sum_{p \in P} w_{i_f j}^p \times C_f = x_j, \ \forall j \in J \tag{9}$$

where C_f is defined as the conversion rate of feedstock type f (gallon/dry ton).

The produced ethanol must be delivered to terminals for blending.

$$\sum_{k \in K} s_{jk} = x_j, \ \forall j \in J \tag{10}$$

No ethanol can be produced from a location *j* unless there is a refinery at this location. This is a logic constraint.

$$\sum_{k \in K} s_{jk} \leq CapR_j \times z_j, \ \forall j \in J \tag{11}$$

where $CapR_j$ represents maximum refinery capacity at location *j*.

A capacity constraint sets the maximum fuel production at refinery *j*.

$$x_j \leq CapR_j, \ \forall j \in J \tag{12}$$

3.2.3 Constraints on terminals

The amount of ethanol flowed into a terminal equals the amount of ethanol flowed out to demand cities.

$$\sum_{j \in J} s_{jk} - \sum_{l \in L} t_{kl} = 0, \ \forall k \in K \tag{13}$$

This is a logic constraint on terminals, stating that no ethanol can be shipped out of a terminal unless the terminal is operating.

$$\sum_{l \in L} t_{kl} \leq CapT_k \times y_k, \ \forall k \in K \tag{14}$$

where $CapT_k$ represents maximum refinery capacity at location *j* (gallon). This capacity can be considered as the maximum service capacity of the terminal.

3.2.4 Constraints on demand cities

All ethanol demand at each city must be satisfied.

$$\sum_{k \in K} t_{kl} = Demand_l, \ \forall l \in L \tag{15}$$

3.2.5 Non-negativity and binary restrictions

Location decision variables z_j and y_k are binary. In addition, all decision variables are non-negative.

4 Case Study Result Analysis

4.1 Baseline Results

The deterministic model was programmed in the mathematical programming language AMPL (Fourer et al. 2003) and solved by the commercial optimization solver package CPLEX on Dell Precision Model 650 Workstation with 4 GB RAM and Dual-Xeon 3.06 GHz processor under Windows XP environment. All numerical experiments were solved within 1000 CPU seconds. The baseline scenario considers the near-term (see Table 1) technology performance and the total in-state cellulosic ethanol demand is 200 million gallons (Fuel Delivery Temperature Study 2008).

The optimal system layout is presented in Fig. 4a, in which four refineries are built with three different sizes geographically dispersed and two fuel terminals co-located with two of the refineries, which mainly serve their respective demand markets.

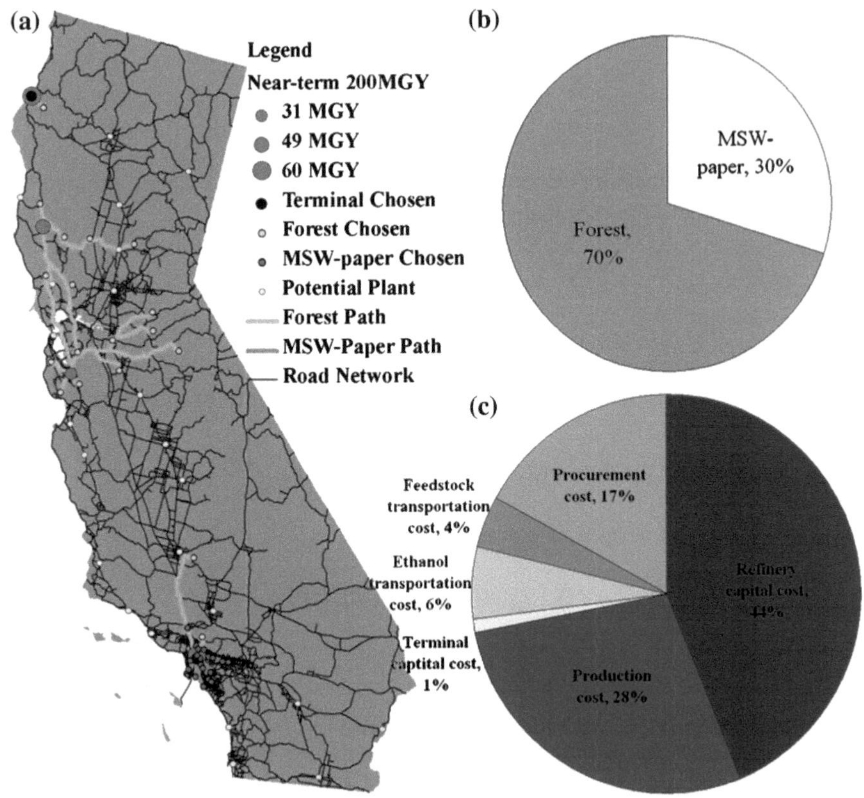

Fig. 4 Results of the model based on near-term technology

The optimal feedstock procurement portfolio suggests using only two types of feedstock resources illustrated in Fig. 4b. In particular, the forest residues are mainly consumed by the refinery in the North of the state and the MSW paper are mainly consumed by the refinery located in the Los Angeles area. This procurement strategy is resulting from tradeoffs between procurement cost, conversion rate, and cost in the feedstock delivery associated with each type of feedstock. The forest residue outranks other types, mainly due to the high conversion rate and overall low procurement cost. The MSW is vast in that region and proximate to the refinery.

The total system cost breaks down to six cost components as shown in Fig. 4c. The refinery capital cost is the major cost driver. The transportation cost including feedstock delivery cost and fuel distribution cost is substantial, accounting for 10 % of the total cost, which justifies the use of systems approach. The resulting average delivered cost is about $1.86 per gallon.

4.2 Impacts of Advancement of Conversion Technology

This section presents the analysis results of the impacts of advancing conversion technology (i.e., mid-term (2015–2025) in Table 2) on system planning and economic efficiency. The resulting optimal system layout indicates a centralized location pattern as shown in Fig. 5a, in which only two refineries with full capacity and two terminals are placed. More MSW (including paper and yard) is used for the refinery in the south of the state. This system configuration takes the advantage of improved economy of scale but raise the transportation cost as indicated in Fig. 5c. The average delivered fuel cost is lowered to $1 per gallon.

4.3 Demand Fluctuations

The cellulosic biofuel demands may fluctuate in response to the future under renewable energy regulations (Congress 2007; California Bioenergy Interagency Working Group 2006). This section aims to examine how significant the demand variations between 100 and 900 MGY can impact the system planning and resulting delivered cost. A set of demand sensitivity analyses is conducted in a range between 100 and 900 MGY.

The system layouts with demands of 200, 500, and 900 MGY as examples are plotted in Fig. 6. In general, increasing demand leads the refineries and terminals to be located nearer to feedstock resources and major demand markets, respectively. Furthermore, Fig. 6 implies that an optimal system planning is not a process of simply adding additional facilities to the existing system as the demand increases; instead, it needs a re-optimization of the entire system in order to better system economics.

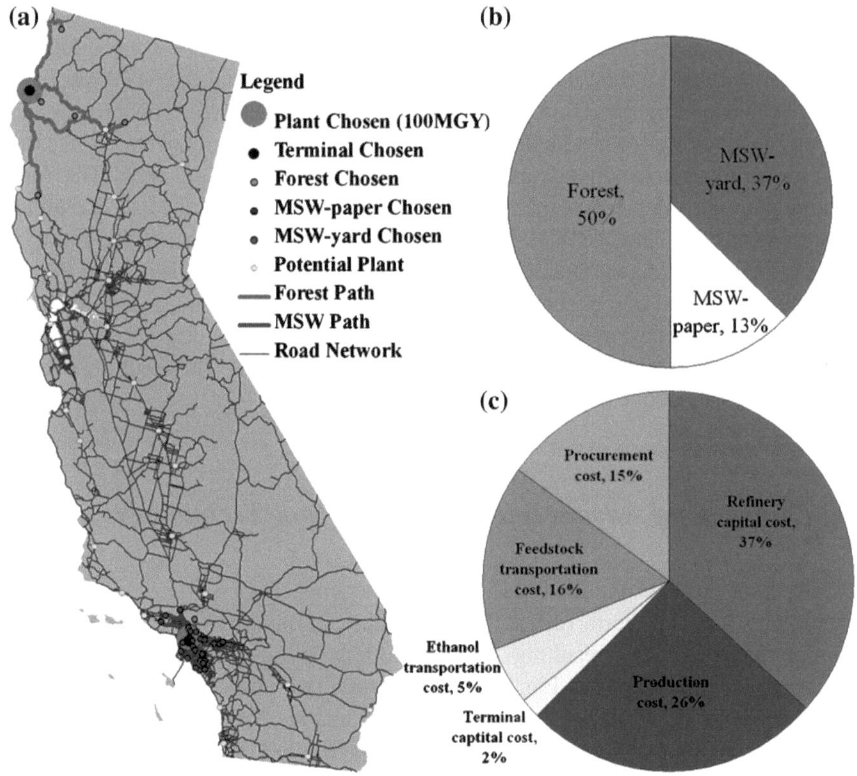

Fig. 5 Results of the model based on mid-term technology

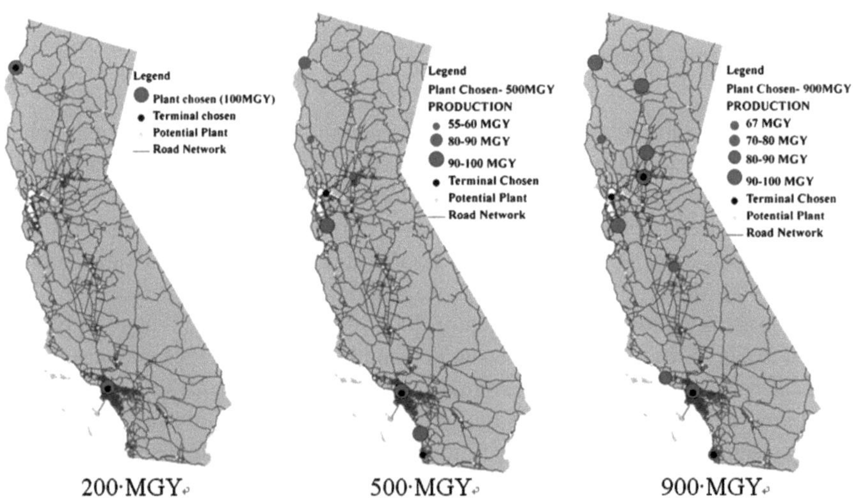

Fig. 6 System layouts under different demand rates

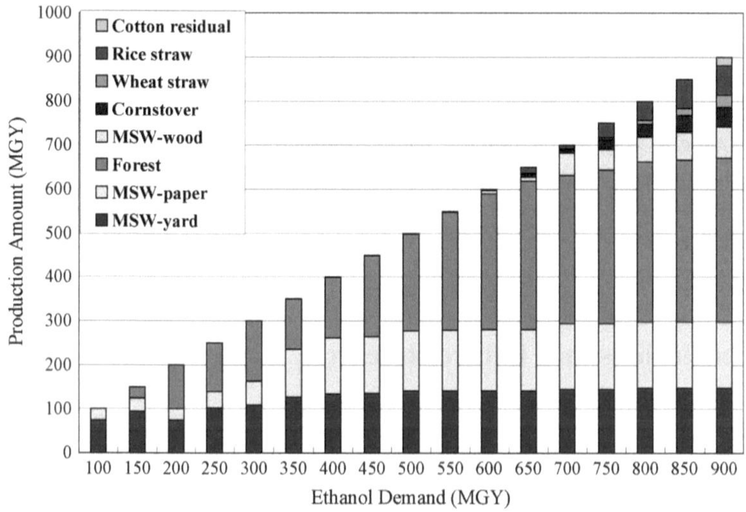

Fig. 7 Feedstock procurement strategies under different demand rates

The feedstock portfolio varies with different demand rates as shown in Fig. 7, which is a compromising result of procurement cost, geographic distribution, moisture content, and resource availability. From the figure, it is clear that the MSW yard, MSW paper, and forests are the primary feedstock resources, which combined

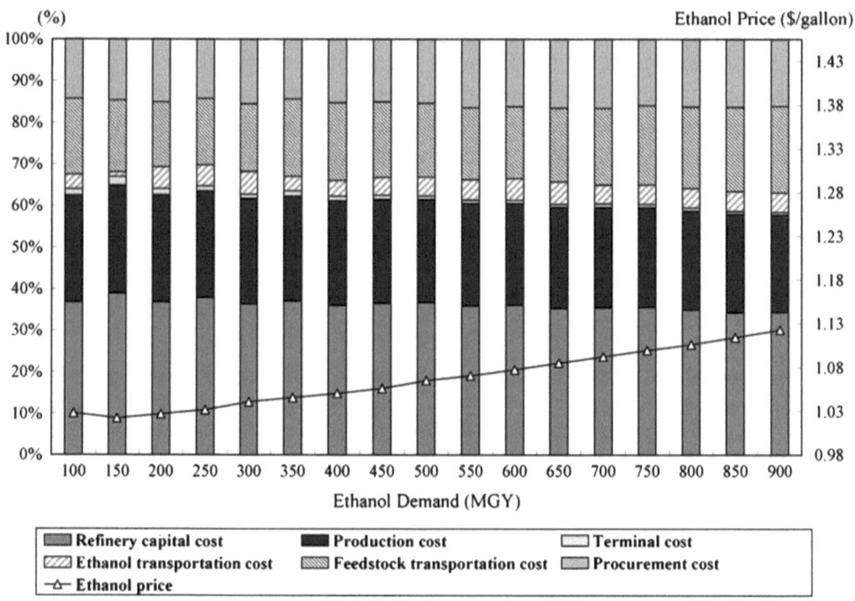

Fig. 8 Breakdowns of the total cost under different demand rates

support at least 75 % of in-state production. They share similarities in terms of low procurement cost and high conversion rate.

The breakdowns of the total cost vary with the demand rates, as plotted in Fig. 8 based on the midterm technology performance. The percent of each cost component in the total cost stays stable and the average delivered cost represented by the line in the figure increases from \$1 to \$1.13 per gallon, which imply that the state has sufficient feedstock resources and intelligent system planning can help maintain cost effectiveness.

5 Conclusions and Future Research

To achieve higher share of in-state biofuel consumption in California, this study has explored the economic potentials of producing the cellulosic ethanol based on biowaste feedstock including crop residues, forest residues, and municipal wastes. We presented a deterministic optimization model for achieving the best system cost effectiveness, with a primary focus on the analysis of economic impacts of geographic distributions, biofuel conversion technology advancements, and demand fluctuations.

The average delivered ethanol cost is \$1.86 per gallon based on the projected 200 MGY in 2015 and near-term conversion technology performance. The forest residues and MSW paper are identified as the primary feedstock resources. These two major resources were also concluded by Leighty (2010), Parker et al. (2007, 2010), while producing ethanol via LCE technology. The advancement in the conversion technology can significantly reduce the delivered fuel cost and alter the system layout by mid-term (2025) with a cost in a range of \$1-1.1 per gallon. Through the analysis of the demand fluctuations, we understand that by using systems approach, the delivered cost can be maintained low.

Some immediate extensions to this study include designing nonlinear cost functions of production and procurement costs to allow for more economic realistic assumptions. Also, it is more realistic to include the layer of pre-processing of different types of feedstock for biofuel conversion. The location and sizing of the involved facilities will be adding another layer of trade-off into the integrated supply chain design.

Acknowledgments This work was completed when authors were at the University of California, Davis. The authors were grateful to Chevron Technology Ventures (LLC) for funding this research. Special gratitude was delivered to Dr. Yueyue Fan for advising this work. We also thank researchers (especially Prof. Joan Ogden, Prof. Bryan Jenkins, and Dr. Nathan Parker) in STEPS program at UC Davis for their generous data support.

References

Acharya AM, Ekşioğlu SD, Petrolia D (2008) In-bound supply chain design for biomass-to-ethanol industry: a study of Mississippi. In: The Proceedings of institute of industrial engineers conference, pp 1296–1301

Awudu I, Zhang J (2012) Uncertainties and sustainability concepts in biofuel supply chain management: a review. Renew Sustain Energy Rev 16(2):1359–1368

Bai Y, Hwang T, Kang S, Ouyang Y (2011) Biofuel refinery location and supply chain planning under traffic congestion. Transp Res Part B Methodol 45(1):162–175

California Bioenergy Interagency Working Group (2006) Bioenergy action plan for California. http://www.energy.ca.gov/2006publications/CEC-600-2006-010/CEC-600-2006-010.PDF

Chen C-W, Fan Y (2012) Bioethanol supply chain system planning under supply and demand uncertainties. Transp Res Part E Logist Transp Rev 48(1):150–164

Cundiff JS, Dias N, Sherali HD (1997) A linear programming approach for designing a herbaceous biomass delivery system. Bioresour Technol 59(1):47–55

Ekşioğlu SD, Acharya A, Leightley LE, Arora S (2009) Analyzing the design and management of biomass-to-biorefinery supply chain. Comput Ind Eng 57(4):1342–1352

Ekşioğlu S, Li S, Zhang S, Sokhansanj S, Petrolia D (2010) Analyzing impact of intermodal facilities on design and management of biofuel supply chain. J Transp Res Board 2191:144–151

Farrell AE, Plevin RJ, Turner BT, Jones AD, O'Hare M, Kammen DM (2006) Ethanol can contribute to energy and environmental goals. Science 311(5760):506–508

Fourer R, Gay DM, Kernighan BW (2003) AMPL: a modeling language for mathematical programming. Thomson, Pacific Grove

Freppaz D, Minciardi R, Robba M, Rovatti M, Sacile R, Taramasso A (2004) Optimizing forest biomass exploitation for energy supply at a regional level. Biomass Bioenergy 26(1):15–25

Fuel Delivery Temperature Study (2008) Annual California county consumption. http://www.energy.ca.gov/transportation/fuel_delivery_temperature_study/documents/2008-06-05_workshop/2008-06-06_ANNUAL_CALIFORNIA_CONSUMPTION.PDF

Hill J, Nelson S, Tilman D, Polasky S, Tiffany D (2006) Environmental, economic, and energetic costs and benefits of biodiesel and ethanol biofuels. Proc Natl Acad Sci 103(30):11206–11210

Huang Y, Chen C-W, Fan Y (2010) Multistage optimization of the supply chains of biofuels. Transp Res Part E Logist Transp Rev 46(6):820–830

Huang Y, Fan Y, Chen C.-W (2014) An integrated biofuel supply chain to cope with feedstock seasonality and uncertainty. Transp Sci

Jenkins B, Dempster P, Gildart M, Kaffka S (2007) California biomass and biofuels production potential (Draft). California Energy Commission

Leighty WW (2010) Deep reductions in greenhouse gas emissions from the California transportation sector: dynamics in vehicle fleet and energy supply transitions to achieve 80 % reduction in emissions from 1990 levels by 2050, transportation technology and policy. University of California, Davis, CA, p 329

Melo MT, Nickel S, Saldanha-da-Gama F (2009) Facility location and supply chain management —a review. Eur J Oper Res 196(2):401–412

Parker N (2007) Optimizing the design of biomass hydrogen supply chains using real-world spatial distributions: a case study using California rice straw. Master's thesis, University of California, Davis, CA

Parker N, Tittmann P, Hart Q, Lay M, Cunningham J, Jenkins B (2007) Strategic development of bioenergy in the western states development of supply scenarios linked to policy recommendations, task 3: spatial analysis and supply curve development. Western Governors' Association

Parker N, Hart Q, Tittmann P, Murphy C, Lay M, Nelson R, Skog K, Gray E, Schmidt A, Jenkins B (2010) National biorefinery siting model: spatial analysis and supply curve development. Western Governors' Association

Perlack RD, Wright LL, Turhollow AF, Graham RL, Stokes BJ, Erbach DC (2005) Biomass as Feedstock for a Bioenergy and Bioproducts Industry: The Technical Feasibility of a Billion-Ton Annual Supply. Oak Ridge National Laboratory, Oak Ridge, TN

Program B (2008) Uniform-format feedstock supply system design for lignocellulosic biomass. Idaho National Laboratory, Idaho Falls, ID

Rentizelas AA, Tatsiopoulos IP, Tolis A (2009) An optimization model for multi-biomass tri-generation energy supply. Biomass Bioenergy 33(2):223–233

Reynolds RE (2000) The current fuel ethanol industry—transportation, distribution, and technical considerations. Downstream Alternatives Inc., Marketing

U.S. Congress (2007) The energy independence and security act: renewable fuels standard. http://www.gpo.gov/fdsys/pkg/PLAW-110publ140/html/PLAW-110publ140.htm

Collaborative Railway Transportation Strategy to Increasing Imports of Russian Wood for the Finnish Forest Energy Industry

Teijo Palander, Jukka Malinen and Kalle Kärhä

Abstract In this research, the logistics alternatives of Finnish forest industry under increasing imports of Russian wood are discussed. The paper show that the transportation strategy of a third party logistic provider (3PLs) used in the railway has large implications for the logistics of imported wood from Russia. In the primary data used in this research, the basic scenario described the railway transportation strategy of 3PLs at 2011, before Russia's World Trade Organization (WTO) membership. The alternative scenarios described the strategies of 3PLs for changed wood transportation and storing needs of imported Russian wood under WTO. In the alternative scenarios also the railway wagon rotation was adjusted to reach the global optimum strategy for more efficient logistics of plants. After adjusting merely 3PLs' wagon traffic flow, the global optimum was not reached, since the reduction of empty wagons caused severe shortages of wood flow and later a need to cut down storage inventory at districts and plants increasing total logistics costs. As the conclusion, the model based on the dynamic multiple objective linear programming (DMOLP) is recommended to the supply chain parties, as the adaptation of Russian wood import to the global railway transportation environment would require collaborative logistics and integrated optimizations of wood and the wagon-traffic flows to be carried out.

Keywords 3PLs · Global wood procurement logistics · Railway transportation strategy · Russian export tariffs · World trade organization (WTO)

T. Palander (✉) · J. Malinen · K. Kärhä
Faculty of Science and Forestry, University of Eastern Finland, P.O. Box 111, 80101 Joensuu, Finland
e-mail: teijo.s.palander@uef.fi

S.D. Eksioglu et al. (eds.), *Handbook of Bioenergy*,
Energy Systems, DOI 10.1007/978-3-319-20092-7_4

1 Introduction

1.1 Background

On several occasions over the last several years, the European Commission (EC) and Russia have been negotiating over Russia's interests in joining the World Trade Organization (WTO). Since July 2006, Russia imposed export taxes on raw materials which increased the tariffs on wood exports from four euro to 15 euro per a cubic meter (m^3). To date, the wood tariff dispute has not been a particularly major issue on the EU scale (European Commission, 2008). However, the tariff hikes caused national problems in Russia's neighboring countries, such as Finland, due to the significant increase in price of imported wood from Russia. The increase in raw material costs has considerable consequence to Finnish forestry, since Finnish forest industry has many industrial facilities near the Russian border. The main wood assortments imported from Russia to Finland in 2011 were birch pulpwood (39.4 %) and chips (33.7 %). The chips were mainly produced from coniferous species (Anon 2012c).

The continual dispute between Finland and Russia over Russia's tariffs on wood exports may have been a reflection of the difficulties that the EU has experienced in dealing with Russia's global economic policy (Piermartini 2004; Anon 2007; Parker and Thornton 2007; Thee 2009). Nevertheless, until 2009 a large share of wood raw material used by Finnish forest industries (25 %, 15 mill. m^3) was imported from Russia, which decreased remarkably and was about 4 mill. m^3 in 2011 (Palander and Vesa 2012). Fortunately, the tariff rise which was expected to take place in 2011 was delayed by Russia for the country's joining of WTO. It seems probable that Russian wood import to Finland would reach 10 mill. m^3 in 2014. In practice, imports started to rise significantly during 2013.

1.2 Rail Transportation Planning in Wood Procurement Logistics

Global wood procurement problems of Finnish forest industry have been managed at the national level (Iikkanen and Mukula 2010; Palander and Vesa 2012). There have also been several railway transportation strategies since the tariff changes have been implemented in 2011. In strategic planning, the predictions of the effects of the tariff increase for the Finnish forest industry have been conducted with the help of static annual optimization until 2015, 2020 or 2030 (Hetemäki and Hänninen 2009; Iikkanen and Mukula 2010; Reini et al. 2010). However, the assessment of the affects of tariff reduction due to Russian WTO membership on wood import from Russia have not been conducted using dynamic optimization for monthly procurement chains of Russian wood, starting from purchase of wood and ending to rail transportation to Finnish plants.

In the wood procurement, as well as in many other branches of the global raw-material logistics, different adjustments take place in the planning and control: New orders arrive, some of the existing orders are completed, some orders may be modified, wood deliveries can be delayed etc. The planning environment is further complicated by significant sequence-dependent set up times among different wood qualities. The supply chain parties should be able to consider these dynamic changes and update the logistics schedule. However, several wood procurement problems are still solved manually by experienced managers. This easily results in procurement schedules which do not meet the optimal service solutions of plants (Palander and Vesa 2012).

In Finland, VR Group (state owned company offering railway services) has basically a monopoly at the railway traffic, and VR Group provides third-party logistic services (3PLs') for industrial customers. 3PLs' share of rail transportation has increased about ten % during five-year period and it represents the majority (75 %) of all wood raw material imported from Russia (Iikkanen and Mukula 2010; Iikkanen and Sirkiä 2011). Finnish forest industry companies estimate that imports of wood from Russia could even double within a few years. In the railway transportation, there will seemingly be delivering problems in the wood procurement logistics. This implies that 3PLs' are required to put into practice different global logistics decisions and service strategies in order to satisfy the global customer requirements. Once the collaborative railway transportation strategy was again possible to implement between 3PLs and customers in 2013, it would have been useful to fulfill the customers' wood services by a multiple objective optimization approach for rail transportation and storing of Russian wood. Then the supply chain parties could also implement integrated information and communication systems (ICT) to enhance information flow across them, and to support a better collaboration and logistics management.

1.3 Dynamics of Rail Transportation

The model presented in this research is based on monthly procurement scheduling of one year in a global wood procurement environment, where Russian wood imported to Finland is transported by rail (Fig. 1). The railway transportation strategy is designed using an order-driven policy of plants to minimize total logistics costs and railway transportation time. Wood flows are also adapted to varying wagon rotation restrictions, which was 4.5 days on average in 2011 (Iikkanen and Sirkiä 2011). In practice, 3PLs controls wagon rotation and determines it for monthly planning periods depending on operating wagon capacity (m^3), transportation speed (kmh^{-1}) and transportation distance (km).

In order to maintain efficient wagon and wood flows, plant (customer) orders are transformed into the sequence dependent functions of wood procurement chains

Fig. 1 Railway wagons are loaded from timber truck in a station (photo: Metsäteho)

adapting inventories and buffer storages in the dynamic model (Fig. 2). A challenging task is to optimize wagon and wood flows so that the release dates are fulfilled, due dates met and the total costs minimized. While a long scheduling horizon contains multiple decision-making objectives, the scheduling of wagon and wood chains is too difficult task to conduct manually. A strong demand exists among managers for good dynamic optimization theory which can be utilized in ICT systems without any significant deviation from global optimality (Palander 2011).

Several efficient models can be applied in ICT systems in order to solve raw material procurement problems. Since the publication of the classical work of Hadley (1962), there has been a continuous effort to develop mathematical formulations for various types of logistics operations, e.g., Bengston (1966). One of the early continues-time formulations for Swedish wood-flow problem was presented by Carlsson and Rönnqvist (1999). At the time, also a useful review of models and several efficient techniques for wood flow optimization was published and can be found in Palander (1998). The wood-flow approach has been found useful for modelling practice in the wood procurement research context in Finland (e.g. Palander and Väätäinen 2005; Palander et al. 2005; Iikkanen et al. 2010; Palander 2011; Palander and Vesa 2012; Palander and Voutilainen 2013). The Dynamic

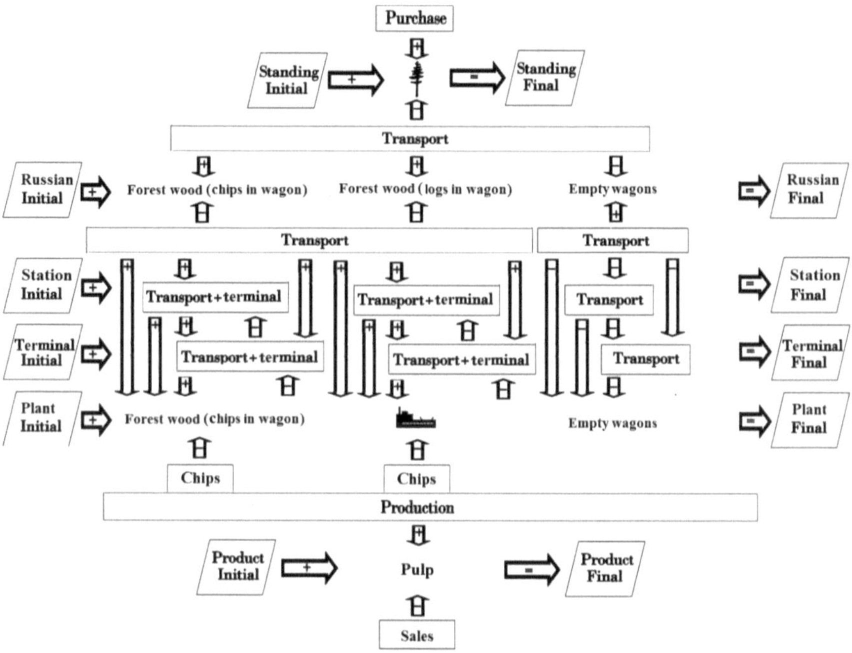

Fig. 2 A model of the dynamics of wood and wagon flows that includes resource inventories: *Vertical arrows* represent sequence-dependent effects for the system; *horizontal arrows* represent time-dependent effects for the system. *Arrows* labeled with a + represent inputs to a component of the system; *arrows* labeled with a—represent withdrawals from a component

Multiple Objective Linear Programming (DMOLP) has been used in wood-flow models. According to Palander (1998), DMOLP can be applied for the optimization of transportation schedules subjected to storing alternatives of forest industry.

In this research, we discuss developing a DMOLP model to the global railway transportation in respect to the different strategic alternatives of wood import companies (3PLs') under the increase in Russian wood import. The focus is set upon the prediction of rail transportation and storing plans (railway transportation strategy) for the south-eastern Finland in 2013. A continuous-time DMOLP-formulation is used to model the planning problem of the supply chain parties considering various sourcing strategies for assumed Russian wood increase in global wood supply. We are going to take into account the significant change in logistics environment of Finnish plants since late 2012 in order to analyze the logic of the DMOLP approach which, as such, may be highly dependent on the initial conditions of the optimization models which describe the pre-2012 environment. Based on the optimization results, conclusions are presented by giving multiple choice criteria to choose the supply chain parties' efficient strategy from the railway transportation alternatives.

2 Dynamic Multiple Objective Linear Programming—DMOLP

2.1 *Mathematical Formulation of Model*

There are three main goals in constructing the optimal wood procurement plan in the global-scale and long-term system (Fig. 2). The first goal is to fulfill the plant orders, i.e., the procurement functions should be completed on time. The monetary goal is to minimize the total logistics costs. A good quality of operations is also an important goal. Nevertheless, the objectives are in most cases contradictory: in order to be able to maintain the confirmed service level (e.g. due dates) it is often necessary to use more quality changes than required. For example, plant managers' scheduling task for 1 year involves a continuous balancing between the three goals, e.g., quality changes can be compensated by paying additional procurement costs or by importing more Russian wood (Palander and Vesa 2012).

The most important scheduling properties, from the practical rail transportation point of view (3PLs'), are that an efficient delivering sequence is guaranteed. Therefore, the scheduled wagon and wood mixes, multiple border stations and Finnish plants, intermediate storage times, and functional transition times are the sub-properties which must be included to the optimization model. As the additional property of the system, the imports were described as wood storages at the railway stations of Finland and Russia (Figs. 4 and 5). Particularly, it is crucial to join system dynamics to the properties of the model, in which scheduling task takes place during procurement periods (Palander 1998). In order to solve the scheduling problem, the multiple-objective model by Palander (2011) can be applied in the planning system of the supply chain parties to solve our research problem. In mathematical terms:

$$\text{Minimize } Z = w_1 D_1 + w_2 D_2 + w_3 D_3 + w_4 D_4 \tag{1}$$

where Z is the optimum weighted sum of D_g, which represents the deviations (€, min) above and below the decision-maker's goals of the problem, the parameter w_g is a positive weight that reflects the decision-maker's preferences regarding to the relative importance of each objective, and G is the number deviations of decision-maker's goals (1, …, g, …, 4). Here the decision-maker regards the penalization of all deviations as equally important and each deviation is turned into a percentage value away its target level. Thus all deviations are measured in the same units using percentage normalization.

Equation (1) is subject to the following restrictions:
Operating costs

$$\begin{aligned}
&\sum_{i=1}^{I}\sum_{j=1}^{J}\sum_{t=1}^{T}(cl_{ijt}L_{ijt} \times (1+p/26)^{t}) \\
&+\sum_{i=1}^{I}\sum_{j=1}^{J}\sum_{k=1}^{K}\sum_{t=1}^{T}(cy_{ijkt}Y_{ijkt} \times (1+p/26)^{t}) \\
&+\sum_{i=1}^{I}\sum_{j=1}^{J}\sum_{t=1}^{T}(cx_{ijt}X_{ijt} \times (1+p/26)^{t}) \\
&+\sum_{w=1}^{W}\sum_{j=1}^{J}\sum_{t=1}^{T}(ct_{wjt}XT_{wjt} \times (1+p/26)^{t}) \\
&+\sum_{i=1}^{I}\sum_{k=1}^{K}\sum_{t=1}^{T}(cm_{ikt}M_{ikt} \times (1+p/26)^{t}) \\
&+D_1 - D_2 = C_{\mathrm{tot}}
\end{aligned}$$

where C_{tot} is the total operating costs (€) of supply chain, L_{ijt} is the volume (m^3) of wood assortment i purchased and transported to border station j during period t, Y_{ijkt} is the volume (m^3) of wood assortment i transported from railway station j to railway station k during period t, X_{ijt} is the volume (m^3) of wood assortment i in inventories of railway station j in the end of period t, XT_{wjt} is the volume (m^3) of empty wagon w in inventories of railway station j in the end of period t, M_{ikt} is the volume (m^3) of wood assortment i in plant and terminal inventory k in the end of period t, D_1 and D_2 are the deviation above or below the total costs (€) of supply chain during the decision-making horizon, cl_{ijt} is the purchase and transportation costs (€ m^{-3}) of wood assortment i to border station j during period t, cy_{ijkt} is the transportation cost (€ m^{-3}) of wood assortment i transported from railway station j to railway station k during period t, cx_{ijt} is the inventory cost (€ m^{-3}) of wood assortment i in railway station j during period t, ct_{wjt} is the inventory cost (€ m^{-3}) of empty wagon w in railway station j during period t, and cm_{ikt} is the plant inventory cost (€ m^{-3}) of wood assortment i in plant and terminal k during period t, p is the annual interest rate (%), I is the number of wood assortments ($1, \ldots, i, \ldots, I$), W is the number of and wagons ($1, \ldots, w, \ldots, W$), J is number of railway stations ($1, \ldots, j, \ldots, J$) as point of departure of wood and wagons, K is the number of railway stations ($J + 1, \ldots, k, \ldots, K$) as destination station of wood and wagons, T is the number of periods ($1, \ldots, t, \ldots, T$), and G is the number of deviation variables ($1, \ldots, g, \ldots, 4$, where $g = 1$ and $g = 2$ are used in the present constraint).

Operating time

$$\sum_{i=1}^{I}\sum_{j=1}^{J}\sum_{k=1}^{K}\sum_{t=1}^{T}(cw_{ijkt}Y_{ijkt}\times(1+p/26)^{t})$$
$$+\sum_{w=1}^{W}\sum_{j=1}^{J}\sum_{k=1}^{K}\sum_{t=1}^{T}(ce_{wjkt}YH_{wjkt}\times(1+p/26)^{t})$$
$$+D_3-D_4=T_{\text{tot}}$$

where T_{tot} is the total operating time (min) from the supply chain during the planning horizon, Y_{ijkt} is the volume (m^3) of wood assortment i transported from railway station j to railway station k during period t, cw_{ijkt} is transportation time (min m^{-3}) of wood assortment i transported from railway station j to railway station k during period t, YH_{wjkt} is the volume (m^3) of empty wagon w transported from railway station j to railway station k during period t, ce_{wjkt} is the transportation time (min) of empty wagon w transported from railway station j to railway station k during period t, D_g are the deviations (min) above and below the total transport time from the supply chain during the planning horizon for deviation g, and G is the number of deviation variables (1, …, g, …, 4, where g = 3 and g = 4 in the present constraint).

Balance of empty wagons and loaded wagons

$$YH_{wjkt}-Y_{ijkt}\geq WB$$
$$YH_{wjkt}\geq L$$

where, WB is the minimum volume (m^3) of empty wagon transportation above the wood transportation volume (m^3) from railway transportation network to Russia during the planning horizon. L is the volume (m^3) of wood assortments purchased from Russia and transported to Finland via the border stations during the planning horizon.

Dynamic equations

$$X_{ijt-1}-Y_{ijkt}+L_{ijt}=X_{ijt}$$
$$X_{ijt-1}+Y_{ijkt}-Y_{ijkt}=X_{ijt}$$
$$XT_{wjt-1}+Y_{ijkt}-YH_{wjkt}=XT_{wjt}$$
$$XT_{wjt-1}+YH_{wjkt}-YH_{wjkt}=XT_{wjt}$$
$$M_{ikt-1}+Y_{ijkt}-MH_{ikt}=M_{ikt}$$

where MH_{ikt} is the volume (m^3) of wood assortment i demanded by plant (respective railway station k) during period t

Wood inventories of railway station

$$
\begin{aligned}
&X_{ijt=0} = XI_{ij} \\
&X_{ijt=12} = XB_{ij} \\
&X_{ijt} \leq X\max_{ijt} \\
&X_{ijt} \geq X\min_{ijt}
\end{aligned}
$$

where XI_{ij} is the volume (m^3) of wood assortment i stored by railway station j at the beginning of the planning horizon, and XB_{ij} is the volume (m^3) of wood assortment i stored by railway station j at the end of the planning horizon, $X\max_{ijt}$ and $X\min_{ijt}$ are the maximum and minimum (respectively) wood inventories (m^3) of wood assortment i held in railway station j during period t.

Wagon inventories of railway station

$$
\begin{aligned}
&XT_{wjt=0} = XTI_{wj} \\
&XT_{wjt=12} = XTB_{j} \\
&XT_{wjt} \leq XT\max_{jt} \\
&XT_{wjt} \geq XT\min_{wjt}
\end{aligned}
$$

where XTI_{wj} is the volume (m^3) of wagon w stored by railway station j at the beginning of the planning horizon, and XTB_j is the volume (m^3) of wagons stored by railway station j at the end of the planning horizon, $XT\max_{jt}$ and $XT\min_{wjt}$ are the maximum and minimum wagon inventories (m^3) of wagons or wagon w held in railway station j during the planning horizon and period t, respectively.

Wood inventories of plant

$$
\begin{aligned}
&M_{ikt=0} = MI_{ik} \\
&M_{ikt=12} = MB_{ik} \\
&M_{ikt} \leq M\max_{ikt} \\
&M_{ikt} \geq M\min_{ikt}
\end{aligned}
$$

where $M\max_{ikt}$ and $M\min_{ikt}$ are the maximum and minimum (respectively) volumes (m^3) of wood assortment i held in plant inventory k (railway station) during period t, MI_{ik} is the volume (m^3) of wood assortment i stored by plant k at the beginning of the planning horizon, and MB_{ik} is the volume (m^3) of wood assortment i stored by plant k at the end of the planning horizon.

Wood demands

$$
\begin{aligned}
&MH_{ikt} \leq MH\max_{ikt} \\
&MH_{ikt} \geq MH\min_{ik}
\end{aligned}
$$

where $MH\max_{ikt}$ and $MH\min_{ik}$ are the maximum and minimum (respectively) volumes (m^3) of wood assortment i demanded by plant (respective railway station k) during period t or during the planning horizon.

Purchase and transportation of wood to Finland-Russia border station

$$L_{ijt} \leq L\max_{ijt}$$
$$L_{ijt} \geq L\min_{ijt}$$

where $L\max_{ijt}$ and $L\min_{ijt}$ are the maximum and minimum (respectively) volume (m^3) of wood assortment i purchased from Russia and transported to Finland-Russia border station j during period t.

Rail transport capacities for wagons and wood

$$YH_{wjkt} \leq YH\max_{wjkt}$$
$$YH_{wjkt} \leq YH\max$$
$$Y_{ijkt} \geq Y\min_{ijkt}$$
$$Y_{ijkt} \leq Y\max_{ijkt}$$

where $YH\max_{wjkt}$ is the maximum volume (m^3) of wagon w transported from railway station j to railway station k during period t, $YH\max$ is the maximum volume (m^3) of wagons during the planning horizon, $Y\max_{ijkt}$ and $Y\min_{ijkt}$ are the maximum and minimum (respectively) volumes (m^3) of wood assortment i transported from railway station j to railway station k during period t.

Non-negativity

$$L_{ijt}, X_{ijt}, XT_{wjt}, Y_{ijkt}, YH_{wjkt}, M_{ikt}, MH_{ikt} \geq 0$$

The problem was formulated using the DMOLP subjected to the cost, time, wagon and wood allocation constraints. The dynamics of the model derive from monthly planning periods and the sequential wagon and wood delivering chains without stop periods. The primary average unit cost rates are by 2011 over all scenarios, because the objective of this research was to (1) predict suitable strategy for 2013 and (2) justify the model for adjustment of transportation and storing to increase Russian wood import (Table 1). This was done by analyzing the effects of optimized wagon and wood flows to their total operating costs and time in business and labor units. Wagon turnaround (%) depicted 3PLs' wagon rotation for planning periods in wagon flow, which depends on wagon capacity (m^3), transportation speed (kmh^{-1}) and transportation distance (km) (Iikkanen and Sirkiä 2011).

In wood flow, purchase, importing and transportation unit costs were determined from the materialized average unit costs in the year 2011. The unit costs were applied for wood of wagons that were delivered from border stations to plants. For a long scheduling period the average unit costs of rail transportation can be considered constant in railway transportation strategies (Fig. 3). This assumption can be

Table 1 Annual volumes of wood import (Totals, mill. m^3) from Russia to Finland (2005–2011) and the predictions E1 and E2 for the years 2012 and 2013

Plant	Wood	2005	2007	2010	2011	E1	E2
Plants 1, 2	A	1.6	1.4	0.7	0.5	0.7	1.0_e
	B	1.0 (60)	0.8 (60)	0.5 (66)	0.3 (60)	$0.4\ (60)_e$	$0.6\ (60)_e$
	C	0.6 (40)	0.6 (40)	0.2 (34)	0.2 (40)	$0.3\ (40)_e$	$0.4\ (40)_e$
Plants 3, 4	A	1.0	0.8	0.5	0.2	0.3_e	0.4_e
	B	1.0 (100)	0.8 (100)	0.5 (100)	0.2 (100)	$0.3\ (100)_e$	$0.4\ (100)_e$
	C	–, (–)	–, (–)	–, (–)	–, (–)	$–, (–)_e$	$–, (–)_e$
Totals		18.8	11.8 (6.9)	7.0 (3.5)	6.5 (3.5)	9.6_e	12.0_e

A Russian wood; *B* Birch chips, *C* Pine chips, Wood share is depicted in parentheses, per cent of used wood, %, *e* estimate

used as real costs have been deflated by indexes of traffic costs in Finland, 2011 = 100 (Strandström 2012).

Plant and district terminals' inventory costs were considered to be two euro per m^3. An annual interest rate of five percent was taken in consideration in the unit costs, used as time-varying coefficients in the model (Palander 1995a). In this research, the annual interest rate was constant, which could be changed by managers before time-varying parameters (unit costs) are computed automatically during model formulation (Palander 1995b).

The allocation constraints were formulated as the volume levels (m^3) in accordance with determination of wagon capacity, plants' wood use and possible wood resources. Aside the minimum total logistics costs and operating time, the other main goals were included into the model by computing adjustment parameters of the constraints. The maximum transportation volumes on each station were determined from the materialized wood supply volumes. Plants' inventory levels were set to correspond to the plants, a priori, weeks wood requirement in the beginning

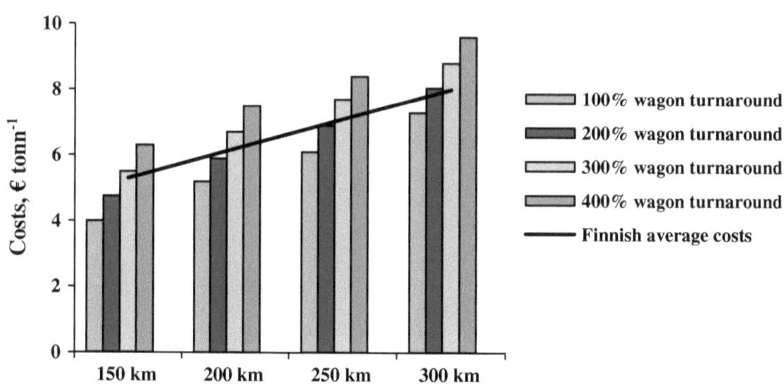

Fig. 3 The unit costs of rail transportation of wood in wood procurement in Finland (Iikkanen and Sirkiä 2011)

of the planning horizon and they were forced to change to a level of plants, a priori, weeks wood requirement in the end of the planning horizon. The periodical plants' inventory levels were scaled in the ratio of the maximum transportation possibilities of the 3PLs. The maximum levels of the districts' inventories of 3PLs were set to correspond to the terminals, a priori, wood volume capacity stored in the area. The technical constraints (e.g. non-negativity) are needed as standards when the model is solved by simplex method.

The modeling methodology was designed to produce a Pareto-optimal result (Dyer et al. 1992; Palander 1998, 2011; Tzeng and Huang 2014). The Pareto frontier is useful in simulations of forest technology: by restricting attention to the set of choices that are Pareto-efficient, a manager can make trade-offs within a non-dominated objective set rather than considering the full range of every parameter.

2.2 Wood Supply Applications in Finland-Russia Railway Network

The research area was formed from the area of the south-eastern Finland, therefore, only the wood resources of the two border stations were considered in the Finland-Russia railway network (Fig. 4). Within this area, three largest forest

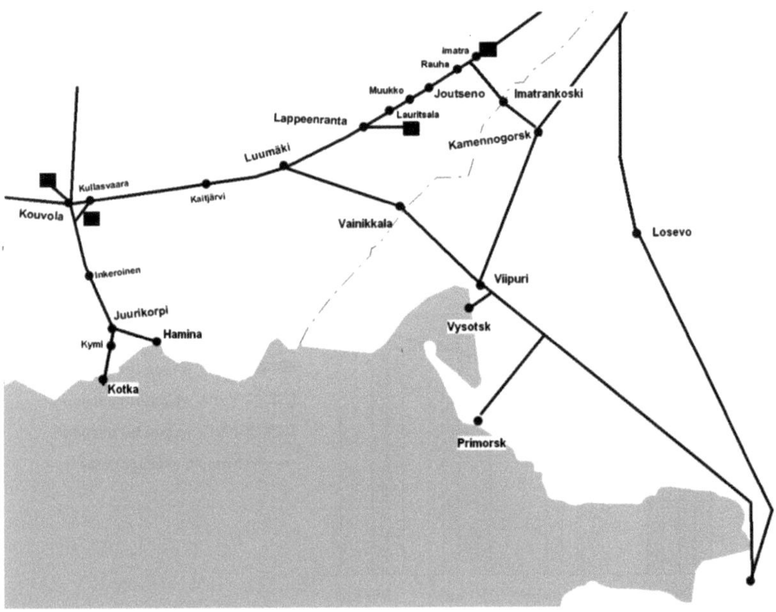

Fig. 4 The research area with energy plants (*black rectangles*) was formed from the area of the south-eastern Finland

products companies have saw mills and pulp and energy plants that use 66 % from imported Russian wood (Anon 2012c). The distances of the border stations and Finnish plants are less than 130 km (Goltsev et al. 2011).

These border stations can be considered as source of the wood resources (2.8 mill. m^3 in 2011) in Finnish forest industry. Within the research area Russian wood supply was 1.9 mill. m^3 (Anon 2012c). The initial energy wood supply of this research (0.6 mill. m^3) was allocated by wagons by trying to satisfy the wood demand of four energy plants (Fig. 5). The primary average wagon turnaround rates for 88 000 wagons are considered to be under 200 % by 2011 over our applications (Iikkanen and Sirkiä 2011; Anon 2012e). The average transportation speed of Russian wagons was 60 kmh^{-1}.

The strategic plans of this study were made for years 2011, 2012 and 2013. We assume that the annual delivery of Russian wood to the production would be increased by 100 %, compared to the delivering volume of 2011. Table 1 summarizes the changes of Russian wood import resulting from the restructuring programme implemented during 2009–2011. The investment growth slowed to a half in 2011 and businesses of the Finnish forest industry were adversely affected by rising cost levels of Russian wood. At the beginning of the programme, global financial crises also affected capacity reductions (2009), which can be found from the annual reports of Finnish forest industry's and Finnish Forest Research Institute

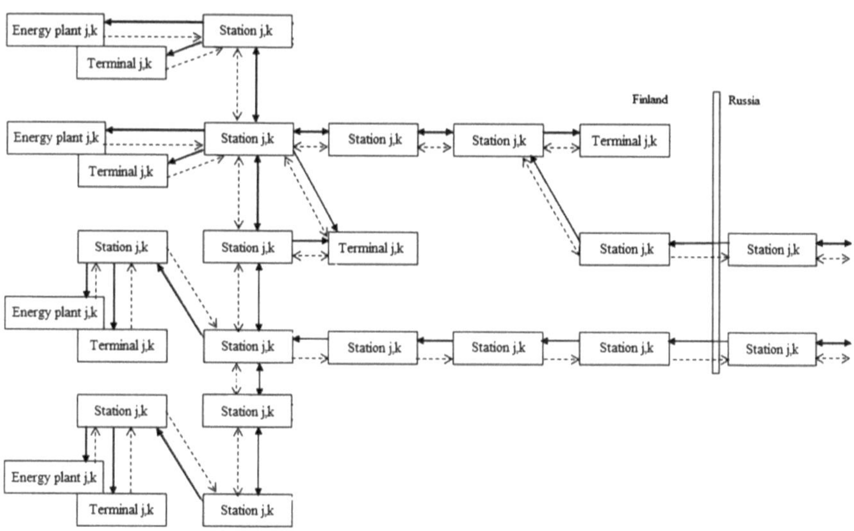

Fig. 5 General network of rail transport and traffic direction of energy wood wagons: Loaded wagon (*solid arrow*) and empty wagons (*dashed arrow*). Nodes indexes *j, k* = transportation from station *j* to station *k*, where $j = 1 \ldots, j \ldots J$, as point of departure of wood and wagons, and $k = J + 1 \ldots, k \ldots K$, as destination station of wood and wagons. Wagons are empty or full loaded by wood

(Anon 2012a, b, c). The total railway transportation volumes decreased by approximate 58 % since 2005 mainly as a result of significant Russian tariff rises over the same time period (Tolonen and Koponen 2011).

To illustrate 3PLs' different strategic alternatives, four scenarios are presented. The first scenario, #1, was a thought railway transportation and storing case of a wood import company based on primary problem data, which has a somewhat old logistics environment with wood supply and demand in 2011 (Table 1). However, the problem well describes the situation before the tariff reduction. The second scenario, #2, is the first of thought three strategic alternatives of the 3PLs. These scenarios were formulated by adjusting the first case. Actually wood demands and Russian wood import were iteratively changed during the test in order to reach the global optimum solution to the logistical wood procurement problem. The importance of the different scenarios is illustrated to the 3PLs in the beginning of the year 2013. Otherwise the scenarios had the same data. In each of the scenario, it is assumed that organizations of the supply chain parties can be prepared facing the scheduled planning horizon and hence the first wood mix to plants can be scheduled at time = 0.

The railway transportation and storing were analyzed using optimization. In the optimization scheduling algorithm was implemented using C programming language and the program was combined with a user interface created with MS Visual Basic of MS Visual Studio version 6.0. The Lindo Api version 5.0 with standard settings was used as the LP-solver. The optimization runs were computed using the experiments on a 2393 MHz x86 Family 6 Model 15 Stepping 10 GenuineIntel—laptop workstation with 4 GB RAM and Windows XP Professional operating system.

We applied DMOLP methodology, which is based on the proven theory to produce global optimum solution or infeasible solution (Taha 2010). Through the usage of optimization and operation analyses, conclusions could be made for the cost-efficient railway transportation strategies. They are based on Pareto optimal solution runs of the sensitivity analysis of the scenario #1. In each strategy, the global optimum solution was the most efficient solution for the problem in question (Hadley 1962; Dykstra 1984). Pareto optimality originates with the concept of efficiency in economics, and has been applied to various problems in management. Pareto optimality is a concept in multi-objective optimization that allows for the optimization of a vector of multiple objectives, enabling tradeoffs among optimal combinations of multiple objectives to be evaluated. Therefore, the results of the solution runs were also used to present the impacts of an increase in Russian wood import on storing and the wagon rotation. The differences of the strategic alternatives as well as the technical performance are discussed during evaluation. For the consideration of the performance, it is assumed that clear and acceptable differences in solution patterns (alternatives) are indicators of good quality of the methodology (Hadley 1962). If the differences are reasonable, they also reveal the importance of the system as a tool to support global strategic planning of the supply chain parties.

3 Results and Discussion

3.1 *Impacts of Railway Transportation Strategy on Supply Chain*

The impacts were revealed as the railway transportation alternatives to the plants, which were optimized before and after the increase in Russian energy wood supply using actual Russian wagons. Based on the DMOLP results of the scenario #1, successful optimal alternatives were feasible to reach in the south-eastern Finland in 2011 (Fig. 6). This indicates that the logistics were in balance to satisfy the wood demands of the plants in 3PLs' railway transportation strategy before the wood import rise in 2012. The solved alternatives are useful to consider with the Finnish energy wood procurement structures (Table 1).

After adjusting merely 3PLs' wagon traffic flow (wagon turnaround), the global optimum for the logistics problem could not be found in the single objective test. The reduction of empty wagons or operating time caused severe logistics shortages in wood flow and a need to cut down storing at terminals of districts and plants, which increased operating costs. Therefore, the supply chain parties should collaborate and select one global strategic alternative using DMOLP.

Traditionally, global wood procurement problems in Finland have been solved by purchasing more wood from Russia. This strategic choice was an easy and a cost-efficient alternative to adjust to diverse production situations. Therefore, it was also the most reasonable of railway transportation strategies to implement and to increase business in Finland for about 15 years until 2006. Between 2007 and 2011, the strategy was not available during the "threat of tariff rise" situation (Palander and Vesa 2012). Since 2012, the traditional strategy has been available and therefore, the suggestions about the efficient railway transportation strategy for the year 2013 were drawn from its impacts on global energy wood supply chain (Fig. 7).

First the minimum operating costs were determined by increasing wood supply from 100 to 200 % for the scenario #2, #3, and #4 (AW1, AW2 and AW3), respectively. Then, the railway transportation strategy was adjusted in the scenarios

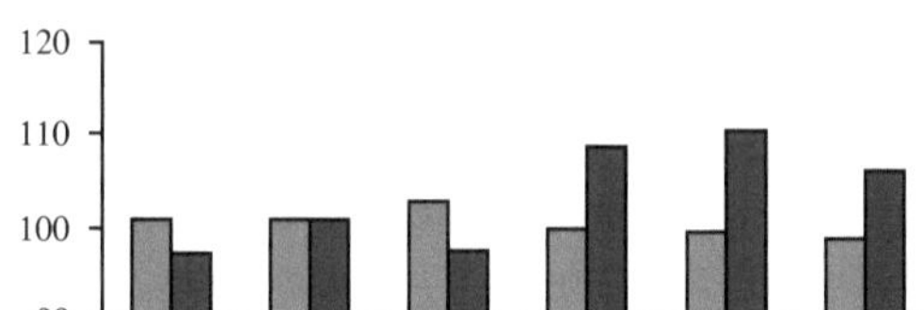

Wagon turnaround, % m^3	100	110	130	240	260	270
Total operating costs, % €	100.9	100.9	102.7	100	99.6	99
Total operating time, % min	97.1	100.9	97.7	108.8	110.5	106.2

Fig. 6 Wagon turnarounds (*horizontal line*) and objective values (*vertical line*) for wood railway transportation from Russia were optimized using the model of 2011, i.e., the model was formulated in accordance with energy wood procurement of plants in the year 2011

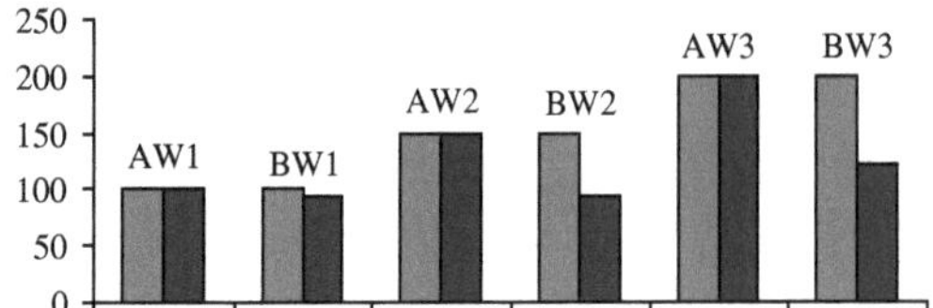

	AW1	BW1	AW2	BW2	AW3	BW3
Wagon turnaround, % m^3	170	210	260	180	190	140
Total operating costs, % €	100	100	150	150	200	200
Total operating time, % min	100	94.9	150	94.9	200	122.3

Fig. 7 Wagon turnarounds (*horizontal line*) and objective values (*vertical bars*) for railway transportation strategy: AW1, AW2, and AW3, supply chains that include Russian wood supply rates of 100, 150, and 200 %, respectively; BW1, BW2, and BW3, supply chains with minimum operating time (*right bar*) that include wood supply rates of 100, 150, and 200 %, respectively

by updating the wagon turnaround coefficients of Russian wagons according to the minimum total operating time (BW1, BW2 and BW3). Using the annual volumes of energy wood import from Russia to Finland in the Table 1, we found, that Russian wood import increased from 6.2 mill. m^3 at 2011 to 12.0 mill. m^3 at 2013. If the Russian wood import had been at the level of 2011, the decision-making situation for the railway transportation strategy of the year 2013 would have been in accordance with the supply chains that include wood supply rates of 100 % (BW1) in the Fig. 7. It also describes—in balance—strategy that should have been decided in 2012. Based on the results of the scenarios 100, 150 and 200 %, we can draw the suggestion that the wagon turnaround could be cut down from 260 to 180 % in 2012 and from 190 to 140 % in 2013, if Russian wood supply increases.

3.2 Supply Chain Parties' Decision-Making Criteria

In this research, it was assumed that Russian energy wood import would rise up to 100 % after the tariff reduce from 15 € m^{-3} to 4 € m^{-3} in 2012. To determine ranges of the decision alternatives for the efficient railway transportation strategy, sensitivity analyses were made by DMOLP for the scenarios (Table 2). The ranges of the decision-making criteria describe possible optimal choices for the supply chain parties, e.g., to what extent is performance enhanced by matching railway transportation strategies to global logistics management.

Table 2 also indicates possible changes in the global supply chain until the end of the year 2013. The results can be justified as useful when the predicted adjustments of the logistics occur after the effects of the Russia's WTO membership on the global decision-making environment. Until the end of the year 2011, the 3PLs cut down Russian energy wood supply in the plants (Table 1). At the end of the year 2012 and during the year 2013, Russia's WTO membership caused operations restoration of the supply chain parties in Finland, which are predicted in this research. It seems that the managers' strategic decisions can be made as

Table 2 Ranges of objective values for globally optimal non-dominated decision alternatives for supply chain management of Russian wood

Scenario		Total costs			Total time	
	I, % €	Goal, % €	D, % €	I, % min	Goal, % min	D, % min
A1	80.1	100.0	0.3	8.6	100.0	16.2
A2	124.9	150.1	0.6	58.8	122.2	54.9
A3	94.4	199.7	0.3	152.0	149.9	0.1
B1	80.1	100.0	0.3	37.7	100.0	33.1
B2	123.8	150.1	0.6	80.2	122.2	56.5
B3	94.7	199.7	0.3	168.9	149.9	0.1

I Increase, *D* Decrease; *A1*, *A2*, and *A3* supply chains that include a wagon turnaround of 100 % and Russian wood supply rates of 100, 150, and 200 %, respectively; *B1*, *B2*, and *B3* supply chains that include a wagon turnaround of 200 % and Russian wood supply rates of 100, 150, and 200 %, respectively

collaborative decisions using DMOLP without a lack of reactions which could cause logistics problems for Finnish forest industry.

Concerning the predictions of the railway transportation strategies, the concluding chain is logical and it illustrates how logistics information can be produced efficiently for the supply chain parties by using optimization results as a part of the adjustment method. The results suggest that the role of the information logistics could be increased as forest industry may try to adjust their rail transportation and storing with respect to potential adjustments in global decision-making environment. Table 2 indicates that the decision support systems could be developed particularly for increasing quality of decision alternatives to be available for decision makers in strategic planning in the future. In practice, the logistic parties' decisions are actually made for short-term planning, due to changes been made without considering the strategic plans as the collaborative manner. In order to increase efficiency in planning and decision making, the right planning methods should be returned to the systems in accordance with the theory of the decision making. As suggested in previous studies (Palander and Vesa 2012), the supply chain parties could increase their development contributions to the operation research and actually take a look at what good it has done related to this global logistics decision-making process and how to improve its efficacy.

The WTO membership was launched just few months prior to the optimization runs of this research by 2012. Therefore, the impacts of the increase in Russia's energy wood import were examined at a relatively high theoretical and abstraction level, which stroke through especially in the determination of used basic data in the logistics problem of the 3PLs in the south-eastern Finland. Obviously, the volumes of the standardized models depart from the real volumes. However, the problem was out of interest, because the purpose was to model their natural variation in the forest industry. Thus, the range figures can be used as predictions for collaborative strategic planning purposes. In addition, optimization found the most cost-efficient railway transportation strategy of the 3PLs that also satisfies criteria of the other supply chain parties.

Russia represents a small share of the world supply of wood. Thus, the overall welfare effect of implementing a tax on wood has been negative in Russia (Piermartini 2004). Recently, Russian tariffs on coniferous wood have fallen the most: in practice by 50 %. Cheaper timber tariffs on birch could also boost the entire forest industry in Finland. In addition to wood imports, Russia's WTO membership will mean easier access for Finnish companies to Russian markets. For example, highly processed tissue and carton products are in high demand in Russia and lower customs charges mean narrower price differences. This will give the supply chain parties better chance to adjust wagon and wood traffic to the global decision-making environment. The hope of easier market access has already increased wagon traffic to Russia by Finnish forest industry.

3.3 Impacts of Wood Supply on Storing and Wagon Transportation

To increase the quality of decision alternatives in decision-making group, the structure of wood storing can be considered more informatively. Here, study on the impacts of Russian energy wood import in the south-eastern Finland was carried out by considering carefully plant and district terminals' storing in the railway network (Fig. 5). The changes in 3PTs' storing plans were computed from the strategic railway transportation plans in the scenarios #2, #3 and #4. We calculated the impacts on the storing by comparing the monthly wood inventories. Figure 8 depicts changes of the storing as terminal volumes. Wood is transported by truck from terminals to production during the same month.

No difference was found in the strategies between the results of the model and the 3PLs' real life storing strategy in research region (Iikkanen and Sirkiä 2011). Therefore, the authors do not suggest additional storage capacity for district terminals. In this respect, the results supported the suggestions for there being enough periodical tolerance in wood inventories for coping with the energy wood supply chains at the procurement levels of 2013. However, in order to find the global optimum strategy for the year 2013 for the railway transportation in the decision-making environments of rising Russian wood export, additional strategic changes can be required in order to restructure storing processes in plant and district terminals.

In practice, inefficient storing solutions are often implemented within the supply chain parties. At the plants of this research the storing volumes have, in fact, gone greatly up in the case of 200 % for imported energy wood is compared to that of 100 % (Fig. 8). The storing of the plant terminals increased of 33 %. Furthermore, the storing in the terminals was not in balance to satisfy 3PLs' operational activity in wood supply of 200 %. To balance terminal's wood flow, the supply chain parties should form a decision-making group that makes collaborative decision (see Palander et al. 2005). This kind of efficiency analysis is useful, because 60 % of district terminals have been withdrawn from use during past decade in Finland.

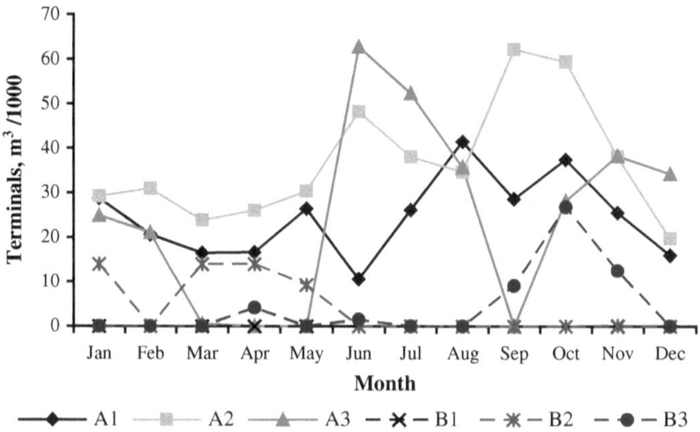

Fig. 8 Storing in the plant terminals (A1, A2, A3) and district terminals (B1, B2, B3) in the supply chain plans that include a wagon turnaround of 180 % and Russian energy wood supply rates of 100, 150, and 200 %, respectively

The target state for 2018 is a rail network consisting of 14 terminals and 32 loading points (Iikkanen and Sirkiä 2011).

To date, Russian wagon supply has been used in order to improve the system's balance in the railway transportation network. Here, 3PLs' purchase and renting needs of wagons is depicted more informatively by allowing the over-allocation (turnaround) of wagons from the Finland-Russia border stations to plants and back to Russia in the adjusted models (scenarios #2, #3 and #4) (Fig. 9). In practice, the over-allocation level of the Russian wagons was 180 % (Iikkanen and Sirkiä 2011).

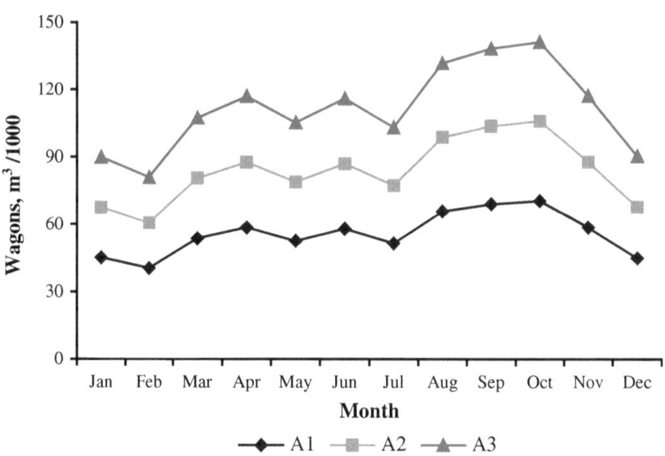

Fig. 9 Backhauling schedule of empty wagons to Russia in the supply chain plans for scenarios A1, A2, and A3, which include a wagon turnaround of 180 % and Russian energy wood supply rates of 100, 150, and 200 %, respectively

The excessive reduction of empty wagons caused severe shortages of wood flow and a need to cut down storing at terminals and plants (Figs. 6 and 7). The results showed that, in theory, by increasing Russian wagon supply to the transportation network, the wagon demand of the 3PLs could be fulfilled in the south-eastern Finland.

The wagon over-allocation rate is within the figures published in the National Transport Agency for 2011 (Iikkanen and Sirkiä 2011). However, it should be beard in mind that the 3PLs' used only Russian wagons and wagon backhauling to Russia is allocated in the scenarios. Finnish wagons (10 364) could also be used in wagon rotation (Anon 2012d). Then the combined wood wagon capacity could be 24 % higher in the allocation case of domestic and Russian wagons. From a technical point of view of modeling, in order to reach the global optimum solution of the balanced strategy, we can use adaptable parameters to adjust the system (Palander 1995b). For example, we can adjust initial conditions of the model to reflect the new wagon data. If the system was in balance, the adjustment option during the decision support process would not need to be used.

3.4 Technical Performance of the Model

The reformed wood-flow model proved to be more efficient than the model developed by Iikkanen et al. (2010) or Palander and Vesa (2012) due to the enhanced model allowing multiple objectives to be taken into consideration. The wood-flow model by Palander and Vesa (2012) did not sufficiently consider storing and rail transportation of Russian wagons and wood. The original model allocated wood from the Finland-Russia border stations and ensured that the best possible wood delivering alternative were in the model. This, however, does not ensure that wagon rotation is in balance. This was important because energy wood import increase resulted in the radical increase of the operating time of delivered wood. Therefore, some of the delivering alternatives could presumably reduce rather than increase the efficiency of railway transportation strategy. This was avoided by the use of the methodology of this research, because the integrated wagon and wood flow model allocated Russian wood supply from the Finland-Russia border stations either by train or truck or by their combination based on the multiple decision-making objectives.

The test results of the optimization run for the scenarios were good. For example, the sum of the LP iterations was 1837 and the duration of the solution procedure 9 CPU-seconds for the scenarios #2. The results show that the methodology performed well due to the decision support system allows efficient modules to adjustment, data management, modeling, optimization and interpretation of results. For example, during a planning process, a high level parameterization and automatic generation of the LP–matrix was found useful. The results of this study are in line with research results of previous wood procurement optimization studies of Finland (e.g. Palander and Väätäinen 2005; Palander et al. 2005; Palander 2011;

Palander and Vesa 2012; Palander and Voutilainen 2013). Research to further improve the solution quality is not necessary, as the optimization guarantees a global optimal solution within the normal computational possibilities. The reached multiple-objective function values are also reasonable, because the results presented in the previous sections have shown that the differences between decision alternatives are clear and acceptable.

The model was primarily applied for adjusting wood procurement logistics in the south-eastern Finland. However, the theory and applications can be developed even for larger industrial environments with the presence of precedence constraints. In that work, the model is an adaptable instrument and the DMOLP method allows large decision-making problems to be solved efficiently. For example, the methodology solved the dynamic problem using multiple criteria, which is important to the systems used for collaborative scheduling tasks of the supply chain parties. In this respect, the methodology proved to be more efficient than the methods used in the similar studies in Finland (Hetemäki and Hänninen 2009; Iikkanen et al. 2010; Iikkanen and Mukula 2010; Reini et al. 2010). Therefore, the methodology has great potential for significant improvements of decision-making efficiency under increasing logistics of similar global decision-making situations.

The scheduling task is actually solved using linear goal programming technique. The controversies surrounding linear goal programming mostly come from misconceptions about the principle of satisfying solution which underlies goal programming theories. It is almost impossible for the decision maker to achieve ideal goals without the expense of other goals in optimization of multiple goals. In this sense, the efficiency of solutions is problem dependent and user dependent. Therefore, in this study, an appropriate set of solutions was prepared by the modeler for the decision maker. If the decision maker sets unreasonable targets or assigns incorrect weights and/or priorities, a DMOLP solution cannot provide the best available or efficient solution. Therefore, the limitations of DMOLP method are due mainly to errors of its users, not to the rationale behind theories. However, there is no doubt that further developments of DMOLP theories are no longer needed.

3.5 *Pareto Optimal Solution of the Model*

In the present in-depth case study Pareto optimal approach was used during collaboration process. According to literature, the basic concept of goal programming is that whether goals are attainable or not, an objective will be stated in which optimization gives a result which comes as close as possible to the desired goals (satisfied solution) (Dyer et al. 1992; Palander 1998; Tzeng and Huang 2014). Based on the theory, goal programming technique yields non-dominated solutions if the goal point is chosen in the feasible domain. It has been proved that when the goals are feasible, the solution given by the technique is efficient, but the efficiency of the provided solution when the goals are not feasible remains an open problem. Major advantage of goal programming is that there always exists a solution to the problem, provided that it has feasible region and this is because of the inclusion of the deviational variables.

In the present approach, four different deviations of the objectives are considered in the DMOLP framework. They are combined with a normalized weighting approach, i.e., the weighted sum of the deviations are optimized. The choice of normalization method can affect the optimal solution found and hence care was taken in choosing a method that reflected the preferences and interests of the decision maker. In this study, the weighting space was explored by the modeler. The percentage normalization method was selected, as the decision maker could not determine range of values from the target value for the total operating time. The decision maker also preferred the presentation of a small number of different solutions given in terms of their achievements of goals and decision variable values. In that sense the percentage normalization method is also an efficient approach.

It is well known that optimal solution depends very strongly on the weights chosen and that one can find an assignment of weight which declares almost any feasible solution as optimal. Thus, Pareto approach is preferable, especially because LP model is applied. The results of the present in-depth case study illustrate a simple case of minimizing two objectives simultaneously (Total operating costs, Total operating time). Results could be presented using a figure with the solid line indicating the Pareto optimal frontier (whereby any improvement with respect to Total operating costs comes at the expense of Total operating time) (Palander 1998). Each point along that frontier represents a unique model parameterization and/or model structure, so Pareto optimality identifies multiple Pareto optimal solutions. In the present study, value points of the objectives are presented in Table 2. The objective function contributions were multiplied by 100 to give percentages, as the weights in the function were divided by the target levels. Through Pareto procedure a decision maker is able to consider differences among the non-dominated multiple optimal solutions that optimize varying combinations of logistics operating cost and time objectives.

3.6 Future Extensions of Railway Transportation Model

In the present article, the logistics railway transportation alternatives of Finnish forest industry under increasing imports of Russian wood are discussed. The DMOLP is used to model the railway transportation from Russia to the energy plants in the south-eastern Finland. The objective of this handbook is to present recent advancements on optimization and simulation models and tools developed to optimize the economic, environmental and social impact of bioenergy. This article is focused on consideration of these aspects (economy, environment and social group) during optimization of impact of bioenergy. Outsourcing can be considered as a social aspect. Outsourcing of logistics operations to 3PLs has been a common approach to organize plant's raw material procurement in global production business. In Finland VR group has recently provided logistics services to energy plants (customers). They form a group of logistic parties which manages a bioenergy chain. The results show that tradeoffs among the different objectives of logistics parties should be considered, when no clear winner among them could be identified.

In the present article, tradeoffs were calculated by constructing an additive objective function where the different objectives were weighted appropriately, according to their relative importance and percentage normalization. In practice, tradeoffs can be found on the efficient frontier that is a practical tool supporting decision making. Instead of complex definition of optimality in the context of multi-objective optimization, logistics parties receive a set of efficient solutions, and they need to select one of them, as decision-making group's compromise solution.

The shortcoming of the present methodology is a traditional win-lose situation (Tzeng and Huang 2014). Hence, a win-win strategy is a focus for future research. Under the win-lose strategy the logistics parties can optimize its system only via its given resources and restricted capabilities. However, outsourcing aims to the win-win situation, where managers are interested in creating added value to logistic parties. As an extension for the presented methodology logistics parties should be interested in achieving better parameters with ideal modeling through re-allocating limited resources. Even more advanced extension with modeling formulates aspiration points using expanded competence set. However, based on preliminary tests (Palander 1998), with this extension computational complexity increases too much for practical applications of forest energy industry.

4 Conclusions

In this research, the DMOLP is used to model impacts of the decrease in Russian wood tariffs and the increase in Russian energy wood imports on the wood procurement logistics of the plants in the south-eastern Finland. If the energy wood import from Russia increases, the 3PLs can implement integrated strategic plans to continue logistics operations with more efficient wood supply chains. The increase in Russian wood supply necessitates changes in the railway transportation strategy as well as in the collaborative planning and decision support system. Then Russian wood import growth may fasten, and transportations of the 3PLs can be positively affected by faster wagon rotation and balanced wood storing levels.

Further studies would be needed to increase the efficiency of the supply chains of Russian energy wood import, e.g., at what point should various sourcing criteria and decisions be rethought in global light of changing decision making environment. This preliminary research could serve as a guideline for further optimizations and discussions on the adjustment of the railway transportation strategies in the global environments. Lessons have been learnt and the integrated optimization model (DMOLP) presented in this paper is recommended to the supply chain parties. Furthermore, the adaptation of the railway transportation to the complicated environment would require carry out collaborative logistics planning in the supply chain parties' group to be concluded.

Appendix

Nomenclature

ce_{wjkt}	The transportation time (min) of empty wagon *w* transported from railway station *j* to railway station *k* during period *t*
cl_{ijt}	The purchase and transportation costs (€ m^{-3}) of wood assortment *i* to border station *j* during period *t*
cm_{ikt}	The plant inventory cost (€ m^{-3}) of wood assortment *i* in plant and terminal *k* during period *t*
ct_{wjt}	The inventory cost (€ m^{-3}) of empty wagon *w* in railway station *j* during period *t*
cx_{ijt}	The inventory cost (€ m^{-3}) of wood assortment *i* in railway station *j* during period *t*
cy_{ijkt}	The transportation cost (€ m^{-3}) of wood assortment *i* transported from railway station *j* to railway station *k* during period *t*
cw_{ijkt}	The transportation time (min m^{-3}) of wood assortment *i* transported from railway station *j* to railway station *k* during period *t*
C_{tot}	Total operating costs (€) of the supply chain during the planning horizon
D_g	Deviation (€) above and below the decision-makers' goals defined by the study problem
G	Number of deviations from the decision-maker's goals (1, …, *g*, …, 4)
I	Number of wood assortments (*1*, …, *i*, …, *I*)
J	Number of railway stations (*1*, …, *j*, …, *J*) as point of departure of wood and wagons
K	Number of railway stations (*J* + *1*, …, *k*, …, *K*) as destination station of wood and wagons
L	The volume (m^3) of wood assortments purchased from Russia and transported to Finland via the border stations during the planning horizon
L_{ijt}	The volume (m^3) of wood assortment *i* purchased and transported to border station *j* during period *t*
$Lmax_{ijt}$	Maximum volume (m^3) of wood assortment *i* purchased from Russia and transported to Finland-Russia border station *j* during period *t*
$Lmin_{ijt}$	Minimum volume (m^3) of wood assortment *i* purchased from Russia and transported to Finland-Russia border station *j* during period *t*
M_{ikt}	The volume (m^3) of wood assortment *i* in plant and terminal inventory *k* in the end of period *t*
MB_{ik}	The volume (m^3) of wood assortment *i* stored by plant *k* at the end of the planning horizon
MI_{ik}	The volume (m^3) of wood assortment *i* stored by plant *k* at the beginning of the planning horizon

MH_{ikt}	The volume (m^3) of wood assortment *i* demanded by plant (respective railway station *k*) during period *t*
$MHmax_{ikt}$	Maximum volume (m^3) of wood assortment *i* demanded by plant *k* during period *t*
$MHmin_{ik}$	Minimum volume (m^3) of wood assortment *i* demanded by plant *k* during the planning horizon
$Mmax_{ikt}$	Maximum volume (m^3) of wood assortment *i* held in plant inventory *k* (railway station) during period *t*
$Mmin_{ikt}$	Minimum volume (m^3) of wood assortment *i* held in plant inventory *k* (railway station) during period *t*
p	Annual interest rate (%)
T	Number of periods (1, …, *t*, …, *T*)
T_{tot}	The total operating time (min) from the supply chain during the planning horizon
W	Number of and wagons (*1*, …, *w*, …, *W)*
w_g	Positive weight that reflect the decision-maker's preferences regarding the relative importance of objective *g*
WB	Minimum volume (m^3) of empty wagon transportation above the wood transportation volume (m^3) from railway transportation network to Russia during the planning horizon
X_{ijt}	The volume (m^3) of wood assortment *i* in inventories of railway station *j* in the end of period *t*
XB_{ij}	The volume (m^3) of wood assortment *i* stored by railway station *j* at the end of the planning horizon
XI_{ij}	The volume (m^3) of wood assortment *i* stored by railway station *j* at the beginning of the planning horizon
$Xmax_{ijt}$	The maximum wood inventories (m^3) of wood assortment *i* held in railway station *j* during period *t*
$Xmin_{ijt}$	The minimum wood inventories (m^3) of wood assortment *i* held in railway station *j* during period *t*
XT_{wjt}	The volume (m^3) of empty wagon *w* in inventories of railway station *j* in the end of period *t*
XTB_j	The volume (m^3) of wagons stored by railway station *j* at the end of the planning horizon
XTI_{wj}	The volume (m^3) of wagon *w* stored by railway station *j* at the beginning of the planning horizon
$XTmax_{jt}$	Maximum wagon inventories (m^3) of wagons held in railway station *j* during period *t*
$XTmin_{wjt}$	Minimum wagon inventories (m^3) of wagon *w* held in railway station *j* during period *t*
Y_{ijkt}	The volume (m^3) of wood assortment *i* transported from railway station *j* to railway station *k* during period *t*
YH_{wjkt}	The volume (m^3) of empty wagon *w* transported from railway station *j* to railway station *k* during period *t*

YHmax Maximum volume (m^3) of wagons the during the planning horizon
$YHmax_{wjkt}$ Maximum volume (m^3) of wagon w transported from railway station j to railway station k during period t
$Ymax_{ijkt}$ Is the maximum volume (m^3) of wood assortment i transported from railway station j to railway station k during period t
$Ymin_{ijkt}$ Is the minimum volume (m^3) of wood assortment i transported from railway station j to railway station k during period t
Z The optimum weighted sum of $w_g D_g$ during the planning horizon

References

Anon (2007) World markets. Russian Log Export Tax. Equity Research Industry Update. Canadian Imperial Bank of Commerce (CIBC), March 13, 2007

Anon (2012a) Annual report of StoraEnso, 2011

Anon (2012b) Annual report of UPM-Kymmene, 2011

Anon (2012c) Forest statistic year book 2012. Finnish Forest Research Institute. http://www.metla.fi/metinfo/tilasto/julkaisut/metsatilastollinenvsk/

Anon (2012d) Annual report of VR Group, 2011

Anon (2012e) Finnish railway statistics, 2011. Statistics of the Finnish Transport Agency 6

Bengston SJ (1966) A mathematical model for optimum timber allocation. M.S. thesis, Oregon State University. Corvallis

Carlsson D, Rönnqvist M (1999) Wood flow problems in Swedish forestry. Skog Forskning Rep 1:26–30

Dyer J, Fishburn P, Steuer R, Wallenius J, Zionts S (1992) Multiple criteria decision-making, multiattribute utility theory—the next ten years. Manage Sci 38:645–654

Dykstra DP (1984) Mathematical programming for natural resource management. McGraw-Hill Book Company, New York, p 318. ISBN 0-07-018552-2

European Commission (2008) The raw materials initiative—meeting a critical needs for growth and jobs in Europe. Communication from the Commission to the European Parliament and the Council. http://ec.europa.eu/enterprise/newsroom/cf/document.cfm?action=display&doc_id=894&userservice_id=1&request.id=0

Goltsev V, Tolonen T, Syunev V, Dahlin B, Gerasimov Y (2011) Wood harvesting and logistics in Russia—focus on research and business opportunities. Working papers of the Finnish Forest Research Institute, vol 210, pp 0–157

Hadley G (1962) Linear programming. Addison-Wesley Publishing Company. Inc., Reading

Hetemäki L, Hänninen R (2009) Arvio Suomen puunjalostuksen tuotannosta ja puunkäytöstä vuosina 2015 ja 2020. Working papers of the Finnish forest research institute, p 122 (In Finnish with English Summary)

Iikkanen P, Keskinen S, Korpilahti A, Räsänen T, Sirkiä A (2010) A national optimisation model for raw wood streams. Research reports of the Finnish Transport Agency, vol 29, pp 0–30 (In Finnish with English Summary)

Iikkanen P, Mukula M (2010) Rail network freight traffic forecast 2030. Research reports of the Finnish Transport Agency, vol 37, pp 0–60 (In Finnish with English Summary)

Iikkanen P, Sirkiä A (2011) Development of the railway raw wood terminal and loading point network. Study covering all forms of transport. Research reports of the Finnish Transport Agency, vol 31, pp 0–50 (In Finnish with English Summary)

Palander TS (1995a) A dynamic analysis of interest rate and logging factor for reducing saw timber procurement costs. Int J For Eng 7:29–40
Palander T (1995b) Local factors and time-variable parameters in tactical planning models: a tool for adaptive timber procurement planning. Scand J For Res 10:370–382
Palander T (1998) Tactical models of wood-procurement teams for geographically decentralized group decision making. D.Sc. (Agr. and For.) thesis. University of Joensuu, Faculty of Forest Sciences
Palander T, Väätäinen J (2005) Impacts of inter-enterprise collaboration and Backhauling on wood procurement in Finland. Scand J For Res 20:177–183
Palander T, Kainulainen J, Koskinen RA (2005) Computer supported group decision-making system for timber procurement planning in Finland. Scand J For Res 20:514–520
Palander T (2011) Modelling renewable supply chain for electricity generation with forest, fossil, and wood waste fuels. Energy 36:5984–5993
Palander T, Vesa L (2012) Potential methods of adjustment to declining imports of Russian roundwood for the Finnish pulp and paper industry. Int J Logist Manage 22(2):222–241
Palander T, Voutilainen J (2013) Modelling fuel terminals for supplying a combined heat and power (CHP) plant with forest biomass in Finland. Biosyst Eng 114:135–145
Parker E, Thornton J (2007) Fiscal Centralisation and decentralisation in Russia and China. Comp Econ Stud 49:514–542
Piermartini R (2004) The role of export taxes in the field of primary commodities. World Trade Organization, pp 1–24. http://www.wto.org/english/res_e/booksp_e/discussion_papers4_e.pdf
Reini K, Törmä H, Mäkinen J (2010) Massa- ja paperiteollisuuden supistumisen ja tulevaisuudenkuvien aluetaloudelliset vaikutukset. University of Helsinki, Rularia-institute. Report, vol 50, pp 0–38 (In Finnish). http://www.helsinki.fi/ruralia
Strandström M (2012) Harvesting and long-distance transportation. Metsätehon katsaus, vol 48 (In Finnish with English Summary). http://www.metsateho.fi
Taha HA (2010) Operations research: an introduction, 9th edn. Prentice Hall, Upper Saddle River
Thee KW (2009) The Indonesian wood products industry. J Asia Pac Econ 14(2):138–149
Tolonen T, Koponen M (2011) Havupuun tuonti Venäjältä Suomeen ja tuontihavupuun kasvinterveystarkastukset vuosina 2005–2009. Working papers of the Finnish Forest Research Institute, vol 197, pp 0–32 (In Finnish with English Summary)
Tzeng GH, Huang JJ (2014) Fuzzy multiple objective decision making. Taylor & Francis Group, Chapman and Hall CRC Press, New York, 322 p. ISBN 978-1-4665-5461-0

Gis-Based Methodology for Optimum Location of Biomass Extraction Plants and Power Plants Using Both Logistic Criteria and Agricultural Suitability Criteria

L. Morales-Rincón, A. Martínez, F.B. Avila-Díaz, J.R. Acero, E.F. Castillo-Monroy and Ariel Uribe-Rodríguez

Abstract A GIS-based methodology to identify the optimal locations for biomass extraction plants and biomass power plants is presented. Both agricultural land suitability criteria and logistic criteria were taken into account to select the optimal locations. Agricultural land suitability criteria were included as several independent variables of edaphic and climate conditions. A generalized additive model (GAM) was developed for estimating crop yield by using those edaphic and climate independent variables in potential zones where crops of interest are not currently grown, planted or seeded. Logistic criteria were incorporated in the model via network analysis of the available roads for the accessibility of each zone. Using a saturation approach of candidate locations, it was possible to generate a ranked list of sites for the project development. This list can be sent as input to an energy supply chain optimization model

Keywords Biomass residues · Crop yield model · Road network · GIS · Plant location · Logistic criteria · Agricultural land suitability

Highlights

- The proposed method provides a systematic, quantitative approach to select the best sites for the development of biomass extraction plants and biomass power plants.
- Using this methodology it is possible to systematically manage the data required to build a crop yield surface over areas where the crops of interest are not currently grown, planted or seeded.
- The use of the proposed methodology allows deriving input data for an energy supply chain optimization model.

L. Morales-Rincón · A. Martínez · F.B. Avila-Díaz · J.R. Acero · E.F. Castillo-Monroy · A. Uribe-Rodríguez (✉)
Instituto Colombiano del Petróleo, Piedecuesta, Santander, Colombia
e-mail: ariel.uribe@ecopetrol.com.co

S.D. Eksioglu et al. (eds.), *Handbook of Bioenergy*,
Energy Systems, DOI 10.1007/978-3-319-20092-7_5

1 Introduction

Biomass energy exploitation through direct combustion processes can be directed to thermal energy or electricity production. The selected locations of bioenergy plants and biomass sources play an important role in the decision making process about plant sizes and type of energy to generate in a given area: Thermal or electrical energy (Freppaz et al. 2004).

Thermal energy as a final product has the disadvantage of requiring demand for this type of energy very near its production site.

On the other hand, biomass power plants require a sustainable supply of biomass in order to keep plant's full capacity in use and cheap operation expenditure per kWh of generated electricity (Panichelli and Gnansounou 2008). Besides, the profitability of these plants is strongly influenced by the biomass cost per ton at the plant gate (Panichelli and Gnansounou 2008).

Transportation costs are of paramount importance for any energy project, and this is especially true for biomass, because of its low energy density (low heating value) compared to fossil fuels (Searcy et al. 2007). Consequently, establishment of biomass-based facilities in suitable locations so as to minimize transportation costs is one of the key issues for their profitability (Sultana and Kumar 2012).

The rather dispersed geographical distribution of biomass potential has raised the interest of researchers in using Geographical Information Systems (GIS) for the evaluation of biomass supply from current crops, estimation of transportation costs to existing power plants, and site selection for energy crop developments (Voivontas et al. 2001).

The major requirement for bioenergy production is land; consequently the competition with food production is an important aspect in determining the global bioenergy potential (Fischer and Schrattenholzer 2001). Not surprisingly, the use of agricultural residues for energy production has become a focal point of various studies, not only to avoid the competition with food production, since the residues could come from food crops; but also to give a double use to the soil: main crop (for production of food or liquid biofuels) and production of crop residues for power generation and other valuable applications (Voivontas et al. 2001; Yu et al. 2012).

In 2010, *Unidad de Planeación Minero Energética*—UPME—published an atlas of Colombia, with energy potential data of residual biomass, using the information reported by the Colombian administrative units in 2007 for the crops harvested that year. For the first time in this country, biomass residues and their energy potential were estimated in 2010 UPME's work (UPME et al. 2010). UPME also developed a model for estimating agricultural yield of the species of interest, in other potential zones different from the ones where the crops already existed (UPME et al. 2010). This model was based on an average value for each influencing edaphic and climate variable; annual or monthly variation of these conditions was not considered by UPME for yield model development or yield prediction. Accessibility of biomass residues required for the development of bioenergy power projects was also disregarded by UPME.

In the following approach, potential locations suitability is determined for new biomass extraction plants and new biomass power plants, by both availability of residual biomass and accessibility to biomass-generating lands. A methodology has been developed, not only to incorporate the effects of annual and monthly climate variability in the crop yield estimation of current and new potential zones, but also to consider accessibility criteria and to select a ranked list of optimal locations for the project. Since, most of the variables involved in this methodology have an intrinsic spatial nature, both GIS and geostatistics are used in this analysis.

2 Methodology Description

2.1 Assessing Current Biomass Availability

Different information sources for current biomass availability were evaluated. The highest spatial resolution of the available information was the administrative unit. In order to compile existing information, a residual biomass inventory basic application was created. Using this application, the amount of residual biomass of the studied species was calculated by taking into account the tons of main product in each administrative unit, hectares of land used by these crops, residues factors for each species and residues availability for alternative uses in the considered areas.

The areas with the highest density of residual biomass were identified using the "Kernel Density" function implemented in ArcGIS, which is based on the quadratic kernel function described in Silverman (1986). This step allowed for the identification of areas that were selected for subsequent analysis.

2.2 Biomass Production Estimation in Potential Zones for Crop Development

Since the current distribution and amount of biomass may not be enough to assure projects' sustainability, it is necessary to identify potential zones for crop development considering the species to be used for biomass production.

Potential areas were selected by identifying zones that met edaphic and climate conditions requirements for crop development. Zone suitability is rated according to main product yield, which is estimated using a statistical model, previously adjusted as explained in section 2.2.1, based on edaphic and climate conditions and average yield reported in 2007–2009 by administrative units where the crops were harvested those years.

Once production is calculated, residual biomass for each byproduct and its availability for alternative uses can be estimated.

2.2.1 Crop Yield Model

The statistical model was created using a generalized additive model (GAM) framework. In a model of this type, the traditional linear or nonlinear parametric function, relating response and predictor variables, can be replaced by a non-parametric smoothing function, which provides a flexible method for visualizing relationships when the choice of a parametric function is not obvious (Wullschleger et al. 2010).

The available information to develop the model included digital datasets of georeferenced crops polygons for the studied species, edaphic conditions (including depth, fertility and drainage), and a set of layers of climate variables (Table 1). Independent variables for the model were strategically selected from the available variables. The proposed model could use not only continuous variables but also categorical variables, as explained below in section 3 where the case of study for oil palm is described.

Random point clouds were created within the areas where the crops are established. For each point the independent variables were extracted for the period 2007–2009 as the data set for these years contained simultaneous data of crop yield and climate conditions for all administrative units. This extraction was implemented in ArcGIS model builder. All the information was stored in a geodatabase. SQL (Structured Query Language) queries were used to create a consolidated table that included information about the independent variables and the average yield reported by the administrative unit where each crop is located.

The model was developed by using the statistical software R-Studio®. The statistical library used for this purpose was "mgcv". The consolidated table previously mentioned was used as input to the "gam" function included in the "mgcv" library.

Table 1 Climate variables available. http://www.cgiar-csi.org/data

Variable	Units	Temporal range
Cloud cover	%	1901–2009
Diurnal temperature range	°C	1901–2009
Annual precipitation	mm/year	1901–2009
Daily mean temperature	°C	1901–2009
Monthly average daily minimum temperature	°C	1901–2009
Monthly average daily maximum temperature	°C	1901–2009
Vapor pressure	Hecta-Pascals	1901–2009
Wet day frequency	Days	1901–2009
Monthly potential evapotranspiration	mm/month	1901–2009
Annual potential evapotranspiration	mm/year	Multi-annual
Monthly actual evapotranspiration	mm/month	Multi-monthly
Annual actual evapotranspiration	mm/month	Multi-annual
Water balance	mm/month	Multi-monthly

The main source of error in this model stems from having to assign an average yield for the administrative unit where the crop is located. Due to this assumption, two crops having significantly different edaphic and climate conditions but located in the same administrative unit were given the same yield.

This assumption made the data dispersion less accurate. However, due to the great number of data records and the spatial representativeness of the georeferenced crop location, it was assumed that the main source of error was generated by those records whose edaphic and climate conditions and yield values differed significantly from the others within the same administrative unit.

In order to identify these records, all of which had to be removed as part of the model construction, a reanalysis procedure was developed. In this procedure, the difference between estimated and actual yield was calculated using the same set of data previously used for the model development. Data with yield differences above 10 % were removed. The resulting data set was then used to develop the final model.

2.2.2 Identification of Potential Development Zones

Ranges of acceptable values for the different variables were established, according to the suitable conditions to develop new crops, for the following variables: temperature, precipitation, depth, drainage, fertility and slope. The same set of variables was averaged over the period 2000–2009, in the entire geographical area of interest, and these values were used to select zones for crop development whose averaged variables met the suitability requirements previously established.

2.2.3 Crop Yield Estimation from Edaphic and Climate Conditions

Random point clouds were created inside the development zones. If each point represents 20 ha, more than 10,000 points would be required to represent all entire zones. To assure reasonable processing time, the number of points was reduced by half. Independent variables values were extracted for the study period using a SQL query and stored in a geodatabase layer. This process used the same methodology described in 2.2.1.

Due to the polynomial nature of the model, it is important to check if the values of the independent variables are within the range of the data used for the model development, in order to ensure proper data interpolation.

The geodatabase layer was linked to R-Studio® using an ODBC connection. Yield prediction was obtained using the "predict" function and the developed model. Then the predicted yield was spatialized using the geographic identifier of each point.

2.3 Prioritization Procedure for Parcel Cells and Power Plant Locations Using Crop Yield and Logistic Criteria

The yield layer was divided into equal-sized cells using the "create fishnet" tool. The area of study was divided into 5 km × 5 km cells. Average yield and available area of the cell were estimated using the "zonal statistics as table" function from the Spatial Analyst extension of ArcGIS, and then the corresponding amount of biomass was calculated. In this step it is assumed that transportation cost per ton of biomass to the center of the cell is uniform within the analysis area.

The main goal of incorporating the logistic analysis was to consider the actual distance of biomass transportation using the available roads network. This analysis was performed using the Network Analyst extension of ArcGIS. In order to use this tool properly, the road layer must meet the connectivity and coherent criteria of a transportation network. The assessment and adjustment of this layer was made by creating a network topology for such a layer and using the tools within ArcGIS for that adjustment. Then, a network dataset was created using the adjusted layer, and the distance was configured like an attribute of cost (impedance).

To develop an analysis based on residual biomass, it was necessary to split the process in two stages. First, the biomass must be transported from the production field to an extraction plant, where the residues are separated, and then from this plant to the facility that will use the biomass residues.

The first problem to be solved was to find the best location for the extraction plants and assign them the closest cells taking into account the available roads. The network dataset, after inclusion of existing-roads layer, was divided into 20 km × 20 km cells; an extraction potential plant was located in the center of each cell. Due to the different types of analysis that can be performed by ArcGIS, it was necessary to define the logistic analysis criteria. For this study, a Location—Allocation analysis was used. It requires the definition of the location of Facilities and Demand points.[1] For this first problem, "Facilities" are the extraction plants (if required, it is possible to establish some of them as compulsory in the solution) whereas "Demand points" are locations where biomass is being generated (cells). Biomass production values were used to assign preference of the demand points. Along with the Location - Allocation analysis, the type of problem to be solved (maximize: coverage, capacitated coverage, attendance, market share; minimize: impedance, facilities, or target market share) had to be selected. It was also necessary to define the number of extraction plants to be selected and the maximum distance beyond which the analysis must be suspended.

[1]A Location—Allocation analysis requires two sets of input data: "Facilities" and "Demand Points". In this ArcGIS tool, "Facilities" are defined as generic production plants and "Demand points" are the customers for its products. For this case the interpretation is taken backwards: "Demand points" will produce biomass and "Facilities" will demand that biomass.

The best locations for the new extraction plants were selected using the Location—Allocation analysis. Available biomass and distance to the parcels were also calculated from this analysis. An indicator (λ) was defined and calculated as a production/distance ratio; λ is directly proportional to the feasibility of each parcel. From the available biomass and applying the residues factor it is possible to calculate the amount of residual biomass that each extraction plant is able to produce.

The second problem consisted in assigning the extraction plants to the power plants to be using the biomass residues based on the shortest transportation distance. In order to achieve that, another Location—Allocation analysis was performed. In this analysis, the potential power plants were defined as Facilities and extraction plants were defined as Demand points. It is possible to define potential locations for power plants, or to use the already established extraction plants' locations as potential locations for the power plants. Biomass demand is used as prioritization factor for the potential power plants. Prioritization factor for the extraction plants was the indicator (λ), calculated as formerly defined for parcels. The Location—Allocation analysis was performed with these settings and adopting criteria similar to the ones used in the previous step. The results of such analysis encompassed the prioritization of the extraction plants and the selection of the potential plants that will use biomass residues.

In this step, there is an origin-destination matrix with distances and assignments that can be used as input to a supply chain optimization model intended to make a final selection based on economic performance indicators.

It is important to point out that for an analysis with more than one type of crop, it is required to estimate the amount of biomass for each type of residue and each crop, and then group the residues that could have the same use. The total amount of residue going into the same plant is what should be set as prioritization factor. If the amount of biomass assigned to each power plant is required to be known, then it is possible to name each element that is associated to a plant and then rerun the analysis. For this approach, the prioritization factor should be set as the amount of biomass produced in the extraction plants.

The methodology is presented in the diagram shown in Fig. 1.

3 Case of Study: Oil Palm Crop (*Elaeis Guineensis*)

The methodology described was implemented in ArcGIS and applied to a Colombian region. The main goal was to identify potential zones for the development of oil palm crops and select the best location for oil extraction plants and power plants using the biomass residues, taking into account the maximum crop yield and logistic criteria.

It is important to point out that this problem concerns to assigning a use to residual biomass, but it is also necessary to realize that the biomass residues are a by-product of an extraction plant.

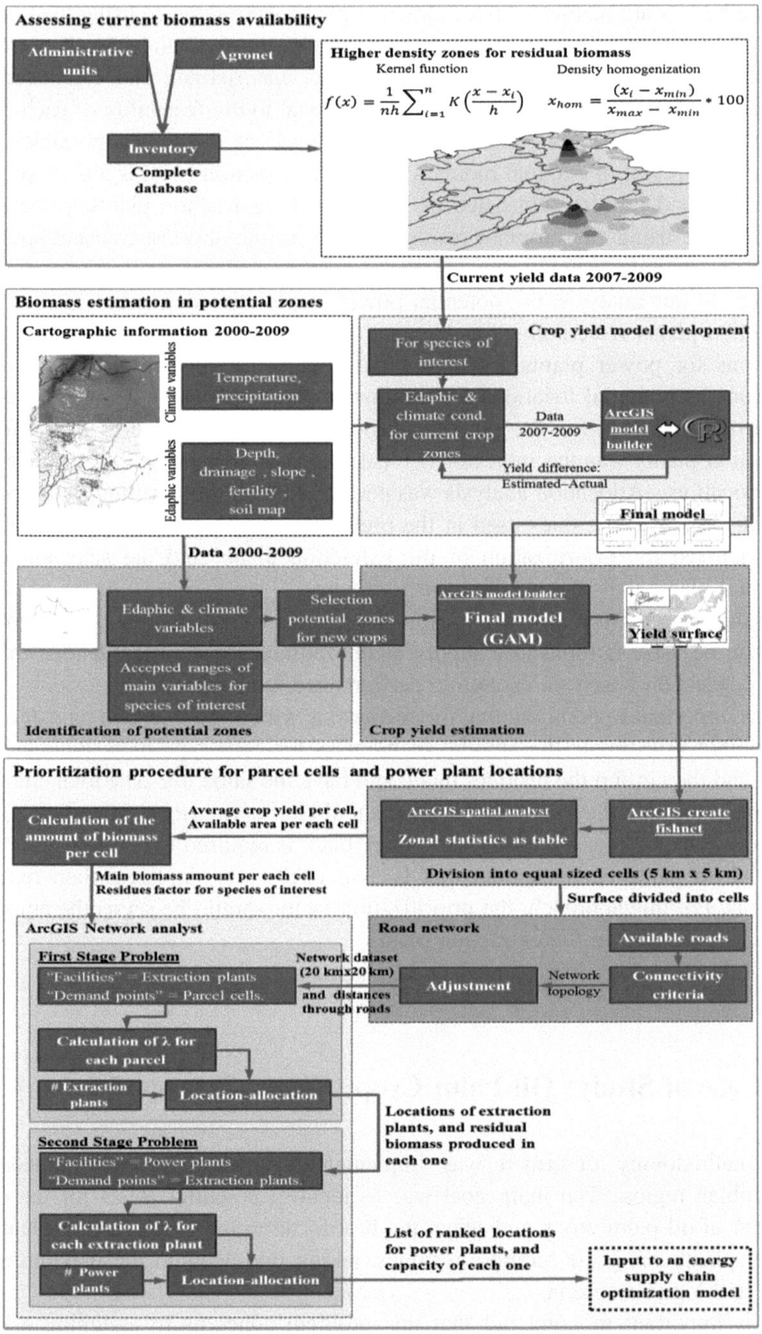

$$f(x) = \frac{1}{nh}\sum_{i=1}^{n} K\left(\frac{x - x_i}{h}\right) \qquad x_{hom} = \frac{(x_i - x_{min})}{x_{max} - x_{min}} * 100$$

Fig. 1 Methodology general scheme

3.1 *Current Availability of Raw Material—Empty Fruit Bunches (EFB) of Oil Palm*

A biomass residue density analysis was performed by using the application developed for the residual biomass inventory. The criterion used for this analysis was the amount of available dry biomass within a 30 km radius. This analysis identified 10 zones, four of which were prioritized (Fig. 2, Table 2). Each zone consists of different administrative units.

Zones presented in Table 2 are the four most productive ones among ten zones where oil palm is produced in Colombia. However crop yield is well established by

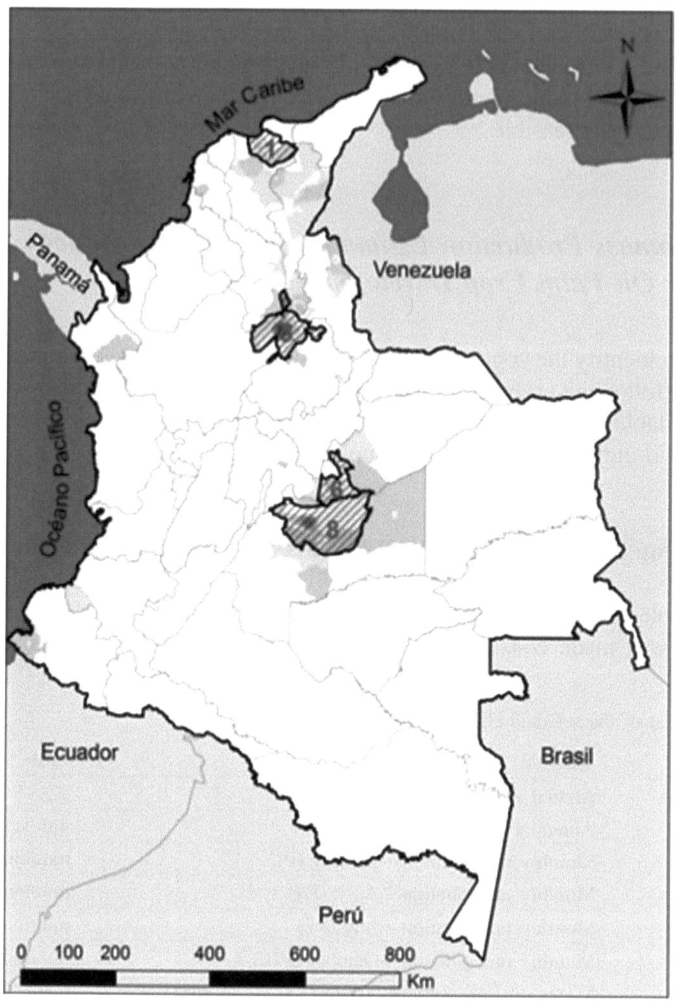

Fig. 2 Location of the zones with the biggest amount of available residual biomass from oil palm

Table 2 Amount of residual biomass of empty fruit bunches of oil palm

Zone ID	Available dry biomass [Thousands of tons/year]
5	84.8
8	74.4
1	30.1
6	30.1

the corresponding administrative units for the ten zones, and this data is collected in the residual biomass inventory basic application.

Despite the greater amount of biomass is located in Zone 5, the region of interest for developing the project is zone 8, which is considered a strategic zone where ECOPETROL has other assets in operation.

Based on the results, we concluded that the current amount of available residual biomass in Zone 8 is not enough for the development of the project.

Therefore it is necessary to evaluate if implementing new oil palm crops, zone 8 would be able to produce the residual biomass in such a distributed pattern to make the energy project feasible.

3.2 *Biomass Production Estimation in Potential Zones for Oil Palm Crop Development*

In order to identify the zones with the highest potential for the development of new crops, the following criteria were used: (1) main product yield and (2) accessibility to the available road network. For estimating crop yield, a statistical model based on edaphic and climate conditions was applied.

3.2.1 Crop Yield Model for Oil Palm (Fresh Fruit Bunches: FFB)

The variables used in the yield model are those presented in Tables 3 and 4. This selection was made considering the relative importance of the variables for the

Table 3 List of the selected climate variables used by the model

	Variable	Unit
Climate	Annual average temperature (Ta)	°C
	Annual average precipitation (Pa)	mm/year
	Monthly precipitation—January (P1)	mm/month
	Monthly precipitation—April (P4)	mm/month
	Monthly precipitation—July (P7)	mm/month
	Monthly precipitation—October (P10)	mm/month
	Maximum soil humidity (Hmax) at 90 cm deep	mass %
	Minimum soil humidity (Hmin) at 90 cm deep	mass %

Table 4 Conversion of categories into alpha-numerical levels for edaphic variables selected by the model

Level	Soil depth (Prof)		Fertility (Fer)		Drainage (Dre)	
1	MUS	Very superficial	MUB	Very low	ED	Very well drained
2	S	Superficial	B	Low	BD	Well drained
3	MOP	Moderately deep	MO	Moderate	MOBD	Moderately well drained
4	P	Deep	MUA	Very high	ID	Imperfectly drained
5	MUP	Very deep			PD	Poorly drained
6					MUPD	Very poorly drained

development of oil palm crops. Since the yield can be specially affected by water availability, by precipitation (rainfall) during some stages of the production cycle, this variable was incorporated with a higher time resolution (annual precipitation, and monthly precipitation of January, April, July and October).

All ten zones where oil palm is produced in Colombia were used to gather edaphic and climate historical information for the crop yield model development stage. Distribution of these ten zones in three great regions is presented in Fig. 3. The statistics of the extracted data, which included 425,439 records, are presented in Table 5.

The generalized additive model (GAM) for estimation of oil palm (FFB) productivity was developed in the statistical software R-Studio®. The statistical library used for this purpose was "mgcv".

The GAM corresponds to a sum of non-linear relationships, derived from applying smoothing function on the dataset of parametric variables (continuous variables) and categorical variables (defined as non-continuous variables). These datasets of independent variables were used as input for the model developed in R-Studio®, which used the structure presented in Eq. (1). Then, Eq. (1) represents the mathematical expression of productivity model for oil palm (FFB), with predicting or independent variables that can be classified in two groups. The first one corresponds to continuous variables such as annual temperature, annual precipitation, soil humidity and monthly rainfall (January, April, July and October). The second group considers the independent edaphic variables: Soil depth, fertility and drainage, with categories defined for each one as shown in Table 4.

$$\begin{aligned} Prod = {} & s(T_a) + s(P_a) + s(H_{\max}) + s(H_{\min}) + s(P_1) + s(P_4) + \cdots \\ & + s(P_7) + s(P_{10}) + Prof + Fer + Dre \end{aligned} \tag{1}$$

where:

Prod Oil palm (FFB) yield or productivity (dependent variable [Ton/ha-year])

S() Represents smooth functions for non-lineal additive adjustment

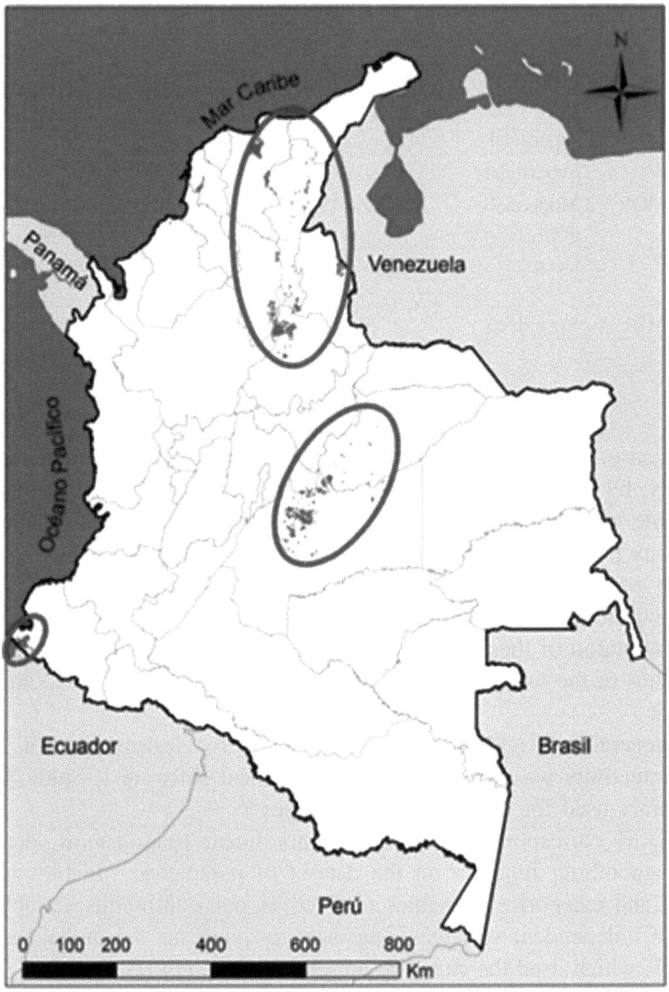

Fig. 3 Location of the current crop lands for oil palm

Table 5 Statistical parameters of the data used for model development

	T_a^*	H_{max}^*	H_{min}^*	P_a^*	P_1^*	P_4^*	P_7^*	P_{10}^*	Prod
Min.	177	72	16	8395	1	264	302	860	10.23
1st Qu.	257	100	39	24497	85	1977	1356	2970	14.88
Median	264	100	52	28690	260	2504	2867	3499	16.27
Mean	264	99	51	28454	517	2774	2814	3782	15.82
3rd Qu.	278	100	58	33708	556	3081	4036	4949	16.74
Max.	289	100	100	52119	4104	7739	9051	7459	19.02

*Multiplied by 10 in order to be easily stored in raster format

T_a Annual average temperature (independent and continuous variable [°C])
P_a Annual average precipitation (independent and continuous variable [mm/year])
H_{max} Maximum soil humidity (independent and continuous variable [%])
H_{min} Minimum soil humidity (independent and continuous variable [%])
P_1 Monthly precipitation—January (independent and continuous variable [mm/month])
P_4 Monthly precipitation—April (independent and continuous variable [mm/month])
P_7 Monthly precipitation—July (independent and continuous variable [mm/month])
P_{10} Monthly precipitation—October (independent and continuous variable [mm/month])
Prof Soil Depth (independent and categorical variable)
Dre Soil Drainage (independent and categorical variable)
Fer Soil fertility (independent and categorical variable)

Mathematical expression of the crop yield model for oil palm (FFB) was defined as shown above in Eq. (1). When parametric and categorical variables are combined, recommended error function for data distribution is the type "Gaussian". Then, independent variables data (400,000 edaphic-climatic records) was loaded for this preliminary model development. It is noteworthy that before loading data into software, null values and records with missing information were eliminated from data set of the continuous variables and categories of the non-continuous edaphic variables were converted to alpha-numeric levels as shown in Table 4.

With all input data loaded, GAM model was executed using R-Studio®. After an analysis of variance, the results obtained from the first execution of the model are shown below:

```
summary(Oil palm)
  Family: gaussian
  Link function: identity
  Formula:
   Prod ~ s(Ta) + s(Pa) + s(Hmax) + s(Hmin) + s(P1) + s(P4) + s(P7)
     + s(P10) + Prof + Fer + Dre
```

Parametric coefficients:

	Estimate	Std. Error	t value	Pr(> |t|)	
(Intercept)	16.025698	0.007407	2163.68	<2e-16	***
ProfMUP	−0.443831	0.430829	−1.03	0.303	
ProfMUS	1.192869	0.010124	117.82	<2e-16	***
ProfP	−0.730176	0.008165	−89.43	<2e-16	***
ProfS	−0.566664	0.008372	−67.69	<2e-16	***
FerMO	−0.070455	0.006671	−10.56	<2e-16	***
FerMUA	−0.203836	0.010071	−20.24	<2e-16	***
FerMUB	−1.121413	0.010358	−108.27	<2e-16	***
DreED	0.218235	0.024467	8.92	<2e-16	***
DreID	0.341391	0.007492	45.57	<2e-16	***
DreMOBD	0.313108	0.010011	31.28	<2e-16	***
DreMUPD	−0.505361	0.012013	−42.07	<2e-16	***
DrePD	0.941905	0.008435	111.67	<2e-16	***
–					

Signif. codes: 0 ‘***’ 0.001 ‘**’ 0.01 ‘*’ 0.05 ‘.’ 0.1 ‘’ 1

Approximate significance of smooth terms:

	Edf	Ref.df	F	p-value	
s(Ta)	8.992	9.000	1474.0	<2e-16	***
s(Pa)	8.998	9.000	2988.4	<2e-16	***
s(Hmax)	8.963	8.999	686.3	<2e-16	***
s(Hmin)	8.991	9.000	2902.3	<2e-16	***
s(P1)	8.998	9.000	1624.0	<2e-16	***
s(P4)	8.993	9.000	1561.3	<2e-16	***
s(P7)	9.000	9.000	947.8	<2e-16	***
s(P10)	9.000	9.000	1694.7	<2e-16	***
–					

Signif. codes: 0 ‘***’ 0.001 ‘**’ 0.01 ‘*’ 0.05 ‘.’ 0.1 ‘’ 1
R-sq.(adj) = 0.576 Deviance explained = 57.6 %
GCV score = 1.1119 Scale est. = 1.1116 n = 415439

Analysis of variance showed that non-continuous edaphic variables contribute significantly to the explanation of the model, having a value of statistical probability (p) lower than 0.05, with the exception of the variable "depth", for category "Very deep" (MUP), that took a p-value greater than 0.05. Regarding the statistical significance of continuous variables, all of them obtained a p-value lower than 0.05. Moreover, a greater contribution of some of the continuous variables for explaining the model can be confirmed for those variables having a greater value of F. Annual precipitation and minimum soil humidity have the greatest values of F; annual temperature can also be considered a significant variable. Variables such as maximum soil humidity and precipitation for the month 7 (July) have the lowest contribution for the explanation of crop yield model (lowest values of F).

After completing the analysis of variance of the additive model for oil palm (FFB) productivity, the explained variance for the whole model was estimated, including the percentage of contribution of both edaphic and climate variables. Thus, the estimation of the explained variance for this first execution of the model was 57.6 %, too low for considering the model a good approximation. The main source of error in the model stems from assigning, in the input information, the reported annual average yield of the administrative unit to all the crops harvested in that administrative unit in 2007–2009. This assumption was made because there is not more detailed information in a consolidated official database for the crop yields or productivities, per hectare-year of oil palm (FFB).

Therefore, a second analysis of the initial data used in the model development was carried out.

This second analysis consisted in:

- Determining the yield prediction values with the model as developed in the first analysis
- Calculation of the absolute differences between predicted value and observed real value
- If difference of productivities was greater than 1 FFB ton/ha-year, the corresponding data set was removed,

Thus, a higher concentration of adjusted data to the model was guaranteed and part of uncertainty of the response variable was eliminated.

Data cleaning by removing these differences involve a reduction from 415,000 records of initial data to 312,000 records in the final input data for model development. After the input data was refined, a second execution of model was carried out with the final 312,000 records as input data. Results are shown below.

```
summary(Oil palm 2)
   Family: gaussian
   Link function: identity

   Formula:

    Prod ~ s(Ta) + s(Pa) + s(Hmax) + s(Hmin) + s(P1) + s(P4) + s(P7)
       + s(P10) + Prof + Fer + Dre
```

Parametric coefficients:

	Estimate	Std. Error	t value	Pr(> |t|)	
(Intercept)	16.208354	0.003829	4232.517	<2e-16	***
ProfMUP	−0.567989	0.184296	−3.082	0.00206	**
ProfMUS	0.952838	0.006676	142.724	<2e-16	***
ProfP	−0.532010	0.004434	−119.991	<2e-16	***
ProfS	−0.534454	0.004413	−121.116	<2e-16	***
FerMO	−0.132347	0.003446	−38.409	<2e-16	***
FerMUA	−0.074119	0.006512	−11.382	<2e-16	***
FerMUB	−1.029908	0.006494	−158.592	<2e-16	***
DreED	0.137661	0.011679	11.788	<2e-16	***
DreID	0.383900	0.004288	89.525	<2e-16	***
DreMOBD	0.096980	0.005793	16.742	<2e-16	***
DreMUPD	−0.834221	0.008459	−98.625	<2e-16	***
DrePD	0.713796	0.005527	129.153	<2e-16	***
–					

Signif. codes: 0 '***' 0.001 '**' 0.01 '*' 0.05 '.' 0.1 '' 1

Approximate significance of smooth terms:

	edf	Ref.df	F	p-value	
s(Ta)	9.000	9.000	5288	<2e-16	***
s(Pa)	8.999	9.000	10811	<2e-16	***
s(Hmax)	8.989	8.999	1968	<2e-16	***
s(Hmin)	9.000	9.000	11324	<2e-16	***
s(P1)	8.997	9.000	2402	<2e-16	***
s(P4)	9.000	9.000	3011	<2e-16	***
s(P7)	9.000	9.000	3290	<2e-16	***
s(P10)	9.000	9.000	2598	<2e-16	***
–					

Signif. codes: 0 '***' 0.001 '**' 0.01 '*' 0.05 '.' 0.1 '' 1
R-sq.(adj) = 0.87 Deviance explained = 87 %
GCV score = 0.2034 Scale est. = 0.20334 n = 312102

The results obtained from this second execution show an increase of the explained variance of the model from 57.6 % to 87 %. Besides, the variables: Minimum soil humidity, annual precipitation and annual temperature are still having the greatest contributions (Greater values of F).

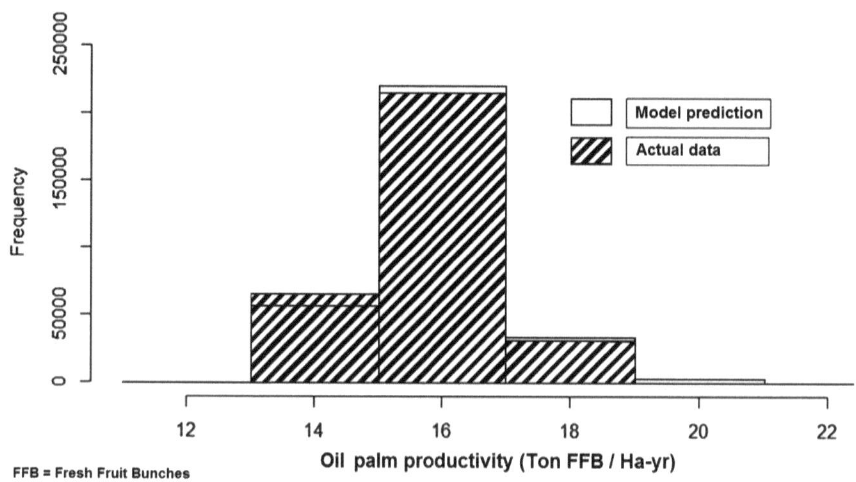

Fig. 4 Reproducibility of Data Oil palm (FFB) Productivity Model

The above explained variance obtained for oil palm (FFB) productivity model (about 87 %) can be considered a good approximation. Thus, it is possible to forecast oil palm (FFB) productivity with good accuracy.

Figure 4 shows a hatched histogram for real values of oil palm (FFB) productivities and an unfilled histogram for predicted oil palm (FFB) productivities from the developed crop yield model. As shown in Fig. 4, the range of real productivity data is almost completely covered by the response surface of the predicted productivity data. This indicates that the edaphic and climate variables selected as input data explain very well the oil palm (FFB) productivity behavior.

3.2.2 Identification of Potential Developing Zones for Oil Palm

Ranges for the more influential variables on oil palm (*Elaeis Guineensis*) productivity were properly defined, as depicted in Table 6.

Categorical variables Soil depth, fertility and drainage are selected in all their categories for not closing the possibilities to very narrow areas but having a wide region to generate the prediction of yield surface. As mentioned previously, the zone of strategic interest is the zone 8 shown in Fig. 2.

3.2.3 Yield Estimation for Oil Palm (FFB) in Potential Zones

The point clouds generated for each yield prediction zone are shown in Fig. 5, while the corresponding statistics of the extracted data are presented in Table 7.

Table 6 Criteria used for identifying potential development zones for oil palm

Variable		Variable range
Precipitation (mm/year)		1000–4000
Temperature (annual average)	Min °C	21
	Max °C	31
Soil depth		Moderately deep, Deep, and very deep
Soil fertility		Low, moderate, high, very high
Soil drainage		Imperfectly drained, well drained, moderately well drained, poorly drained
Slope		<8 %

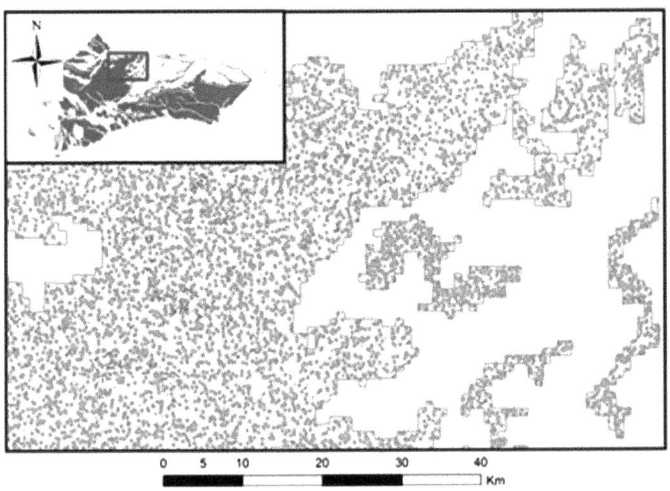

Fig. 5 Point clouds for crop yield prediction zones

Table 7 Statistic parameters of the data used for crop yield prediction

	H^*_{max}	H^*_{min}	T^*_a	P^*_a	P^*_1	P^*_4	P^*_7	P^*_{10}
Min.	100	35	128	15780	106	719	783	1155
1st Qu.	100	49	260	28195	308	2356	661	2923
Median	100	53	262	28647	439	2721	915	95
Mean	100	54	258	29704	462	2721	900	3243
3rd Qu.	100	58	264	29581	589	2900	9	3400
Max.	100	95	272	50476	1469	5154	549	5974

*Multiplied by 10 in order to be easily stored in raster format

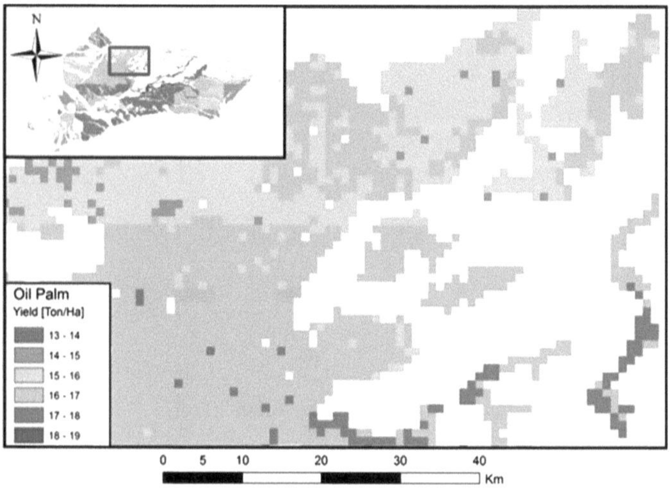

Fig. 6 Oil palm (FFB) yield map [Ton/ha per year]

Due to the polynomial nature of the model, it was necessary to verify if the values of the independent variables were within the range of the data used for the model development, in order to ensure data interpolation. This verification will be carried out below.

Yield prediction was performed and the results were spatially distributed using the geographic identifier of each point. Figure 6 presents the yield layer created from the geographic distributed points using Kriging interpolation method.

3.2.4 Analysis of Residual Values: Edaphic Variables

The behavior of residual values for each independent variable in the model allows identifying which ones have greater variability and significance to the predictive statistical model. In this particular case of oil palm (FFB) model, the Fig. 7 presents the residual values (Y-axis), for the variable "depth" (Prof) and its categories (MUS, S, MOP, P, MUP). In contrast with other categories, a high statistical significance (high absolute residual value) is found for the category "MUS" (very superficial). Surprisingly, the category "MUS" also has a low variability and a minimum contribution to the uncertainty of the model; this is shown in Fig. 7 by the very close spacing among the dashed lines and solid line of the residual value for category "MUS".

On the other hand, category "MUP" (very deep) has a lower absolute residual value than category "MUS". Accordingly, category "MUP" has a lower statistical significance to explain the model. A wide spacing among the dashed lines and solid line shows that category "MUP" has a high variability and therefore a high contribution to the model error. Category "MUP" also has a low availability of data

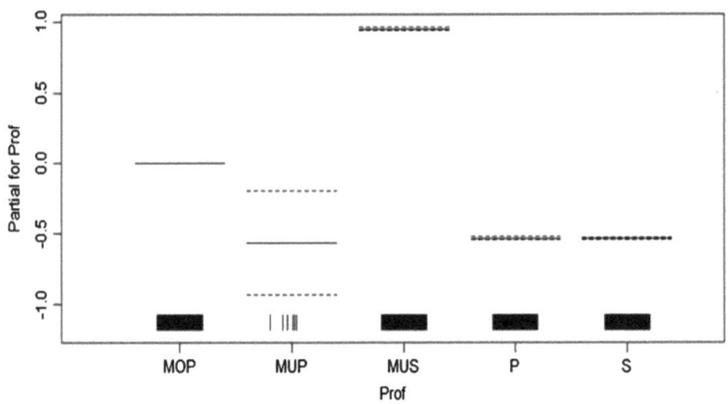

Fig. 7 Variation of residual values for the variable "depth" (Prof)

compared with the other categories. This is shown in Fig. 7 by the wide spacing among the vertical short lines, which do not make up a solid bar, like in the other categories. This low availability of data makes difficult to get a high precision in category "MUP".

In the case of the variable "drainage" (Dre), Fig. 8 shows that all categories belonging to this variable have a low variability in the subsets of data and therefore a minimum contribution to model uncertainty.

According to Fig. 8, the categories of greater significance or contribution for model explanation correspond to "PD" (poorly drained) and "MUPD" (very poorly drained) with the highest absolute residual values. Other categories have residual drainage values very near to zero, thus its contribution to significance level on global productivity model is reduced.

Finally, Fig. 9 shows the residual values for soil fertility categories. All categories have a low variability in the data and therefore lower contribution on global model error. Besides, category "MUB" (very low fertility) has the greatest significance and therefore the highest contribution for explaining the additive model

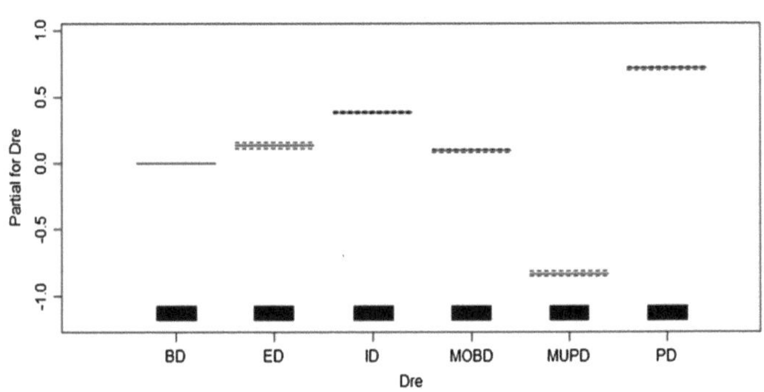

Fig. 8 Variability of residual values for "Drainage" variable

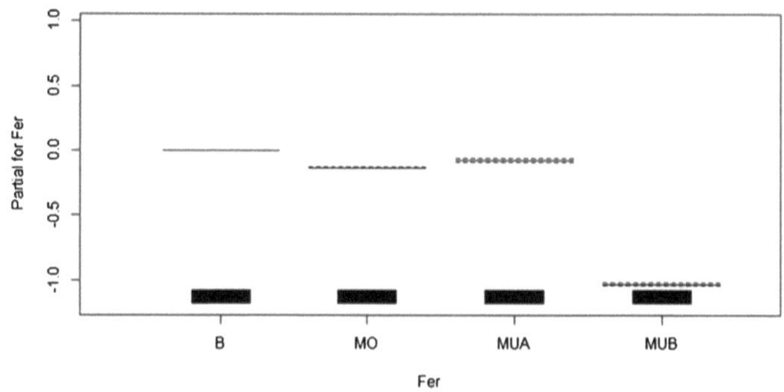

Fig. 9 Variation of residual values for "Fertility" variable

compared with other soil fertility categories, which has absolute residual values very close to zero.

In summary, the residual values analysis for edaphic variables allowed to infer that extreme edaphic conditions, particularly very superficial soils, poorly drained soils and soils with very low fertility, have a high statistical significance in order to explain the additive model and therefore present a greater impact on the estimation to oil palm (FFB) productivity.

3.2.5 Analysis of Residuals: Climate Variables

The behavior of residual values for parametric variables (climate conditions) is presented below. The residual values for continuous variables make up a continuous curve. Figure 10 presents the curve of residuals for the variable "maximum soil humidity" (Hmax), which is associated with the amount of water retained in the soil as consequence of water balance between the natural precipitation, runoff and evapotranspiration. According to Fig. 10, the residual values near to zero indicate a

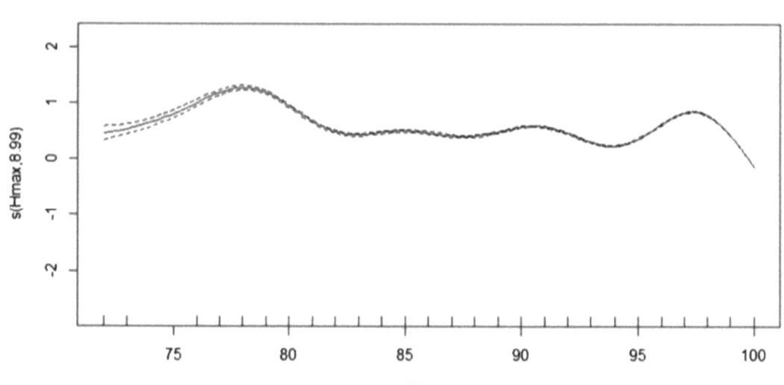

Fig. 10 Variation of residual values for "maximum soil humidity" variable

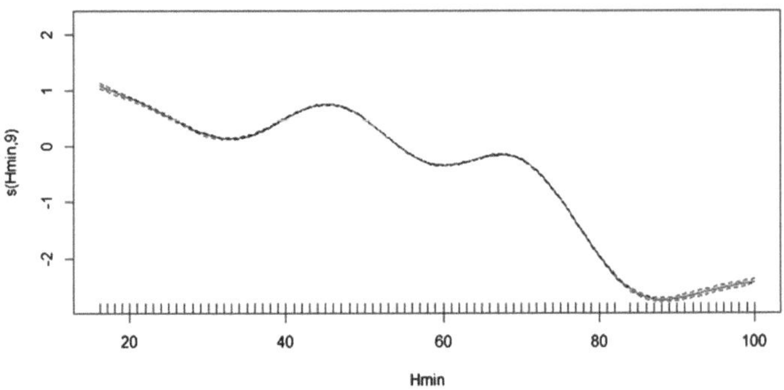

Fig. 11 Variation of residual values for "Minimum soil humidity" variable

partial influence on the explanation of the productivity model; however the data set has a minimal statistical variability (dotted lines very close to the solid curve of residual data). An almost parallel curve of the residuals for the variable Hmax means that oil palm (FFB) productivity is not very influenced by this variable.

On the other hand, the residual values for the variable "minimum soil humidity" (Hmin) are shown on Fig. 11. This variable has a high significance to explain productivity model because its residual values range from 1 to −2. Thus, minimum water content in soil has also a greater impact on oil palm (FFB) productivity than other climate variables such as temperature and precipitation. Additionally, the minimum amount of water in the soil (variable Hmin) is directly related to the continuous variable "precipitation" and to the categorical variable "drainage". A lower precipitation and a very well drained ("ED") soil will have a lower Hmin.

Figure 11 shows a clear trend of Hmin residuals. That trend means a negative impact of minimum soil humidity on the oil palm (FFB) productivity: For higher minimum soil humidity values, oil palm (FFB) productivity decreases. Sensibility of oil palm (FFB) productivity to changes of Hmin is greater than to changes of Hmax.

The other variable with a high significance for explaining the productivity model corresponds to the annual precipitation. Its residual values can be observed in Fig. 12 (values between 1 and −2). Residuals represent a sinusoidal curve and indicates a high significance or impact on productivity model. Specifically, water precipitation between 1000 and 3500 mm/year, shown in Fig. 12, has a low statistical variability in the data set, which lead to minor error or uncertainty on productivity model for the oil palm crop

Another important aspect of the climate conditions that affects the productivity of oil palm (FFB) is related with the annual distribution of rainfall. This phenomenon is evidenced by the behavior of residual values for monthly rainfall variable, particularly in the fourth month (April) as shown in Fig. 13. Rainfall in April between 200 and 500 mm/month[2] has a negative impact on the productivity

[2]In Fig. 13 values are multiplied by 10.

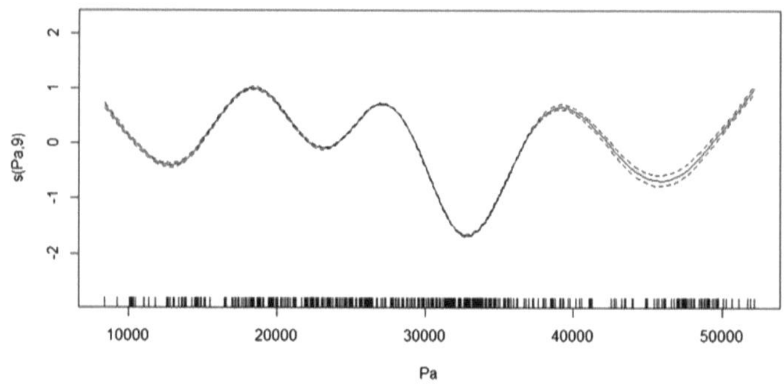

Fig. 12 Variation of residual values for "Annual Precipitation" variable

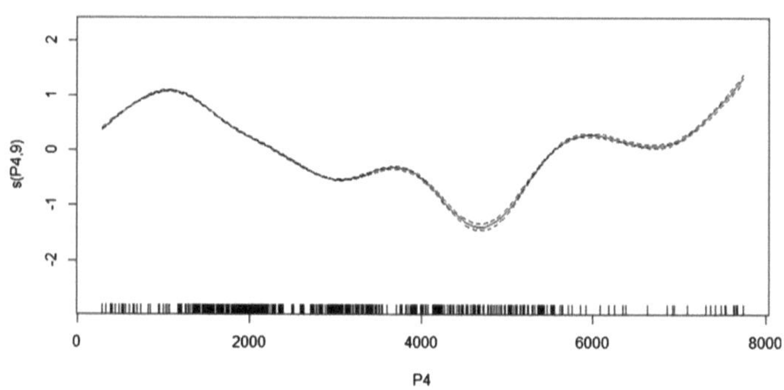

Fig. 13 Variation of residual values for "monthly precipitation—April" variable

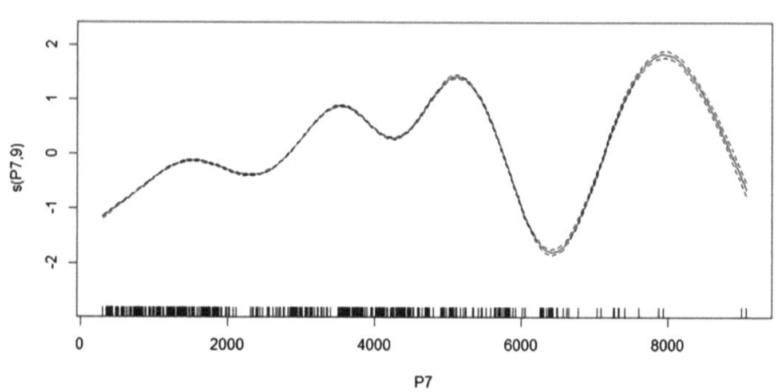

Fig. 14 Variation of residual values for "monthly precipitation—July" variable

of oil palm (FFB). For July period (precipitation of the month 7) the productivity responds positively to an increase in rainfall until 500 mm/month. For greater values of rainfall the impact becomes negative on productivity (see Fig. 14). All the above should be interpreted with some uncertainty and caution, since the general

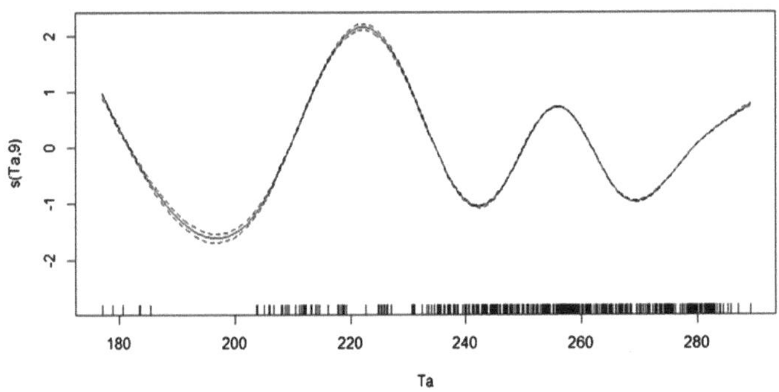

Fig. 15 Variation of residual values for "average temperature" variable

behavior of residual values for both variable, annual rainfall and monthly rainfall (April-July), describes a sinusoidal curve, therefore does not follow a clear trend about the positive or negative incidence of rainfall and its influence on predicting oil palm (FFB) productivity.

Finally, Fig. 15 shows the behavior of residual values for annual average temperature variable, which constitutes the third most significant aspect for description of productivity model. The behavior for temperature residual values follows a sinusoidal curve, which don't allow identifying if this variable can affect positively or negatively the model prediction. Nevertheless this behavior of residual temperature (values different from zero and ranging between 0.5 and -1) indicates a high significance in order to explain the overall productivity model.

3.2.6 Productivity Prediction Model for Palm Oil (FFB)

Residual analysis of the additive oil palm (FFB) productivity model for each independent variable and an explained variance for global model around 87 % suggest that the model represents a good statistical approach for estimating oil palm (FFB) productivity under a defined set of soil and climate conditions. However, the applicability of the model is subject to the domain of variables considered in its development; in other words, the data set of soil and climate conditions over which a forecast is desired must belong to the domain of real data set used for model construction, outside this model's domain, a suitable reproducibility of productivity is not guaranteed.

Accordingly, Fig. 16 shows the domain for the variable "minimum soil humidity". The data frequency used in the model construction extends from 1 to 10 %[3] (unfilled bars); while the minimum soil humidity data used for productivity forecasting (solid-filled bars) ranges from 3 to 8 %. It means that the domain for

[3]In Fig. 16, values are multiplied by 10.

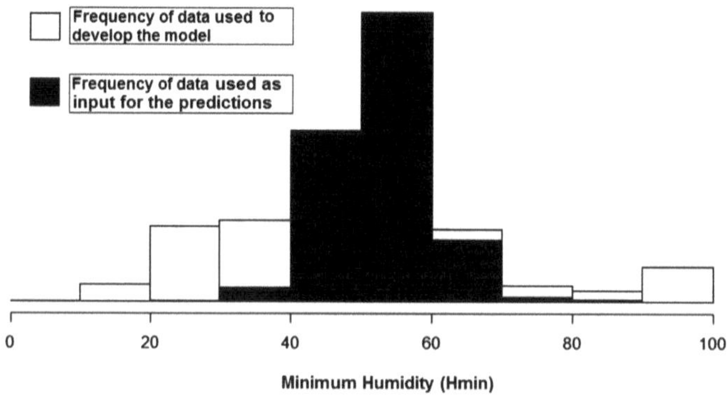

Fig. 16 Frequency data "Minimum soil humidity"—oil palm (FFB)

predictions is contained within the domain used for model construction. Therefore the applicability of the additive prediction model for oil palm (FFB) productivity is validated.

Two other variables of greatest contribution in terms of explaining the productivity model for oil palm (FFB) correspond to annual precipitation and mean annual temperature,[4] whose domains of data used to develop the model are shown in Fig. 17 and Fig. 18 respectively (unfilled bars). Data ranges used for prediction are also shown (solid-filled bars) in these figures. Data domain for prediction is completely contained within the data domain for model development. One more time, developed model shows its suitability for the dataset of edaphic and climate conditions where forecasting is required.

3.2.7 Prioritization Procedure for Parcel Cells and Power Plant Locations Using Crop Yield and Logistic Criteria for Zone Accessibility

In order to apply logistic criteria, it was necessary to group the biomass into point entities. The area of study was divided into 5 km × 5 km cells. Available area for crop development, average yield and amount of main product were calculated for each cell. Additionally, a 20 km × 20 km grid was created for potential location of oil extraction plants. Biomass residues are proportional to the main product amount.

A layer of existing roads was topologically structured to enforce both the connectivity and coherent criteria for a transport network. Distance was set as a cost attribute. Figure 19 presents cells distribution and location of the potential extraction plants.

[4]These values were amplified by a factor 10, in order to avoid errors in the software ArcGIS® which has troubles handling very low values.

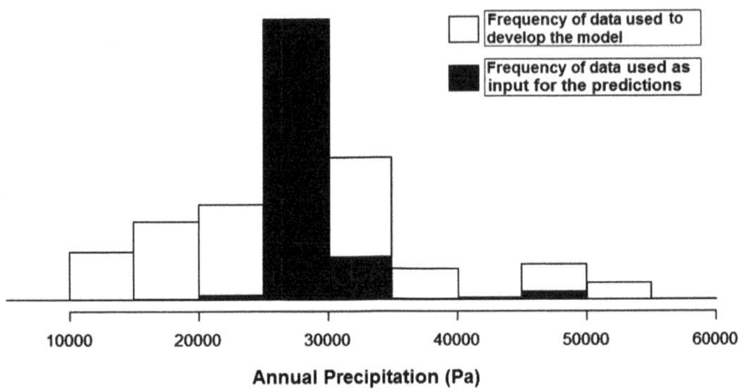

Fig. 17 Frequency data "Annual Precipitation"—oil palm (FFB)

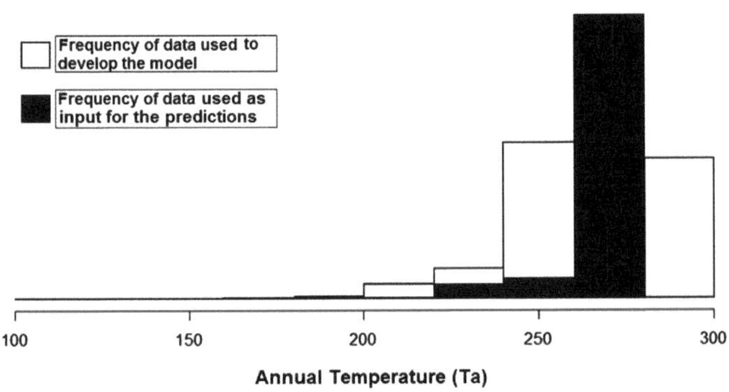

Fig. 18 Frequency data "Annual Temperature"—oil palm (FFB)

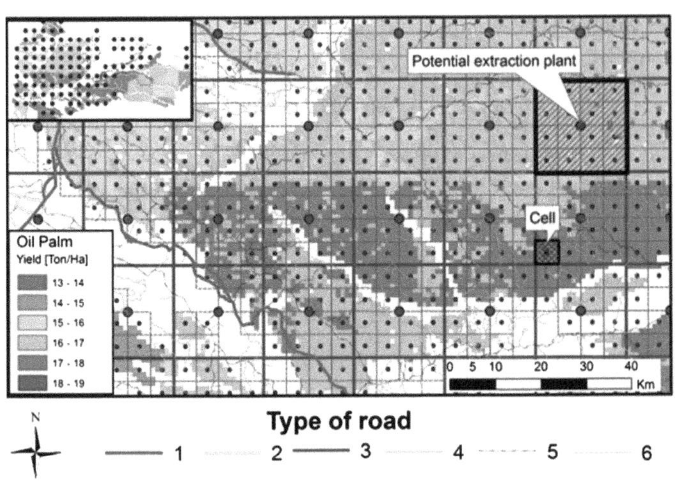

Fig. 19 Distribution of the analysis cells, potential extraction plants and available roads

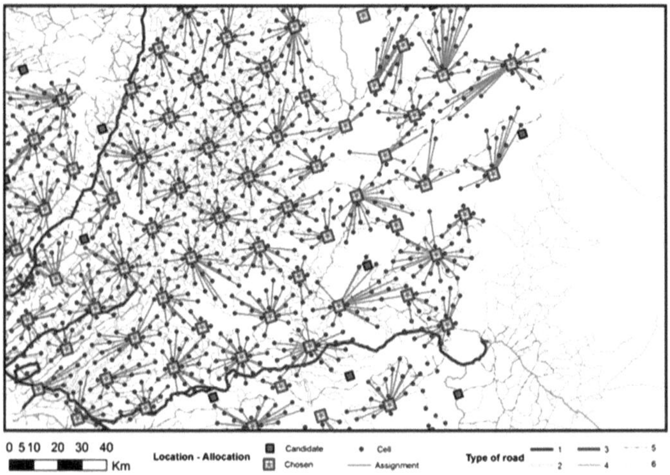

Fig. 20 Location—allocation analysis used for extraction plants prioritization and cells assignation

A total of 1,308 cells suitable for this analysis and 136 potential locations for the extraction plants were created. The east part of the study area was excluded since there were no roads to connect the potential extraction plants to the cells.

A Location—Allocation analysis, in which the total impedance was minimized, was performed in order to select 100 out of the 136 potential extraction plants and a maximum distance of 50 km using the existent road network. Production was established as prioritization factor.

The results of the analysis are presented in Fig. 20. The λ indicator was calculated as the production/distance ratio. This indicator is directly proportional to the feasibility of the extraction plant.

The assignment of extraction plants to the power plants that will use the biomass residues was based on the shortest transportation distance. In order to achieve this, another Location—Allocation analysis was performed. In this analysis, the potential power plants previously selected were defined as facilities and biomass consumption was set as the prioritization factor. The extraction plants were defined as demand points and their prioritization factor was the indicator (λ), as previously defined. The analysis was run using those settings to select three plants for the use of biomass residues with a maximum distance of 100 km.

The results of the analysis are presented in Fig. 21; 72 extraction plants, for which the amount of available biomass and the distance to the plant are known, were selected and the best locations were assigned to the three plants that use the biomass residues.

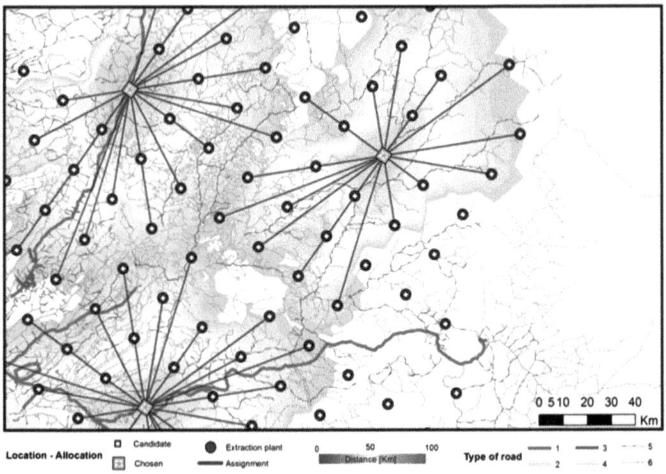

Fig. 21 Location—allocation analysis used for power plants using biomass residues and extraction plants assignation

4 Discussion of Results

The developed yield model explained most of the variance, hence bringing reliability to the ranking of suitable zones. The statistical analysis of the data used for the predictions guaranteed the model applicability under the study area conditions.

The prioritization method used in this approach allowed for an objective land suitability analysis to support the development of new crop areas, using key variables (yield and maximum distance).

Model reliability depends significantly on the information that is used for the model's construction. This is a limitation because the greater the detail of the analysis, the higher the uncertainty. This limitation can be circumvented by gathering information about the yield from the crop currently in production, crop by crop and not as an averaged value for the entire administrative unit where that crop is located.

The model used here considered the availability of water all year long, which made the model more robust, since dry and wet season impacts are then taken into account for yield estimation.

The logistic component was based on a saturation approach, in which potential sites were prioritized according to biomass availability and accessibility.

The analysis included two stages that first prioritized cells and then the extraction plants according to the assigned cells. It is well known that sequential prioritization does not guarantee the global optimal solution and that ArcGIS's heuristic generates an ideal solution but not the global optimal answer. However, this methodology was aimed to generate systematically and objectively a set of alternatives, which included all the data required by a supply chain further analysis.

5 Conclusions

By using the proposed methodology it is possible to make an objective prioritization of the biomass generation places based on primary decision variables (yield and location). This allows the selection, from a large set of candidate zones, of a subset with the best areas for developing new plants, including biomass extraction plants and power plants that use biomass residues. The methodology takes into account different criteria, including yield, amount of biomass, and logistic conditions.

By adopting the approach detailed in the present study, one can derive, in a systematic and straightforward way, all the input data required by the crop-yield-prediction statistical model, while also including the impact of the climate variability on the crop yield.

Finally, it is possible to calculate the maximum capacity for each potential plant and to create an origin-destination matrix between extraction plants and power plants, which can be used as input for a supply chain further analysis.

References

Fischer G, Schrattenholzer L (2001) Global bioenergy potentials through 2050. Biomass Bioenergy 20:151–159

Freppaz D, Minciardi R, Robba M, Rovatti M, Sacile R, Taramasso A (2004) Optimizing forest biomass exploitation for energy supply at a regional level. Biomass Bioenergy 26:15–25

Panichelli L, Gnansounou E (2008) GIS-based approach for defining bioenergy facilities location: a case study in Northern Spain based on marginal delivery costs and resources competition between facilities. Biomass Bioenergy 32:289–300

Searcy E, Flynn P, Ghafoori E, Kumar A (2007) The relative cost of biomass energy transport. Appl Biochem Biotechnol 137–140:639–652

Silverman B (1986) Density estimation for statistics and data analysis. Chapman and Hall, New York

Sultana A, Kumar A (2012) Optimal siting and size of bioenergy facilities using geographic information system. Appl Energy 94:192–201

UPME, et.al (2010) Atlas del Potencial Energético de la Biomasa Residual en Colombia. UPME, Bogotá

Voivontas D, Assimacopoulos D, Koukios EG (2001) Assessment of biomass potential for power production: a GIS based method. Biomass Bioenergy 20:101–112

Wullschleger SD, Davis EB, Borsuk ME, Gunderson CA, Lynd LR (2010) Biomass Production in Switchgrass across the United States: database description and determinants of yield. Agron J. 102:1158–1168

Yu H, Wang Q, Ileleji KE, Yu C, Luo Z, Cen K et al (2012) Design and analysis of geographic distribution of biomass power plant and satellite storages in China. Part 1: straight-line delivery. Biomass Bioenergy 46:773–784

Supply Chain Network Model for Biodiesel Production via Wastewaters from Paper and Pulp Companies

Sushil R. Poudel, Mohammad Marufuzzaman, Sandra Duni Ekşioğlu, Marta AmirSadeghi and Todd French

Abstract This study presents a mathematical model that aids the design and management of a biofuel supply chain network based on the treatment of wastewater sludge generated at Pulp and Paper plants. The model presented here analyzes the supply chain performance under different biomass conversion processes, different requirements for plant location and capacities. We present a case study using data from the states of Alabama, Louisiana and Mississippi. The modes of transportation considered in this analysis are truck, rail and barge. The model identifies the location of biocrude plants, the amounts of glycerol and wood chip to purchase, the amounts of products to deliver, the amount of biodiesel to produce, the modes of transportation and the suppliers to use in order to minimize the system wide cost. Numerical analysis identifies Lincoln, Mississippi and Naheola, Alabama as two potential locations for biocrude plants.

1 Introduction

USA is a big consumer of fossil fuels. In 2011, over 4 billion barrels of crude oil were imported, and over $460 billion were spend on petroleum products. This spending represents 0.8 % of the $60 trillion Gross Domestic Product (GDP) ECODAD

S.R. Poudel (✉) · M. Marufuzzaman
Department of Industrial and System Engineering, Mississippi State University, MS 39759 Starkville, USA
e-mail: srp224@msstate.edu

S.D. Ekşioğlu
Department of Industrial Engineering, Clemson University, 134 Freeman Hall, SC 29634 Clemson, USA

M. AmirSadeghi · T. French
Dave C. Swalm School of Chemical Engineering, Mississippi State University, MS 39762 Starkville, USA

S.D. Eksioglu et al. (eds.), *Handbook of Bioenergy*,
Energy Systems, DOI 10.1007/978-3-319-20092-7_6

(2012). While reducing fuel consumption may not be feasible, using local resources to produce renewable energy is an option worth investigating. Investing in production of bioenergy would not only result in an increase of domestic supply for fuels, but would also result in the creation of new jobs mainly in rural USA. The reality is that in the recent years we have seen an increasing interest on bioenergy production.

Biofuels can be produced using agricultural products (such as, corn, soybean) and agricultural residues (such as, corn stover, wheat straws), forest products and forest residues. First generation biofuels are produced using agricultural products, whereas second generation biofuels are produced using agricultural and forest residues and wastes. For example, wastewaters from paper and pulp mills can be used to cultivate activated sludge from municipal wastewater system along with lignocellulosic sugars to generate biocrude, a lipid feedstock that can be further processed to produce biodiesel (Upadhyaya et al. 2013). This production will contribute to meet the goals set by the Renewable Fuel Standards (RFS) of producing over 1 billion gallons per year in the next few years United States Department of Agriculture (2014).

The pulp and paper (P&P) industry is one of the largest users of water resources in the USA. This industry uses on the average 54 cubic meters of water per metric ton of finished product. Approximately 85 % of the water is used in their fabrication processes. Pulping is the first of the processes used in paper fabrication. During this process, organic waste, in the form of wood waste, is disposed to waters. Next, the mechanical and the chemical pulping processes generate organic and inorganic wastes which are as well disposed to waters (Pokhrel and Viraraghavan 2004). About 40–50 % of the original wood material is disposed to water during the process of chemical pulping; and about 5–10 % of the original wood material is disposed to water during the mechanical pulping (Smook 1992). Thus, wastewaters comprise primarily of bark particles, fiber, fiber debris, filler and coating materials (Thompson et al. 2001). Due to the presence of inorganic and recalcitrant organic chemicals such as sodium hydroxide, sodium carbonate, sulphites, adsorbable organic halides, terpenes, phenols, acetone and alcohols; the wastewaters generated do contain high levels of biological oxygen demband (BOD), relatively high levels of chemical oxygen demand (COD), and other organic matters (Upadhyaya et al. 2013). Therefore, this process requires onsite treatment solutions. For this purpose, plants use a sedementation process which removes about 80 % of the suspended solids from wastewaters prior to discharge (Rajvaidya and Markandey 1998). Typically, a plant would dispose wastewaters and wasted fibers on ponds. With time, ponds water holding capacity decreases due to the accumulation of fibers. Plants spend time and money to eliminate the fiber sediments from these ponds.

Because of the growing awareness about the impact of water pollutants on public health, and because of the stringent regulations established by governmental authorities (such as, the US Environmental Protection Agency (EPA)), the P&P industry has invested in developing treatment methods and treatment effluents to comply with the requirements. One of the compliance programs is the "The National Pollutant Discharge Elimination System (NPDES)" which sets restrictions on the chemical composition of wastewaters discharged in the environment (Pokhrel and Viraraghavan 2004). The industry has estimated that the cost of treating wastewaters accounts for about 10 % of the total production cost. This chapter proposes a model which optimizes the system-wide costs of using wastewaters from P&P plants to produce biodiesel. Using wastewaters to produce biodiesel reduces the costs associated with the treatment of wastewaters at P&P plants. It reduces the cost of disposing fibers and eliminates the need of building new ponds. This also reduces the amount of waste disposed in the environment, and increases the amount of biofuels produced.

The particular supply chain network investigated in this chapter focuses on using P&P plants as suppliers of fibers, wood chip plants as suppliers of wood chips, and diesel refineries as suppliers of glycerol. The mix of wasted fibers, wood chips, and glycerol is treated using a consortium of microorganisms in order to produce biocrude based on the technology developed by Dufreche et al. (2007) and later modified by Upadhyaya et al. (2013). Biocrude is produced at a biocrude plant. Such a plant could be co-located with a biodiesel plant to minimize the delivery cost of glycerol; or with a P&P plant to minimize the delivery cost of fibers; or could be centrally located to minimize the total transportation costs. Next, biocrude is delivered to biodiesel refineries where it is further processed to produce biodiesel. Biodiesel is then delivered to the market for consumption. The mathematical model we propose identifies locations for biocrude plants, and identifies a production and transportation plant to meet customer demand at minimum cost. We develop a case study in order to validate the model presented. The case study uses data from the following states: Alabama, Louisiana and Mississippi. The motivation for focusing our study in this region is because a number of P&P plants are located in these three states, which collectively provide about 50 thousand tons of fibers annually. Also, a number of wood chips suppliers are located in this region. Thus, using these local resources to produce biocrude makes economic sense. The model we propose should help decision makers evaluate the supply chain costs associated with biocrude production via wastewaters from P&P plants.

2 Literature Review

In this section we summarize the relevant literature. The model we propose is an extension of the deterministic location-transportation problem. This work contributes to the existing literature on deterministic supply chain models. In particular, this work contributes to the literature on biomass and biofuel supply chain design and management.

The financial practicality of using corn to produce ethanol, and using corn stover for co-firing in power plants has been studied by economists for a long time English et al. (1981). Current research is focused on building models to optimize the performance of biofuels industry through better managing logistics and supply chain activities. The literature presents a number of deterministic models which aid supply chain decisions. For example, Chinese and Meneghetti (2009) propose a mixed-integer linear programming (MILP) model to determine the optimal supply chain design for the delivery of forest residues to biofuel plants. This model determines whether it is beneficial to deliver the residues to a centrally located biofuel production facility; or to preprocess residues at the farm and then deliver the processed biomass to a biofuel plant.

Ekşioğlu et al. (2009) propose an MILP model to support biofuel supply chains design and management decisions. The model determines the size, number and location of the biofuel plants. The goal is to minimize the total system costs. The authors develop a case study. This case study considers the biomass available in Mississippi. They conduct sensitivity analysis in order to evaluate the impact of biofuel conversion rates, and plant capacities on the performance of the supply chain. A model which optimizes the supply chain design for biodiesel production from forest resources has been introduced by Kim et al. (2011a). These authors also use MILP model. The objective is to maximize the profit while fulfilling customer demand. The authors compare different plant capacity options and product specifications. The authors develop a case study using data from the Southeastern region of the USA. Santibanez-Aguilar et al. (2011) develop a multi-objective optimization model to determine the types of biomass feedstock, processing technologies, and product characteristics which maximize supply chain profits, and minimize the environmental impacts of biofuels during their life cycle.

A number of papers focus on modeling uncertainties faced by companies in the forest products industry (Gupta and Maranas (2003); You et al. (2009); M.K. Zanjani et al. (2010)). These studies indicate that it is important to consider uncertainties when solving real life problems. Solving such models provides robust solution. Awudu and Zhang (2012) provide a review of models that consider uncertainty in the bioenergy supply chain. Kim et al. (2011b) propose a two-stage stochastic programming model to design a biofuel supply chain. This model considers the uncertainties related to the price of the product, conversion rate, biofuel demand and biomass supply. The goal is to maximize the expected profits in the supply chain. The authors generate 33 different scenarios by altering the value of these uncertain parameters. Outcomes of this model were the capacity and location of processing plants—the first stage decisions,—and the flow of biomass and biofuels—the recourse decisions. Marufuzzaman et al. (2014) also propose a two-stage stochastic program to model the supply chain for biodiesel production via wastewater treatment. The model captures biomass supply and technology development uncertainties. The model determines a supply chain design which minimizes the delivery costs and the carbon footprint of biodiesel. The authors analyze the impact of carbon regulatory mechanisms in transportation mode selection and facility location decisions. Chen and Fan (2012) present a mixed integer stochastic programming model to identify the least expected system cost under potential future supply and demand uncertainties. The authors introduce a two-staged stochastic programming model that identifies refinery and terminal sizes and locations. A bi-criterion, multi-period, stochastic MILP model is proposed by Gebreslassie et al. (2012) to address the optimal design of biorefinery supply chain under supply and demand uncertainties.

Work by Graham et al. (2000) proposes a GIS to address biomass availability, cost variations, and environmental impacts of switching from conventional crops to energy crops for production of renewable energy. Haddad and Anderson (2008) use a GIS to identify potential locations of biofuel plants which use corn stover for production of bioenergy. Ekşioğlu et al. (2013) develop an integrated optimization-simulation model to optimize the performance of biofuel supply chain. The integrated model uses the output of the deterministic model as an input for the simulation model. The simulation model captures fluctuations in biomass availability, weather conditions, transportation and production capacities. The model captures the trade-offs that exist between transportation, inventory and investment in the supply chain.

The model presented in this chapter is related to the model proposed by Marufuzzaman et al. (2014). Their model was tested on case study which considers the use of wastewater from municipal wastewater treatment plants in Mississippi. The model proposed here extends the research of Marufuzzaman et al. (2014) the following ways: we consider waste is being supplied from P&P, wood chip and biodiesel plants, therefore, the structure of the supply chain is different; we consider a different biomass conversion process, thus, different requirements are set for plant location and capacities; our case study uses a larger data set since we consider the biomass supply and biofuel demand in Alabama, Louisiana and Mississippi; and finally, since we consider a wider geographical region, modes of transportation such as truck, rail and barge are considered for the delivery of products among facilities.

3 Problem Description

The main objective of this study is to build a mathematical model that aids the design and management of a biofuel supply chain network based on the treatment of wastewater sludge generated at P&P plants. In this supply chain, the suppliers of biomass are P&P plants that provide fibers and wastewater sludge, wood chip plants that provide wood chips, and biodiesel plants that provide glycerol. A biocrude plant is where these types of biomass are processed in order to produce biocrude. Biocrude is then delivered to biodiesel plants for further processing to make biodiesel (see Fig. 1). Biodiesel is then delivered to the market for consumption.

About 80 % of the sludge generated at a P&P plant is usable for the production of biocrude. Of this usable amount, 80 % can be converted to glucose. Fibers contain 100 % cellulose, and assuming 80 % conversion rate of cellulose to glucose, the 80 % conversion rate of fibers to glucose is very stable. This source contains a very low amount of lignin. Typically, about 50 % of the wood chips produced at a wood chip plant are usable. The remaining 50 % is water. Out of the usable -dry wood chip- amount 70 % conversion rate of cellulose/hemicellulose to glucose/xylose is typically obtained. A conversion rate of 70 % is observed most of

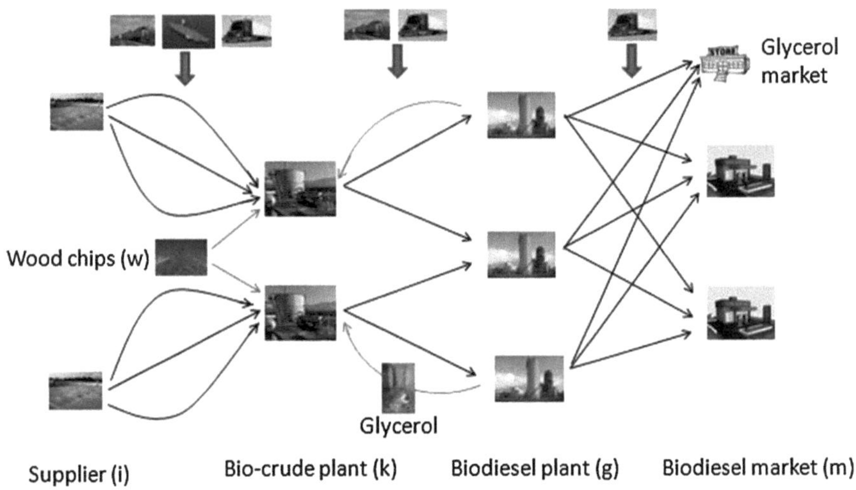

Fig. 1 The supply chain network in support of biocrude production and delivery

the time; however, this conversion rate could fluctuate from 60 to 75 %. In a biocrude plant, the main process is extracting glucose from fibers that exist in the wastewater sludge. In this process, the pulp and paper wastewater is used to cultivate pulp and paper activated sludge microbial community along with extracted glucose from fibers and other carbon sources. In this case, the conversion rate from glucose to oil is 37.8 %. Experimental results indicate that, adding carbon to the culture improves the conversion rate. Carbon sources are either glucose/xylose which can be supplied from wood chip plants; or glycerol which can be supplied from biodiesel refineries. Glycerol is a byproduct of biodiesel refineries. There is a market for high quality (highly concentrated) glycerol, and refineries would typically sell this byproduct. The glucose generated from the treatment of sludge is mixed in a ratio of 1:2 with a carbon source in order to produce oil. The carbon source could be just glycerol; or just glucose/xylose; or a mixture of both. Figure 2 summarizes the process of biocrude production in biocrude plants when using different mixes of carbon sources.

The model we propose in this section is an MILP. This model integrates location, transportation, and production related decisions in a biocrude supply chain network. The network design problem consists of locating biocrude plants among a set of candidate locations, determining the best production and routing decisions among different layers of the supply chain network. The goal is to minimize the overall supply chain costs. Figure 2 presents the structure of the biocrude supply chain network consisting of 2 sludge suppliers, 1 wood chip supplier, 2 potential location of biocrude plants, 3 biodiesel refineries, and 2 customers.

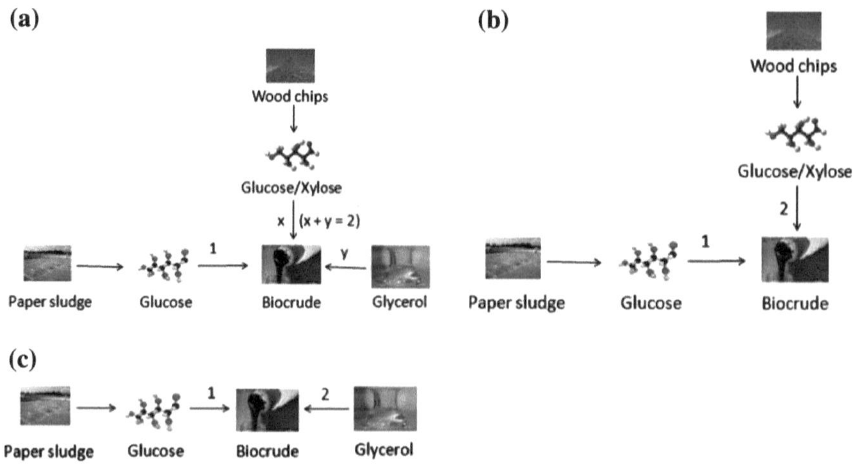

Fig. 2 Mixture ratios of glucose, glucose/xylose, and glycerol at a biocrude plant

4 Problem Formulation

Let $\mathbb{G}(\mathbb{N}, \mathbb{A})$ denote the supply chain network where $\mathbb{N}$ is the set of nodes and $\mathbb{A}$ is the set of arcs. Set $\mathbb{N}$ consists of the set of sludge supplier $\mathcal{I}$, the set of candidate biocrude plant locations $\mathcal{K}$, the set of wood chip suppliers $\mathcal{W}$, the set of biodiesel plants $\mathcal{G}$ and the set of markets $\mathcal{M}$. Each supplier $i \in \mathcal{I}$ produces s_i units of sludge. Each market $m \in \mathcal{M}$ demands d_m gallons of biodiesel. This demand is met by either biodiesel produced in this supply chain, or by substitute products that exist in the market. The price of the substitute product π_m, which is exogenously determined, represents the penalty per unit of unsatisfied demand. This penalty cost also serves as a threshold which indicates that if the unit cost of delivering biofuel through this supply chain exceeds the threshold, then the demand will be satisfied by using a substitute product.

Four different products flow along the arcs of this network: sludge flows from P&P plants to biocrude plants, biocrude flows from biocrude plants to biodiesel plant, glycerol flows from biodiesel plants to biocrude plants, and biodiesel flows from biodiesel plants to markets. There are three types of arcs between sludge suppliers and biocrude plant nodes in $\mathbb{G}(\mathbb{N}, \mathbb{A})$. We use these arcs to represent the fact that three different modes of transportation could be used to ship sludge: truck, train, and barge. The motivation for using these different modes of transportation is the fact that large quantities of sludge would be transported between facilities. Thus, using barge or rail makes economic sense. Typically, P&P plants have access to barge and rail, thus, using these modes of transportation comes without additional infrastructure investments. A single transportation arc exists between each biocrude plant node and biodiesel plant node, and between each biodiesel plant and market node. These arcs represent the use of trucks for shipment of biocrude, glycerol, and biodiesel. Since the quantities shipped between these facilities are rather small, using trucks for these deliveries makes economic sense.

The transportation cost function for trucks, trains, and barges consist of a fixed and a variable component. The fixed cost component mainly includes the annual sales tax, insurance costs, and management costs. These costs are not a function of the distance traveled or quantity shipped. The variable costs are a function of the distance traveled or quantity shipped. These costs include fuel, labor, maintenance, and repair costs. Let v_n be the fixed transportation cost for when using transportation mode $n \in \mathcal{N}$, and let β_n be the variable transportation cost when using transportation mode $n \in \mathcal{N}$. Also consider ξ_n and d_{ij} being the transportation capacity and distance between different nodes of the network respectively. The unit transportation cost (in \$/ton) between any two nodes i and j when using transportation mode n is calculated as follows: $c_{ijn} = \frac{v_n+\beta_n d_{i,j}}{\xi_n}$.

The following MILP model identifies: the location of biocrude plants, the amounts of glycerol and wood chip to purchase, the amounts of products to deliver, the amount of biodiesel to produce, the modes of transportation and the suppliers to use in order to minimizes system wide costs.

$$\text{[BSCM]} \quad \text{Minimize} \quad \sum_{l\in\mathcal{L}}\sum_{k\in\mathcal{K}} \Psi_{lk}Y_{lk} + \sum_{i\in\mathcal{I}}\sum_{k\in\mathcal{K}}\sum_{n\in\mathcal{N}} c_{ikn}X_{ikn} + \sum_{k\in\mathcal{K}}\sum_{g\in\mathcal{G}} c_{kg}X_{kg} +$$

$$\sum_{g\in\mathcal{G}}\sum_{m\in\mathcal{M}} c_{gm}X_{gm} + \sum_{g\in\mathcal{G}}\sum_{k\in\mathcal{K}} c_{gk}W_{gk} + \sum_{w\in\mathcal{W}}\sum_{k\in\mathcal{K}}\sum_{n\in\mathcal{N}} (c_{wkn} + \omega_k)X_{wkn} +$$

$$\sum_{m\in\mathcal{M}} \pi_m U_m + \sum_{l\in\mathcal{L}}\sum_{k\in\mathcal{K}} p_{lk}P_{lk} + \sum_{l\in\mathcal{L}}\sum_{g\in\mathcal{G}} p_{lg}P_{lg} - \sum_{g\in\mathcal{G}} a_g A_g$$

Subject to

$$\sum_{k\in\mathcal{K}}\sum_{n\in\mathcal{N}} X_{ikn} \le s_i \quad \forall i \in \mathcal{I} \tag{1}$$

$$\sum_{i\in\mathcal{I}}\sum_{n\in\mathcal{N}} X_{ikn} \le \sum_{l\in\mathcal{L}} c_{lk}^{cap} Y_{lk} \quad \forall k \in \mathcal{K} \tag{2}$$

$$\phi^1 \sum_{i\in\mathcal{I}}\sum_{n\in\mathcal{N}} X_{ikn} + \phi^2 \sum_{g\in\mathcal{G}} W_{gk} + \phi^3 \sum_{w\in\mathcal{W}}\sum_{n\in\mathcal{N}} X_{wkn} = \sum_{l\in\mathcal{L}} P_{lk} \quad \forall k \in \mathcal{K} \tag{3}$$

$$\phi^1 \sum_{i\in\mathcal{I}}\sum_{n\in\mathcal{N}} X_{ikn} - \frac{1}{2}(\phi^2 \sum_{g\in\mathcal{G}} W_{gk} + \phi^3 \sum_{w\in\mathcal{W}}\sum_{n\in\mathcal{N}} X_{wkn}) = 0 \quad \forall k \in \mathcal{K} \tag{4}$$

$$P_{lk} \le p_{lk}^{cap} Y_{lk} \quad \forall l \in \mathcal{L}, k \in \mathcal{K} \tag{5}$$

$$\sum_{g \in \mathcal{G}} X_{kg} \leq \sum_{l \in \mathcal{L}} P_{lk} \quad \forall k \in \mathcal{K} \tag{6}$$

$$\delta \sum_{k \in \mathcal{K}} X_{kg} = \sum_{l \in \mathcal{L}} P_{lg} \quad \forall g \in \mathcal{G} \tag{7}$$

$$P_{lg} \leq p_{lg}^{cap} \quad \forall l \in \mathcal{L}, g \in \mathcal{G} \tag{8}$$

$$\sum_{m \in \mathcal{M}} X_{gm} \leq \sum_{l \in \mathcal{L}} P_{lg} \quad \forall g \in \mathcal{G} \tag{9}$$

$$\sum_{l \in \mathcal{L}} Y_{lk} \leq 1 \quad \forall k \in \mathcal{K} \tag{10}$$

$$A_g + \sum_{k \in \mathcal{K}} W_{gk} = \zeta \sum_{l \in \mathcal{L}} P_{lg} \quad \forall g \in \mathcal{G} \tag{11}$$

$$\sum_{g \in \mathcal{G}} X_{gm} + U_m = d_m \quad \forall m \in \mathcal{M} \tag{12}$$

$$Y_{lk} \in \{0, 1\} \quad \forall l \in \mathcal{L}, k \in \mathcal{K} \tag{13}$$

$$X_{ikn}, X_{kg}, X_{gm}, X_{wkn}, W_{gk}, P_{lk}, P_{lg}, A_g, U_m \geq 0 \tag{14}$$

Constraints (1) indicate that the amount of fibres delivered by a sludge supplier is limited by its availability. Constraints (2) indicate that the total amount of fibres shipped to a biocrude plant is limited by its capacity. Constraints (3) are the flow conservation constraints at biocrude plants. These constraints set the mass balance between biomass used and biocrude produced using the technology developed by Dufreche et al. (2007) and later modified by Upadhyaya et al. (2013). Constraints (4) ensure that the requirements about the ratio of glucose, glucose/xylose and glycerol of biocrude production process are met. Constraints (5) indicate that the amount of biocrude produced at a biocrude plant is limited by plant capacity. Constraints (6) indicate that the amount of biocrude delivered to a biodiesel plant is limited by the amount of biocrude produced in a biocrude plant. Constraints (7) are the flow conservation constraints for biodiesel plants. Constraints (8) indicate that the amount of biodiesel produced at a biodiesel plant is limited by its available capacity. Constraints (9) indicate that the amount of biodiesel delivered to the market is limited by the amount produced at biodiesel plants. Constraints (10) indicate that at most one biocrude plant of capacity $l \in \mathcal{L}$ is located at a given location $k \in \mathcal{K}$. Constraints (11) ensure that glycerol produced at a biodiesel plant is either sold to the market or to a biocrude plant. Constraints (12) indicate that market demand must be fulfilled either through this supply chain network or through other substitute products available in the market. Finally, (13) are the binary constraints and (14) are the standard non-negativity constraints.

5 Case Study: Data Description

In order to validate our model we develop a case study. This case study is focused in the states of Alabama, Mississippi and Louisiana. Thus, the data collected about biomass supplies, biofuel demands, and costs is specific to these states.

5.1 Biodiesel Demand

We assume that the biodiesel produced will be used within Alabama, Mississippi and Louisiana. We set the overall demand for biodiesel equal to the constant value of 20 million ton. To derive this number, we estimated the total amount of biodiesel that could be produced using the waste generated from the P&P plants in these states. Next, we distributed this demand among all the counties in these states based on their population. So, we used the relative population of a county (as compared to the total population) as the means of distributing the total demand among these counties.

5.2 Facility Locations

Fiber Suppliers: The main sources of fiber in the region are P&P plants. We consider the P&P plants listed in the Primary Forest Products Network (2008). Figure 3a represents the distribution of these plants in the region, and the corresponding amounts of sludge available. The Primary Forest Products Network lists 26 different P&P plants which could supply a total of 55 thousand tons/year of wasted fibers.

Biocrude plants: There are no plants in Alabama, Louisiana and Mississippi that would use wastewaters and fibers as raw materials for production of biocrude. Therefore, we identify a few potential plant locations. We expect that such plants could be located in any county within these three states. Thus, the set of potential plant locations includes 86 counties of Alabama, 63 counties of Mississippi, and 76 counties of Louisiana. The centroid of a county is assumed to be the potential location of a plant. Additionally, this set includes the locations of existing P&P plants, wood chip suppliers, and biodiesel plants since we are interested to evaluate whether co-locating the biocrude plants with these existing facilities makes economic sense.

Wood chip suppliers: There is a total of 55 wood chip suppliers located in Alabama, Mississippi and Louisiana. Of these, 14 are located in Alabama, 34 in Mississippi, and 7 in Louisiana (see Fig. 3d). These locations are collected from the Primary Forest Products Network (2008).

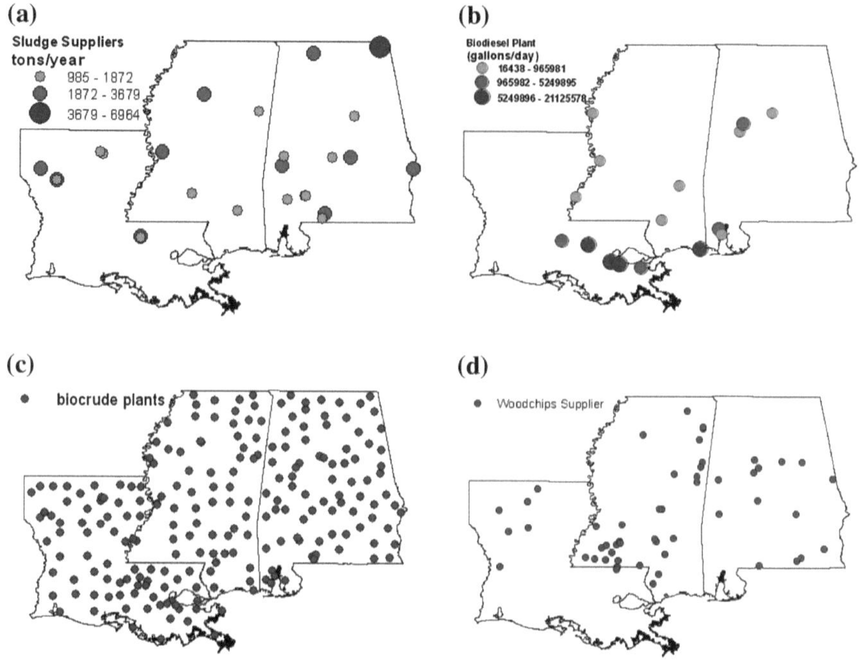

Fig. 3 Locations of suppliers, biocrude and biodiesel plants, **a** sludge suppliers, **b** biodiesel plants, **c** potential biocrude plants, **d** wood chips suppliers

Biodiesel refineries: We consider that the existing petroleum and diesel refineries located in Mississippi, Louisiana and Alabama will be used for further processing biocrude to biodiesel (see Fig. 3b). There is a total of 16 refineries in these states.

5.3 *Costs*

Investment costs: Data about investment costs and operating costs of a biocrude plant was collected from the following research paper (Dufreche et al. 2007) which presents the conversion technology and estimates corresponding costs. In this study we consider three different plant sizes, which produce accordingly. The small-sized plant produced 9000 tons, the medium-sized plant produces 18000 tons, and the large-sized plant produces 35000 tons of biocrude annually. The estimated investment costs are \$5.8 million, \$9.9 million, and \$16.8 million respectively.

We assume that biocrude will displace diesel in diesel refineries, and therefore, no additional investments are required at those plants.

Transportation costs: We consider that three different modes of transportation are available to ship fibers: truck, barge and rail. The associated transportation costs are presented in Table 1. Truck transportation costs for fiber are calculated for a single trailer truck with a load capacity of 30 tons, and average traveling speed of 40 miles per hour. Truck transportation costs for biocrude and biodiesel are calculated per tank (5 Axle 42′) with a load capacity of 23 tons. Barge transportation costs for fibers are calculated for general cargo of a barge with a load capacity of 11000 tons. Rail transportation costs are calculated for the standard commodity group of P&P products (Table 2).

6 Case Study: Discussion of Results

The main goal of our numerical analysis is to determine the structure of the supply chain in support of production and delivery of biodiesel using the technology developed by Dufreche et al. (2007), Upadhyaya et al. (2013). The model identifies locations for biocrude plants, and proposes a production and transportation plan which minimizes system wide costs. Through our numerical analysis we also evaluate the impact of market demand and price for glycerol on the amount of wood chips purchased and biodiesel produced.

The model is coded in GAMS 24.2.1. The solver used is CPLEX 12.6.

6.1 *Experimental Results*

Figure 4 presents the supply chain network for biocrude production and delivery. The model selects two sites to locate the biocrude plants. One large-sized plant is located in Lincoln, Mississippi, and one medium-sized facility is co-located with Georgia-Pacific Corp in Naheola, Alabama. The figure also presents the suppliers of each plant, and transportation modes selected.

Figure 5 represents the distribution of the total supply chain costs. Investment costs count for 72 % of the total costs, purchasing costs count for 12 % of the total costs, transportation cost count for 10 % of the total costs, and production costs count for 7 % of the total costs. This distribution of costs explains the structure of the supply chain in Fig. 4. Due to these high investment costs, only one large- and one medium-sized plants are selected to meet demand at the minimum cost. This strategy minimizes the total investment costs. One of the plants is co-located with a major fiber supplier in order to minimize transportation costs.

Figure 6 summarizes the results of the sensitivity analysis with respect to fiber supply and penalty cost. In these figures, the market price of glycerol is $150/ton. Figure 6a presents the impact that changes on the amount of fiber supply available at each P&P plant have on the amount of biodiesel produced and total supply chain costs. As more fiber becomes available at P&P plants, the supply chain costs decrease and biodiesel production increase. The decrease in costs is mainly due to a

Table 1 Summary of mathematical notations

Notation	Explanation
Sets	
$\mathcal{I}$	Set of sludge suppliers
$\mathcal{K}$	Set of biocrude plants
$\mathcal{G}$	Set of biodiesel plants
$\mathcal{W}$	Set of wood chip suppliers
$\mathcal{M}$	Set of markets
$\mathcal{L}$	Set of production capacities
$\mathcal{N}$	Set of transportation modes
Parameters	
s_i	Amount of biomass available at site $i \in \mathcal{I}$
Ψ_{lk}	Investment cost for biocrude plant of size $l \in \mathcal{L}$
c_{ikn}	Unit cost to transport sludge from sludge supplier $i \in \mathcal{I}$ to biocrude plant $k \in \mathcal{K}$ using ($/ton)transportation mode $n \in \mathcal{N}$
c_{kg}	Unit cost to transport biocrude from biocrude plant $k \in \mathcal{K}$ to biodiesel plant $g \in \mathcal{G}$ ($/ton)
c_{gm}	Unit cost to transport biodiesel from biodiesel plant $g \in \mathcal{G}$ to market $m \in \mathcal{M}$ ($/gallon)
c_{gk}	Unit cost to transport glycerol from biodiesel plant $g \in \mathcal{G}$ to biocrude plant $k \in \mathcal{K}$ ($/gallon)
c_{wkn}	Unit cost to transport wood chip from wood chip supplier $w \in \mathcal{W}$ to biocrude plant $k \in \mathcal{K}$ using transportation mode $n \in \mathcal{N}$ ($/ton)
π_m	Unit penalty cost of not satisfying demand in market $m \in \mathcal{M}$ ($/gallon)
ω_w	Purchasing cost of wood chip at wood chip supplier $w \in \mathcal{W}$ ($/ton)
p_{lg}	unit production cost at biodiesel plant $g \in \mathcal{G}$ of size lß$\mathcal{L}$ ($/gallon)
a_g	Unit cost of glycerol ($/gallon)
c_{lk}^{cap}	Sludge storage/handling capacity of a bio-crude plant of size l at location k (ton)
p_{lk}^{cap}	Biocrude production capacity of a biocrude plant of size l at location k (ton)
p_{lg}^{cap}	Biodiesel production capacity of a biodiesel plant of size l at location g (gallon)
ϕ^1	Conversion rate from sludge to biocrude
ϕ^2	Conversion rate from glycerol to biocrude
ϕ^3	Conversion rate from wood chip to biocrude
ζ	Production rate from biodiesel to glycerol
δ	Conversion rate from biocrude to biodiesel
Variables	
Y_{lk}	1 if a biocrude plant of size l is opened at location k; 0 otherwise
X_{ikn}	Amount of sludge transported from sludge supplier i to biocrude plant k using transportation mode $n \in \mathcal{N}$ (ton)
X_{kg}	Amount of biocrude transported from biocrude plant k to biodiesel plant g (ton)
X_{gm}	Amount of biodiesel transported from biodiesel plant g to market m (gallon)

(continued)

Table 1 (continued)

Notation	Explanation
X_{wkn}	Amount of wood chip transported from wood chip supplier w to plant biocrude k using transportation mode $n \in \mathcal{N}$ (ton)
W_{gk}	Amount of glycerol transported from biodiesel plant g to biocrude plant k (gallon)
P_{lk}	Amount of biocrude produced from biocrude plant k of size l (ton)
P_{lg}	Amount of biodiesel produced from biodiesel plant g of size l (gallon)
U_m	Amount of unsatisfied demand at market m (gallon)
A_g	Amount of glycerol sold by biodiesel plant g (gallon)

Table 2 Summary of transportation costs

Costs	Parameter	Value	Unit
Fixed cost of loading and unloading fiber on trucks	v_1^1	1.2	\$/ton
Fixed cost of loading and unloading fiber on rail	v_2^1	2.5	\$/ton
Fixed cost of loading and unloading fiber on barge	v_3^1	3.5	\$/ton
Variable unit cost for delivering fibers by trucks	β_1^1	0.16	\$/ton/mile
Variable unit cost for delivering fibers by rail	β_2^1	0.05	\$/ton/mile
Variable unit cost for delivering fibers by barge	β_3^1	0.02	\$/ton/mile
Fixed cost of loading and unloading biocrude on trucks	v_1^2	1.2	\$/ton
Variable unit cost of delivering biocrude by trucks	β_1^2	0.02	\$/ton/mile
Fixed cost of loading and unloading glycerol on trucks	v_1^3	0.10	\$/ton
Variable unit cost of delivering glycerol by trucks	β_1^3	0.02	\$/ton/mile
Fixed cost of loading and unloading biodiesel on trucks	v_1^4	0.10	\$/ton
Variable unit cost of delivering biodiesel by trucks	β_1^4	0.01	\$/ton/mile
Fixed cost of loading and unloading wood chips on trucks	v_1^5	1.2	\$/ton
Variable unit cost of delivering wood chips by trucks	β_1^5	0.16	\$/ton/mile

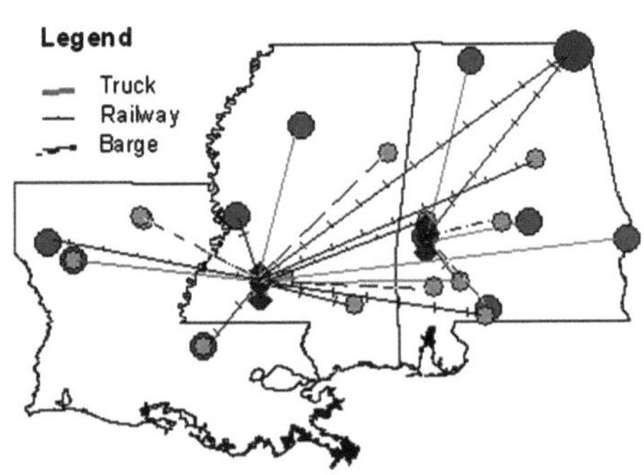

Fig. 4 The supply chain network in support of biocrude production and delivery

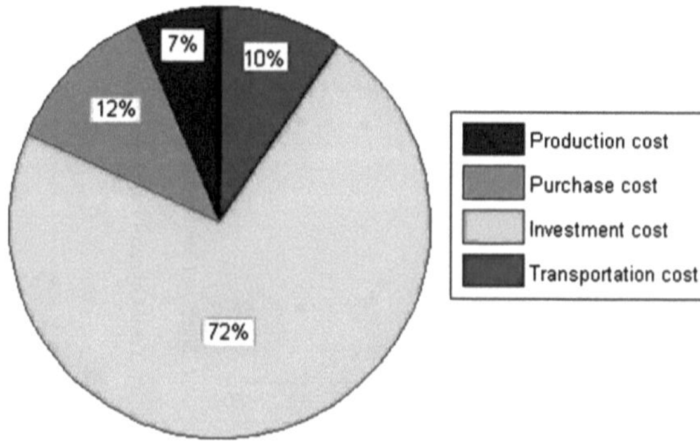

Fig. 5 Distribution of total costs among investment, production, purchase and transportation costs

decrease in transportation costs since a biocrude plant can receive the necessary supplies from plants located nearby. Other plants could gain from the economies of scale that come with using high-volume and cost efficient modes of transportation such as rail and barge.

Figure 6b presents the relationship among an increase in the availability/use of fiber, and the total amount of glycerol and wood chips purchased. Using more of the fiber available impacts the production of biodiesel, and consequently, impacts the use of glycerol and wood chips. However, as the plants use more fiber, the amount of wood chips used increases at a higher rate than the amount of glycerol used. This is mainly due to the limited supply of glycerol.

Figure 6c presents the relationships among penalty costs, biodiesel unit production cost, and biodiesel production quantity. The penalty cost (π_m for $m \in \mathcal{M}$) represents the market price of substitute products. This is called a penalty cost since it represents the lost profit from not satisfying customer demand. Since the objective of this model is to minimize costs, we expect that a customer will purchase the more economical product. Thus, as this penalty cost increases, the competition becomes weaker. A higher penalty cost provides more flexibility to the supply chain to satisfy customer demand. This is the reason why an increase in the penalty cost increases production and consumption of biodiesel. Unit cost also increases in order to satisfy the larger demand. However, this unit cost is smaller than the penalty cost.

Figure 6 represents the relationship among the penalty cost, and the amount of glycerol and wood chips purchased. As the penalty cost increases, the amount of biodiesel produced increases, and consequently, the use of glycerol and wood chips increase.

Figure 7 presents the relationship among the total unit cost, the amount of glycerol sold and glycerol market price. As the market price for glycerol increases, it becomes expensive to purchase and use glycerol. Through our numerical analysis we identified that \$50/ton is a break-even point. For a market price larger than \$50/ton, it is not economical to use glycerol. However, glycerol is a by-product in a biodiesel refinery, so there is an incentive to use it internally and save on transportation and other costs required to deliver this product to the market. We conducted a sensitivity analysis with respect to the amount of glycerol that could be sold in the market.

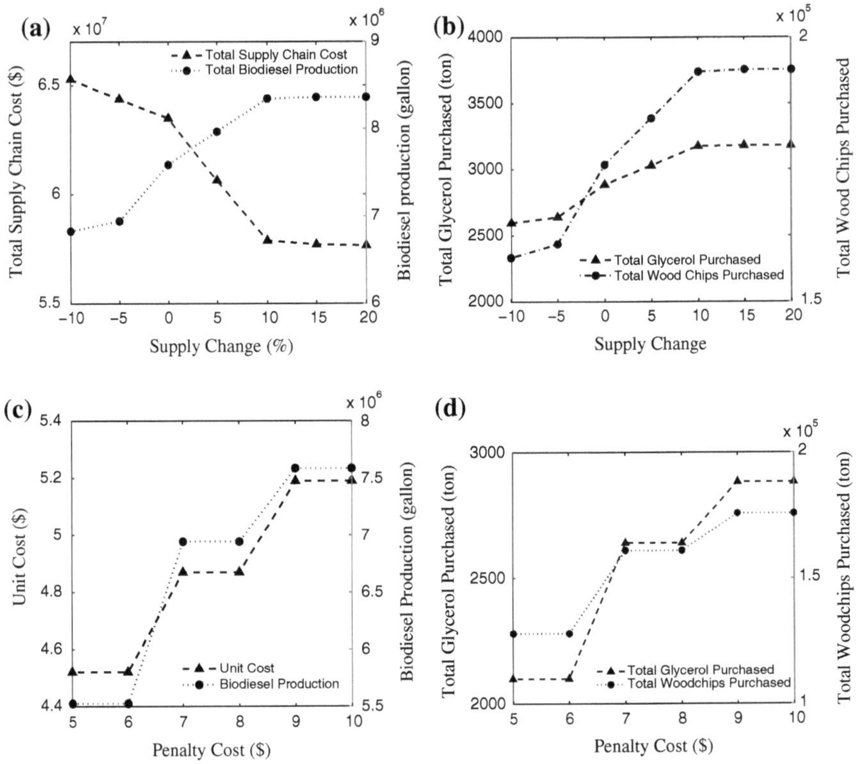

Fig. 6 Sensitivity analysis with respect to fiber supply and penalty cost, **a** relationships among fiber supply, total cost, production quantity, **b** relationships among fiber, glycerol and wood chip supplies, **c** relationships among penalty cost, unit cost, and production quantity, **d** relationships among penalty cost, glycerol purchased, and wood chips purchased

Fig. 7 Relationship among the total unit cost, the amount of glycerol sold and glycerol market price

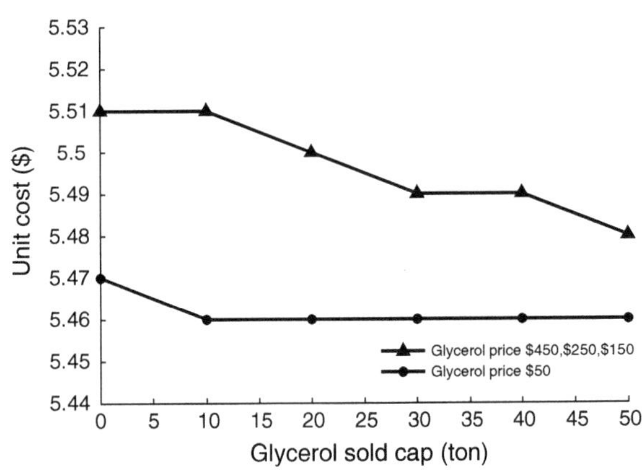

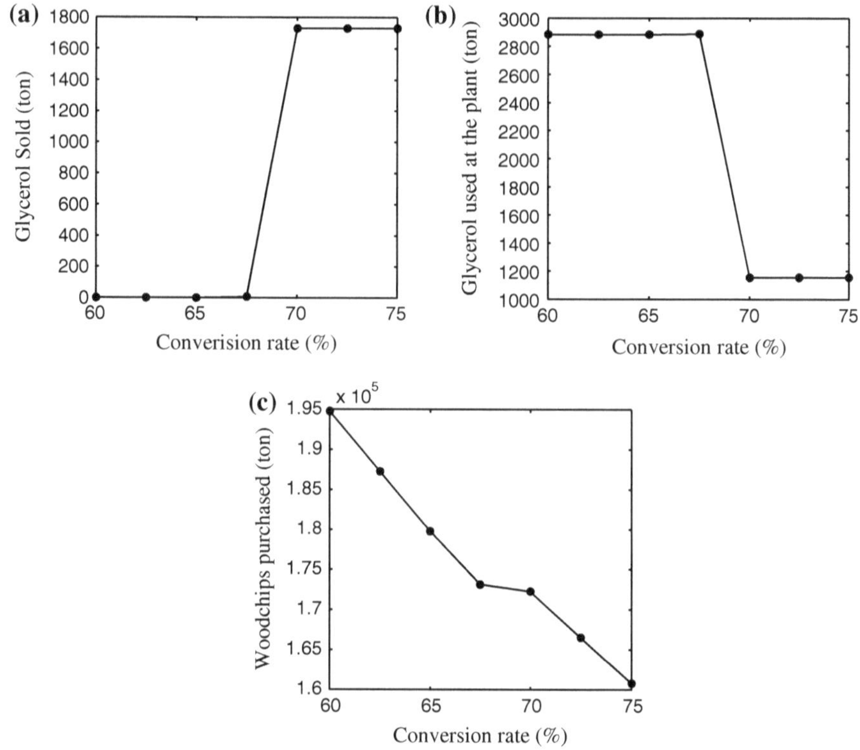

Fig. 8 Analyzing the impacts of conversion rate of wood chips to glucose/xylose on glycerol and wood chips consumption, **a** conversion rate versus the amount of glycerol sold, **b** conversion rate versus the amount of glycerol used, **c** conversion rate versus the amount of wood chips purchased

When the market price is higher than \$50/ton, as the amount of glycerol sold to the market increases, the unit cost of biodiesel decreases. This is because instead of using the highly priced glycerol, the plant uses wood chips which are less expensive.

Figure 8 presents the relationship between the amount of glycerol sold and the conversion rate of wood chips to glucose/xylose. As mentioned in Sect. 3, of the usable -dry wood chip- amount 70 % is glucose/xylose and 30 % is lignin. A conversion rate of 70 % is observed most of the time; however, this conversion rate could fluctuate from 60 to 75 %. These fluctuations do impact the amount of glycerol sold in the market, the amount of glycerol used in a biocrude plant, and the amount of wood chips purchased. Figure 8 shows that the amount of wood chips used and the amount of glycerol sold to market increases with this conversion rate. This is because at a higher conversion rate, more glucose/xylose using the same amount of wood chips, thus less of these resources are needed to maintain production. For the same reason, when the conversion rate is greater than 70 % we see a drop in the use of glycerol.

7 Conclusions

This chapter presents an MIP model in support of designing the supply chain of biodiesel production via wastewaters generated by P&P plants. The technology used to transform wastewaters into biorude, and then biocrude to biodiesel is developed by Dufreche et al. (2007), Upadhyaya et al. (2013). The particular supply chain network investigated here considers P&P plants as suppliers of fibers, wood chip plants as suppliers of wood chips, and biodiesel refineries as suppliers of glycerol. The mix of wasted fibers, wood chips, and glycerol is treated using a consortium of microorganisms in order to produce biocrude. These processes happen at biocrude plants. Such a plant could be co-located with a biodiesel plant to minimize the delivery cost of glycerol; or with a P&P plant to minimize the delivery cost of fibers; or could be centrally located to minimize the total transportation costs. Next, biocrude is delivered to biodiesel refineries where it is further processed to produce biodiesel. Biodiesel is then delivered to the market for consumption.

The mathematical model we propose identifies locations for biocrude plants, and identifies a production and transportation plant to meet customer demand at minimum cost. We develop a case study in order to validate the model presented. The case study uses data from the following states: Alabama, Louisiana and Mississippi. The motivation for focusing our study in this region is because a number of P&P plants are located in these three states, which collectively provide about 50 thousand tons of fibers annually. Also, a number of wood chips suppliers are located in this region. Thus, using these local resources to produce biocrude makes economic sense. The model we propose should help decision makers evaluate the supply chain costs associated with biocrude production via wastewaters from P&P plants.

The numerical analysis using real-life data identifies the following two sites as potential locations for biocrude plants Lincoln, Mississippi and Naheola, Alabama. The Lincoln plant is a large-sized plant, and the Naheola plant is a medium-sized plant. The plant in Naheola is co-located with Georgia-Pacific Corp, a P&P plant and large supplier of fibers. Through the numerical analysis we estimate the unit cost of delivering biodiesel to be \$5.19. Investment costs count for 72 % of the total costs, purchasing costs count for 12 %, transportation costs count for 10 %, and production costs count for 7 % of the total costs. We conducted sensitivity analysis with respect to fiber supply, penalty cost, and conversion rates. The fluctuations of these parameters did impact supply chain costs.

The numerical analysis provides insights about the impact that the market price of glycerol has on the performance of this supply chain. The result indicate that, as the market price for glycerol increases, the amount of glycerol used in the supply chain decreases. Thus, the market price determines the amount of glycerol that can be used. When market price of glycerol is higher than \$50 per gallon, it is not economical to use glycerol in this supply chain.

Acknowledgments This research was partially supported by NSF CAREER Award Nr. 1462420.

References

Awudu I, Zhang J (2012) Uncertainties and sustainability concepts in biofuel supply chain management: a review. Renew Sustain Energy Rev 16(2):1359–1368

Chen CW, Fan Y (2012) Bioethanol supply chain system planning under supply and demand uncertainties. Transp Res Part E: Logistics Transp Rev 48(1):150–164

Chinese D, Meneghetti A (2009) Design of forest biofuel supply chains. Int J Logistics Syst Manage 5(5):525–550

Dufreche S, Hernandez R, French T, Sparks D, Zappi M, Alley E (2007) Extraction of lipids from municipal wastewater plant microorganisms for production of biodiesel. J Am Oil Chem Soc 84(2):181–187

ECODAD (2012) Green life style changes. http://www.greenlifestylechanges.com/how-much-did-the-us-spend-on-imported-oil-in-2011/. Accessed 30 Jan 14

Ekşioğlu SD, Acharya A, Leightley LE, Arora S (2009) Analyzing the design and management of biomass-to-biorefinery supply chain. Comput Ind Eng 57(4):1342–1352

Ekşioğlu SD, Palak G, Mondala A, Greenwood A (2013) Supply chain designs and management for biocrude production via wastewater treatment. Environ Prog Sustain Energy 32(1):139–147

English BC, Short C, Heady EO (1981) The economic feasibility of crop residues as auxiliary fuel in coal-fired power plants. Am J Agric Econ 63(4):636–644

Gebreslassie BH, Yao Y, You F (2012) Multiobjective optimization of hydrocarbon biorefinery supply chain designs under uncertainty. In: 2012 IEEE 51st annual conference on decision and control (CDC). IEEE, pp 5560–5565

Graham RL, English BC, Noon CE (2000) A geographic information system-based modeling system for evaluating the cost of delivered energy crop feedstock. Biomass Bioenergy 18 (4):309–329

Gupta A, Maranas CD (2003) Managing demand uncertainty in supply chain planning. Comput Chem Eng 27(8):1219–1227

Haddad MA, Anderson PF (2008) A gis methodology to identify potential corn stover collection locations. Biomass Bioenergy 32(12):1097–1108

Kim J, Realff MJ, Lee JH (2011a) Optimal design and global sensitivity analysis of biomass supply chain networks for biofuels under uncertainty. Comput Chem Eng 35(9):1738–1751

Kim J, Realff MJ, Lee JH, Whittaker C, Furtner L (2011b) Design of biomass processing network for biofuel production using an milp model. Biomass Bioenergy 35(2):853–871

Marufuzzaman M, Eksioglu SD, Huang Y (2014) Two-stage stochastic programming supply chain model for biodiesel production via wastewater treatment. Comput Oper Res 49:1–17

Zanjani MK, Nourelfath M, Ait-Kadi D (2010) A multi-stage stochastic programming approach for production planning with uncertainty in the quality of raw materials and demand. Int J Prod Res 48(16):4701–4723

Network, Primary Forest Products (2008) Primary forest products network. forestproductlocator. www.forestproductslocator.org. Accessed 03 Oct 14

Pokhrel D, Viraraghavan T (2004) Treatment of pulp and paper mill wastewater: a review. Sci Total Environ 333(1):37–58

Rajvaidya N, Markandey DK (1998) Advances in environmental science and technology: treatment of pulp and paper industrial effluent. A.P.H. Publishing, Ansari Road, New Delhi, India

Santibanez-Aguilar JE, González-Campos JB, Ponce-Ortega JM, Serna-González M, El-Halwagi MM (2011) Optimal planning of a biomass conversion system considering economic and environmental aspects. Ind Eng Chem Res 50(14):8558–8570

Smook GA (1992) Handbook for pulp and paper technologists. Angus Wilde Publications, Vancouver

Thompson G, Swain J, Kay M, Forster CF (2001) The treatment of pulp and paper mill effluent: a review. Bioresour Technol 77(3):275–286

United States Department of Agriculture, Economic Research Service (2014) Findings. USDA. http://www.ers.usda.gov/topics/farm-economy/bioenergy/findings.aspx.VCHxbPldW1V. Accessed 23 Sep 14

Upadhyaya K, Lamichhane A, Mondala R, Hernandez T, French M, Green L, McFarland W Holmes (2013) Biocrude production by activated sludge microbial cultures using pulp and paper wastewaters as fermentation substrate. Environ Technol 34(13–14):2171–2178

You F, Wassick JM, Grossmann IE (2009) Risk management for a global supply chain planning under uncertainty: models and algorithms. AIChE J 55(4):931–946

Decision Support Models for Integrated Design of Bioenergy Supply Chains

Joseph Amundson, Sumesh Sukumara, Jeffrey Seay and Fazleena Badurdeen

Abstract The scarcity of fossil fuels and the environmental implications of their use has drawn increasing attention to the production of bioenergy from nonfood sources. To validate the progressive experimental research in this field, we require a credible tool that can quantify various impacts of potential biorefining processes. This chapter will demonstrate a novel decision support model that can provide comprehensive techno-economic results to various stakeholders. The framework integrates process optimization, supply chain optimization and discrete event simulation (DES) capabilities to provide a comprehensive and multi-disciplinary tool for bioenergy supply chain design following an iterative process. The tool is further enhanced by the incorporation of supply chain risk modeling to capture various uncertainties. A proof of concept case study is presented to illustrate the applicability of this framework to any given geographic region.

1 Introduction

Demand for energy and transportation fuel has increased tremendously and, based on current projections (EIA 2013a), it will continue to rise in the years to come. Currently, a major portion of the demand is met by imported crude oil (EIA 2013b). Reliance on foreign fossil fuel is unsustainable and it is susceptible to economic and political vulnerabilities. The United States Environmental Protection Agency (EPA) in collaboration with refiners, renewable (from plant, animal or waste) fuel producers and various stakeholders developed the Renewable Fuel Standards (RFS) which is

J. Amundson (✉) · F. Badurdeen
Department of Mechanical Engineering, University of Kentucky, Lexington, KY, USA
e-mail: jsamun2@g.uky.edu

S. Sukumara · J. Seay
Department of Chemical and Materials Engineering, University of Kentucky, Lexington, KY, USA

F. Badurdeen
Institute of Sustainable Manufacturing, University of Kentucky, Lexington, KY, USA

S.D. Eksioglu et al. (eds.), *Handbook of Bioenergy*,
Energy Systems, DOI 10.1007/978-3-319-20092-7_7

intended to set production targets based on these growing energy needs. Under the Energy Independence and Security Act (EISA) of 2007, the RFS program was extended which further targeted a production of 36 billion gallons of renewable fuel by 2022 (EPA Office of Transportation and Air Quality (OTAQ) 2007). Presently, under the Clean Air Act, the EPA is required to set annual guidelines for the RFS (EPA 2014). Although these are certainly challenging targets, EISA has created an opportunity for regions rich in biobased resources in the US. However, as it currently stands, only a few countries have made significant progress towards reducing energy dependence on fossil fuel (Batidzirai et al. 2012). The limited progress in biofuels sector has largely been due to the economic uncertainty introduced by biomass feedstock availability and supply chain-related challenges. To advance development of the bioenergy sector, stakeholders such as growers, biorefinery operators, local and regional policymakers and the community at large, require more accurate regional data to assess on the long-term returns of investments in biobased production of power, fuels and chemical products. However, because of the lack of comprehensive models to assess the techno-economic viability as well as the operating risk of utilizing biobased resources, the development of renewable energy processes and infrastructure is limited and has hampered progress towards RFS targets.

Biomass used to produce renewable fuel is mainly classified into four categories: first, second, third and fourth generation feedstock. First generation biomass mainly utilizes the oil and starches present in the food crops whereas, second generation biomass comprises of lignocellulosic plants and wastes. Recently, the focus of research has shifted from first generation feedstocks to lignocellulose based second generation feedstocks (Batidzirai et al. 2012). Presently, researchers around the world are exploring further possibilities of biofuel production by utilizing the third generation microalgae and the fourth generation modified organism-based feedstocks. However, the major portion of economically viable biofuel is still produced by first generation feedstocks, although projections suggest that both first and second generation biofuels will contribute equally by 2022 (Coyle 2010). This is because second generation biofuels have not been able to overcome technological and supply chain issues to prove their viability as a long-term sustainable source of energy. Hence, based on lessons learned from negative impacts of the first generation biofuels, an understanding of the future consequences of the second generation biofuels is needed.

The fundamental disadvantage of biomass as a feedstock is its low energy density when compared to fossil-based resources and, as a result, transportation over long distances is prohibitively expensive. Therefore, supply chain optimization is a critical component in developing economically viable processes utilizing biomass as a feedstock. Additionally, the lack of a flexible framework for determining where and at what capacity biorefineries should be designed has hampered the progress in utilizing these potential resources. A large number of combinations of process configurations as well as possible products arise (Sammons et al. 2007);

different configurations are required based on variables such as the condition of the biomass feedstock available, the quantity of biomass available per acre, the total acres in a given region and the product needs of the surrounding area. All these factors make this a complex optimization problem to solve. Developing specific solutions for this optimization problem that biorefinery operators and other stakeholders can use to identify the optimal product distribution and process configurations and to react to the changes in market prices and environmental targets will be the principle focus of this research.

Rigorous process simulation models are required to validate the utilization of biobased resources in a given region. The potential slate of products generated from a biorefinery is a function of the characteristics of the biomass. Therefore, the economic models generated as an outcome of this research must include input data from simulations of the individual processes to determine which products are the most viable for a given type of biomass feedstock. Although simply generating electricity from biomass is an option, complex molecules like lignin can be converted into high value products by both chemical and biological pathways. While several conversion pathways exist, it is necessary to quantify the long term impacts of any promising conversion process, guiding the investment to the appropriate direction.

A comprehensive and flexible modeling framework is required for the utilization of bio-based resources because the optimum path along with conversion technology will vary from region to region. The developed model must be capable of incorporating fundamental details pertaining to conversion technologies and the supply chain to reflect the effect on the overall economics. The inputs to this model will be information such as the types and available quantities of biomass in the region, details of the regional transportation network and the specific product requirements in terms of fuels, power or chemicals in a region. While there is a need for economic, environmental and societal assessment of bioenergy supply chains, this paper is limited to an economic assessment with some environmental analysis for the processes. When a new technology is proposed, it can be quickly evaluated to determine the economic viability for a given region. The interdisciplinary approach to this methodology is novel because the goal is to provide a modeling and optimization framework that can not only be used to evaluate new technologies by utilizing locally available biomass resources, but also evaluate corresponding resources on a region-by-region basis. This chapter elaborates on the unique manner in which chemical process optimization, supply chain optimization and supply chain simulation models can be linked to create a detailed decision support tool utilizing strengths of each method. By virtue of this contribution, the development of a multidisciplinary tool is described that will guide stakeholders to meet the heightened projections set by environmental agencies in developing energy and fuels from bio-based feedstock. In addition, the developed multidisciplinary framework can be used by various stakeholders such as investors, policy makers, environmentalists and growers to determine relevant impacts of biorefineries based on lignocellulosic and waste biomass.

2 Background

There are numerous process pathways for converting various generations of bio-based resources to biofuels and chemical products. Figure 1 shows a summary of conversion pathways for the first (primarily food-resources based), second (non-food based), third and fourth (micro-organism based) generation biofuels.

When compared to other biofuels, second generation biofuels have a more complex supply chain network. Adding to this complexity, most of the second generation biofuels feedstocks are less energy dense compared with first generation biomass feedstock. Nevertheless, unlike other biofuels, second generation biomass has a distinct advantage due to nonexistence of food competition, flexibility in feedstock supply and the potential to be converted into various marketable products (Naik et al. 2010).

The challenge in developing biorefining processes is to utilize the available renewable resources to determine the optimum product slate and corresponding production rate, in terms of environmental impacts and economic profitability. Looking at the broader prospects from a supply chain perspective, the demand locations for energy and transportation fuel are often far from biomass feedstock sources. Hence, the logistics of transportation of biomass from feed source to biorefinery location and transportation of products to end users must be solved (Faulkner 2012; Amundson 2013; Sukumara et al. 2013). In order to justify the usage of biomass resources, optimization models are required to navigate a path through the matrix, illustrated in Fig. 1, to maximize potential profit and minimize the environmental impacts.

In the last decade, many researchers have made contributions highlighting biofuel conversion, process integration and supply chain logistics for various processes.

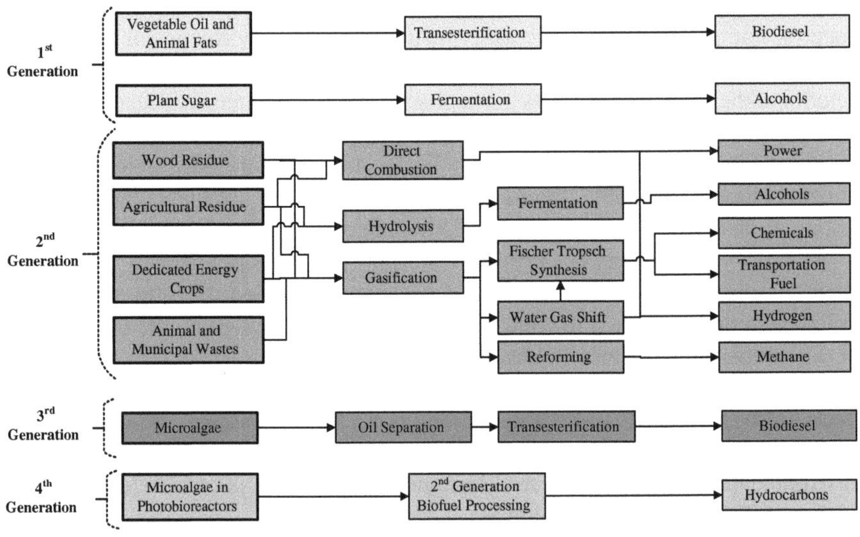

Fig. 1 Pathways for various biorefining processes

A techno-economic analysis of corn stover to produce ethanol was performed by the National Renewable Energy Lab (NREL) (Humbird et al. 2011), expanding on previous work (Aden et al. 2002). This contribution completed a thorough analysis incorporating major fermentation processes and capital cost details. Sammons Jr et al. (2008) showed a framework for assessing profitability of various biofuel production routes by uncoupling process models from the optimization framework for robustness. Another Mixed Integer Linear Programming (MILP) optimization model for strategic planning was demonstrated by Sharma et al. (2011) for a sustainable biorefinery by maximizing stakeholder value. A comprehensive review by Yue et al. (2014) provides a summary of contributions in modelling and optimization, highlighting major challenges in this field. As they point out, most of the existing contributions either focus solely on process optimization while making broad assumptions regarding the transportation logistics and supply chain issues or, vice versa. Hence, future research should be directed towards integrating the process and supply chain models, simultaneously solving for an optimum configuration. A review article on sustainability applications for integrated biorefineries (Azapagic 2014), showed various economic, environmental and social aspects that need to be considered while planning future biorefinery projects. Focusing on sustainability, an optimum planning and site selection model was presented by Santibañez-Aguilar et al. (2013) that incorporated economic, environmental and societal objectives. In the long term such an approach will be of great value as it integrates all the key aspects of sustainability. However, the contribution does not incorporate details of various operation and corresponding variability for the conversion process considered.

This chapter will discuss a novel methodology to encompass all the key aspects, starting with techno-economic feasibility. So far, in the research contributions reviewed related to determining optimal supply chains for biofuel production and distribution, significant voids are observed in capturing details from the process and supply logistics perspectives while accounting for inherent uncertainties. The following section will navigate the various adaptable simulation and optimization models used for developing this extensive framework.

3 Decision Support Framework

Processing of lignocellulosic feedstock into energy and transportation fuel is an energy intensive process. Adding to the complexity, the low energy density of the feedstock demands accurate planning of supply chain logistics. Also, one must be cautious that, while solving this complicated problem, potential impacts are not displaced from one stage to another. For instance, processing of biomass is carbon neutral but uncalculated feed and product transportation planning may increase emissions, leading to an unsustainable supply chain (Azapagic 2014).

Therefore, a multidisciplinary approach is required to determine the effect of conversion technology as well as transportation network for future biorefineries. This research focuses on developing a decision support model that will assist in designing and evaluating various impacts of biofuel supply chains. Preliminary validation of the framework is performed by estimating detailed economic performance for any biorefining process. Multiple concepts from process systems engineering and supply chain optimization are coupled to develop this dynamic optimization framework. The following sections outline the development of the core components of the suggested framework visually presented in Fig. 2. Sukumara et al. (2013) described the process optimization framework based on which the novel methodology is framed that takes into account both the process and transportation effects pertaining to the biomass to bioenergy supply chain. Figure 2 demonstrates

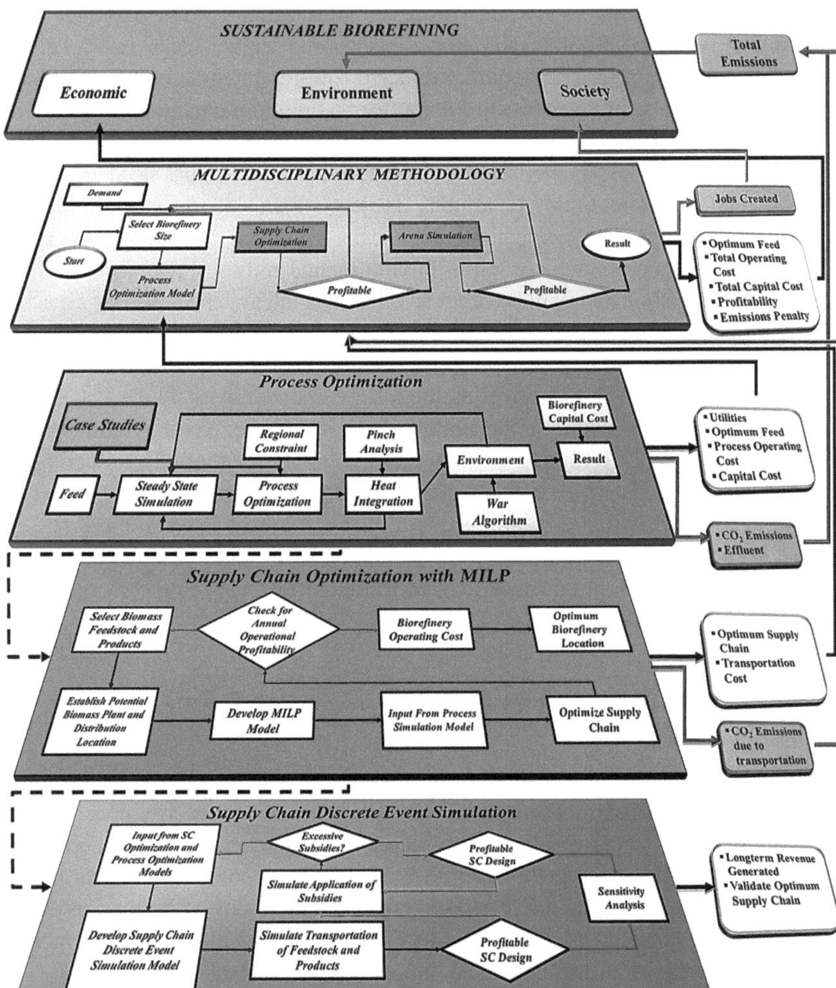

Fig. 2 Multidisciplinary methodology for estimating and validation of various biorefining processes

the detailed scheme for the flow of data in determining the most economically viable process and transportation configuration for potential biorefining in any given region. Note: The proposed methodology shows a unique linking of process optimization, supply chain optimization and discreet event simulation. This innovative approach is one of a kind and is a detailed expansion of previous a contribution (Sukumara et al. 2013). A section of the Fig. 2 (process optimization) has been reproduced with kind permission of Springer Science & Business Media.

3.1 Process Simulation Development

Capturing the details of chemical conversion processes has always been a challenge, especially with non-conventional feedstock, like biomass. Thorough process simulation models need to be formulated to acknowledge the evolving processes. Based on the feedstock available and the product demand in any given region, one or more potential conversion processes are selected. Steady state simulation models are developed using the Aspen Plus® engineering suite. Aspen Plus® is extensively used in process industries (mainly petroleum refining and petrochemical industries) around the world. Inclusion of such a rigorous simulation not only ensures accuracy, but also creates ports for the incorporation of fundamental changes in kinetics, design and utility estimation for existing and fledgling processes. Additionally, useful options such as heat integration and economic assessment are other salient features of this process engineering suite, enabling the framework the quickly assess capital investment requirements and energy saving potential. Most of the current tools lack this approach and hence have to depend on other sources for process details.

3.1.1 Thermochemical Process Simulation

Biomass gasification is a thermochemical process that is used to develop the proposed model. Feed for the gasification is distinguished based on proximate and ultimate analysis, enabling the inclusion of various compositional changes. The feedstock is preprocessed to reduce the size and screened before being sent to the gasifier. The gasification process uses air and steam as inputs to convert biomass into synthesis gas and ash. Some undesired products like tar and sulfur compounds are also produced that must be separated prior to downstream processing. A Water Gas Shift (WGS) reactor is used to enhance the H2:CO ratio to 2:1. The resulting synthesis gas is then split into two streams:

- The first stream is used for power generation
- The second stream is sent to Fisher Tropsch Synthesis (FTS) reactor for hydrocarbon production.

Figure 3 shows details that go into the development of thermochemical process simulation model. The ratio of the splitter can be varied based on local demand and

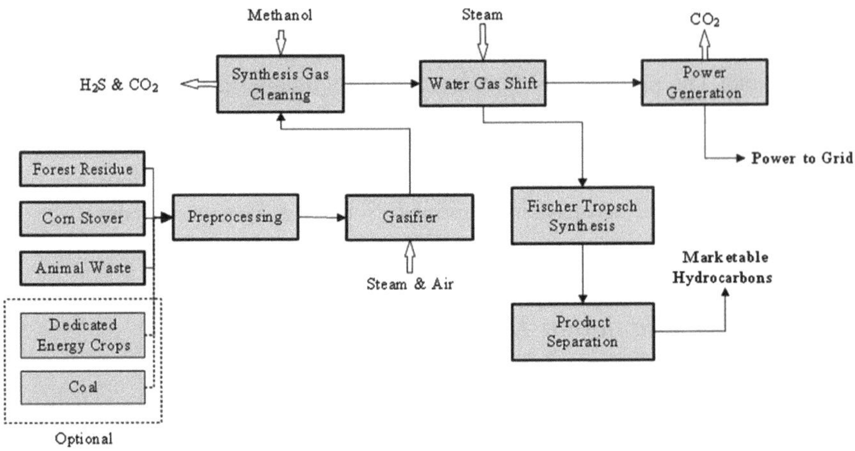

Fig. 3 Thermochemical conversion process of lignocellulosic biomass to power and transportation fuel

potential profit. Also, the WGS and FTS processes can incorporate conversion kinetics based on reaction conditions and catalyst used. Finally, the hydrocarbons produced are separated by distillation to various marketable products.

3.1.2 Biochemical Process Simulation

Fermentation of agricultural residues such as corn stover to ethanol is another process that has been developed. In this case, the biomass is distinguished based on compounds such as lignin, hemicellulose and cellulose composition which is unique for various bio-based residues. Two pretreatment steps are used to prepare the biomass for fermentation:

- Mechanical Pretreatment: feedstock is washed and reduced in size prior to the chemical pretreatment
- Chemical Pretreatment: feedstock is treated with dilute acid to expose the cellulose for downstream saccharification and fermentation processes

Figure 4 shows the major unit operations for a biochemical conversion process. The pH of the treated product is controlled by adding lime before sending it to the enzymatic saccharification and fermentation process. Here, cellulose is broken down into glucose and finally to ethanol by enzymes. Then, the ethanol is concentrated by distillation and passed through a molecular sieve to achieve a concentration of 99.5 %. The remaining effluents which mainly consists of lignin are collected and combusted to produce process heat. Due to the high water utilization, a treatment plant is simulated that cleans waste water before being recirculated. The process developed uses critical thermodynamic details from NREL's previous contribution (Aden et al. 2002). Subsequently, the developed process simulation model is run with user specified material, heat and electricity costs. Based on the

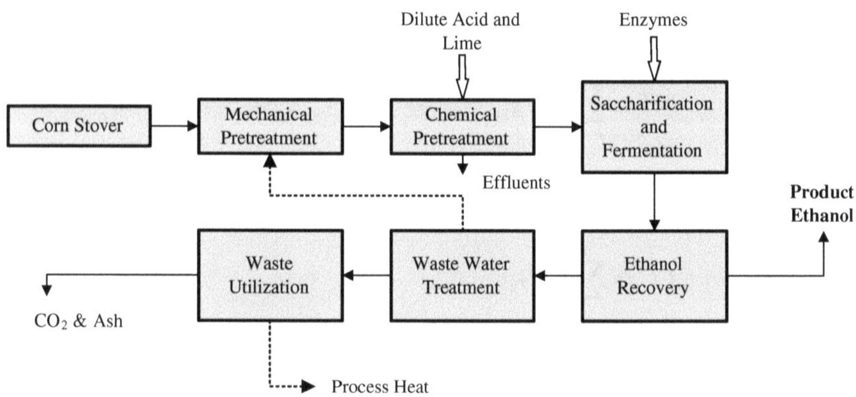

Fig. 4 Biological conversion process for ethanol production

methodology proposed in Fig. 2, the process optimization model is run, starting with the steady state simulation. Based on the user input, the process simulation runs and is optimized for maximum profit or minimum utility consumption as shown in Eqs. 1 and 2, respectively. The model provides an optimal mix of feedstocks as well as an optimal slate of products for the modeled conversion process. Notations used in formulating Eqs. 1 and 2 are detailed in Table 1.

Maximize Profit

$$Profit = Product\ Sales\ Income(PSI) - Biorefinery\ Operating\ Cost(BOC) \quad (1)$$

OR
Minimize Operating Cost

$$BOC = Variable\ Feed\ Cost(VFC) + Chemicals\ Cost(CC) + Utility\ Cost(UC) \quad (2)$$

Table 1 Notations and subscript indices

C_h, B_w, Q_v	Product ($/gal), feed ($/kg) and chemical ($/kg) unit cost, respectively
P_h, F_w, G_v	Product (gal/s), feed (kg/s) and chemical (kg/s) flow rate, respectively
CH, CR, CC	Cost of hot, refrigeration and cold utility, respectively ($/J)
QH_x, QR_y, QC_z	Hot, refrigerant and cold utility consumed (W)
CL, FR, CS	Generated chicken litter, forest residue and corn stover (kg/sec)
A, D, R	Rate of input of CL, FR and CS, respectively (kg/sec)
O_p, O_q, O_r	Non-negative multiplication factor of CL, FR and CS, respectively
a, b, d	Number of product, feedstock and chemicals, respectively
e, f, g	Number of hot, refrigeration and cold utility streams, respectively
h, w, v	Product, feed and chemical indices respectively

where

$$PSI = \sum_{h=1}^{a} C_h P_h,$$

$$VFC = \sum_{w=1}^{b} B_w F_w,$$

$$CC = \sum_{v=1}^{d} Q_v G_v$$

$$UC = CH\sum_{x=1}^{e} QH_x + CR\sum_{y=1}^{f} QR_y + CC\sum_{z=1}^{g} QC_z$$

Subject to:

$$A = CL^* O_p,\ D = FR^* O_q,\ R = CS * O_r$$
$$0.9 \le O_p \le 1,\ 0.9 \le O_q \le 1,\ 0.9 \le O_r \le 1$$

Heat integration is performed by using Aspen Energy Analyzer® to obtain the optimum heat exchanger network, further increasing the operating profit. Using a detailed simulation tool like Aspen Plus® has the following inherent advantages: the detailed results obtained can be used by existing environmental impact assessment tools like Waste Reduction Algorithm (Young et al. 2000) and realistic capital costs can be readily estimated by integrated Aspen Economic Analyzer® for investment analysis. Finally, all the results obtained from process simulation are passed to the supply chain model for further investigation.

3.2 Supply Chain Modeling Approaches

Supply chain design can be a significant challenge to biomass utilization because the supply chain is a complex system; each element affects the others and the choices are numerous: wide variety of usable biomass resources and conversion techniques; the multiplicity of end-use applications for products, many different market segments and stakeholders. In addition, agricultural yields are subject to seasonal fluctuations and vagaries of weather; when feedstocks are perishable, transportation and handling can be costly (Gold and Seuring 2011).

3.2.1 Supply Chain Optimization

Analytical modeling and optimization for supply chains in general is a mature field with an extensive literature (Min and Zhou 2002). Recent reviews have captured advances in supply chain design and logistics issues in biomass systems (Gold and Seuring 2011; Nikolopoulou et al. 2012; Yue et al. 2014). Techno-economic analysis for biomass supply chain design through optimization has also drawn attention. Recent work to identify refinery and biomass source locations, plant sizes, biomass allocations to plants, transport channels, etc., has considered spatially explicit approaches, used multi-period models with some focusing on ethanol production, while others consider biomass-to-heat generation, multiple feedstocks and conversion technologies (Dunnett et al. 2007; Dunnett et al. 2008; Corsano et al. 2011; Akgul et al., 2011; Eksioglu et al. 2009; Zamboni et al. 2009a,b; Aksoy et al. 2011; Kim et al. 2011a, b). While some researchers have expanded the scope to include multiple objectives (Zamboni et al. 2009a, b; Mele et al. 2011; You et al. 2012; Bernardi et al. 2013) the vast majority work to identify optimal supply chains from purely economic perspective, either to minimize costs (Dunnett et al. 2007, 2008; Zamboni et al. 2009a, b; Eksioglu et al. 2009; Aksoy et al. 2011; Akgul et al. 2011; Corsano et al. 2011) or to maximize profits (Kim et al. 2011a, b). While exceptions exist (Dal-Mas et al. 2011; An et al. 2011; Marvin et al. 2012; Gebreslassie et al. 2012) the majority of recent works in biorefinery supply chain design have not considered the uncertainties associated with the supply chain.

By far, the most common means of designing biorefinery supply chains in the literature is mixed integer linear programming (MILP). This method allows for an optimal supply chain to be determined given simplifying assumptions (which may or, may not, correspond to reality). While this method may not provide an optimal solution for all time given supply and demand uncertainties, it provides a static snapshot of what the optimal design may look like, making the MILP an invaluable aspect of the multidisciplinary integrated modeling approach discussed here.

To assess systemic performance and for iterative use with the previously described chemical process simulation and optimization models, a MILP model to optimize total supply chain performance should be formulated by segmenting the biorefinery supply chain into three components: biomass source locations, potential biorefinery locations, and market distribution locations for finished products. Given a selection of feedstock and feedstock supply locations determined via a thorough analysis of Geographical Information System (GIS) data for a given region (the same mix used for optimization and simulation of chemical processes), monthly availability for each supply location must be determined. This can be done by utilizing publically available databases including, for example, those related to agriculture and forest production (NASS and FPR). The strategic supply chain design then accounts for monthly feedstock availability by considering newly available biomass and the aged supply for each month, feedstock, and feedstock location. Other biomass parameters include biomass purchase cost and the amount of biomass shipped per truckload; these parameters can be established based on literature sources. Such parameters with respect to the products are inputs to the model as well. Potential biorefinery siting options should be defined based on relevant regional considerations such as population density, transportation infrastructure access, among others. For product demand, public databases can also be used to determine monthly usage rates.

Such as model was developed by Faulkner (2012) and is used in this section as an example. Given the parameters of the location-allocation problem described, the decision variables included in the model describe the selection of a biorefinery location, the amount of feedstock to be transported from each biomass feedstock location to the selected biorefinery location, and the amount of each product to be transported from biorefinery location to each market distribution locations.

For integrated modeling of biorefinery supply chains, information from chemical process modeling must be included in the design formulation. Costs for heating utilities, cooling utilities, and electricity as well as feedstock requirements for product generation and product yields can be passed from process optimization to improve the supply chain design. In the model described here, the supply chain profit is maximized through the optimization model based on income from the sale of products, product delivery costs, feedstock and operational cost, as well as inventory cost as show in Eq. 3 where the subscript m denotes the month.

$$Total\ Profit = \sum\nolimits_{m=1}^{12} (Income_m - Cost_m) \tag{3}$$

The income for each month depends on the volume of each product generated in that month and product price. The cost associated with the biorefinery supply chain is comprised of monthly biomass purchasing cost, inventory cost, biomass transportation cost, the cost of diesel to transport the biomass, operating cost, product transportation cost, and the cost of diesel to transport product to the market. The objective function for total profit is maximized subject to constraints related to the number of biorefinery locations to be opened, mass flow balances, biorefinery operating requirements, feedstock supply, and product demand as well as relevant non-negativity and binary constraints. Each of the constraint equations and a brief description can be seen in Table 2.

This MILP model development methodology can be applied to any specific region in iteration with the development of chemical process simulation and optimization to yield an optimal supply chain configuration. The optimal supply chain will have one open regional biorefinery location (model can easily be changed if multiple locations are desired) and a set of identified optimal sourcing locations. The model also yields the optimal supply chain costs, income, and, thus, profitability for a given biorefinery technology mix (determined through process simulation and optimization). If a profitable supply chain cannot be established, iteration with process modeling should be done to make changes to the process technologies applied to alter the constituents or magnitudes of the optimal feedstock mix and, therefore, the shape and scope of the designed supply chain. Ideally, iteration of this sort would result in the optimal selection of biorefinery technology for a given region l and, as a result, optimal feedstock requirements for the specified region. A detailed discussion of the application of MILP in the integrated biorefinery supply chain modeling context is provided by Faulkner (2012).

Table 2 Constraint equations used by Faulkner (2012) summarized

Constraint equation	Description
$\sum_{j=1}^{J} P_j = 1$	Ensures only one biorefinery location (j) is selected.
$\sum_{j=1}^{J} X_{fijm} \leq B''_{fim} \quad \forall f, i, m$	The amount of biomass feedstock (f) to be transported from each biomass location (i) to the biorefinery location (j) should not exceed the total amount of biomass available at the supply location for each type of biomass each month (m).
$B_{fim} = B''_{fim} \quad \forall f, i, \ \&m = 1$	For the first month (m = 1) of the planning period, the total amount available of each feedstock type (f) at each biomass location (i) is equal to the new supply.
$B_{fim} + B'_{fim} = B''_{fim} \quad \forall f, i, \&m = 2\ldots12$	For months (m) 2 through 12, the surplus of biomass or aged biomass is considered along with the new supply (*B fim*) for the total available amount.
$E_f\left(B''_{fim-1} - \sum_{j=1}^{J} X_{fijm-1}\right) = B'_{fim} \quad \forall f, i\&m = 2\ldots12$	The amount of aged biomass at biomass location is solely dependent on the amount shipped in the prior month, the total amount in that month and the degradation or erosion factor (E).
$\sum_{i=1}^{I}\sum_{j=1}^{J} X_{fijm} = BN_{fm} \quad \forall f, m$	For all months, the amount of biomass transported is the amount that is needed at the biorefinery for that period.
$\sum_{j=1}^{J}\sum_{k=1}^{K} Y_{pjkm} = PS_{pjm} \quad \forall m, p$	The amount of product to be shipped in a month should be equal to the amount created in that month
$\sum_{j=1}^{J} Y_{pjkm} \leq PD_{pkm} \quad \forall p, k, m$	The amount of product transported to the market location should not exceed the monthly demand for that product.
$\sum_{f=1}^{F}\sum_{i=1}^{I}\sum_{j=1}^{J}\sum_{m=1}^{M} X_{fijm} \leq MP_j \ \forall j$	Ensures biomass feedstock is not sent to a biorefinery location which is not opened. M is an arbitrarily large number.
$\sum_{p=1}^{P}\sum_{j=1}^{J}\sum_{k=1}^{K}\sum_{m=1}^{M} Y_{pjkm} \leq MP_j \ \forall j$	Ensures bioreinfery product is not sent to market locations from a closed location. M is an arbitrarily large number.
$Y_{pjkm}, X_{fijm} \geq 0$	A non-negativity constraint has been added for the amount of products and biomass feedstock to be transported.
$P_j = binary$	Specifies that a binary variable indicates whether the biorefinery is open or not.

3.2.2 Discrete Event Simulation (DES)

Simulation models to study various systems, including supply chains, has been used widely (Tako and Robinson 2012). Some early work in biomass system simulation has been done to analyze biomass harvesting and transportation costs (Mantovani and Gibson 1992). Simulation and optimization models have been used together by De Mol et al. (1997) to analyze all operations from biomass farm to conversion plant within a given network of facilities to assess economic feasibility of the biofuel industry. The simulation tool, Integrated Biomass Supply Analysis and Logistics (IBSAL), has been used to determine processing and transportation related carbon emissions (Sokhansanj et al. 2006) and to study switch grass logistics with several options for collection and transportation (Kumar and Sokhansanj 2007). An Arena© based DES model is used by Zhang et al. (2012) to study a biofuel production supply chain incorporating harvesting/processing, transportation, and onsite storage to assess delivered feedstock cost, energy consumption, and GHG emissions. However, the widespread use of DES models to study biomass supply chains is still limited.

In general, DES models represent reality as a sequence of events that change the state of a system at an instant in time and provide a powerful means of examining operational performance of a system. Here, a DES model is developed to simulate the activities observed in supply chain activities to deliver biomass feedstocks to the biorefinery and products from the biorefinery to demand locations. The events included in the model are represented in the flow chart in Fig. 5.

The MILP model discussed previously is able to determine the optimal supply chain design to source the biomass feedstock; for a selected technology and feedstock portfolio, the previously described chemical process simulation model determines optimal operating costs, product yield, and feedstock requirements. All these models are lacking in their consideration of the time dependent variability inherently associated with agricultural feedstocks and uncertainty related to demand in product markets over the long term. The use of these models together with a DES

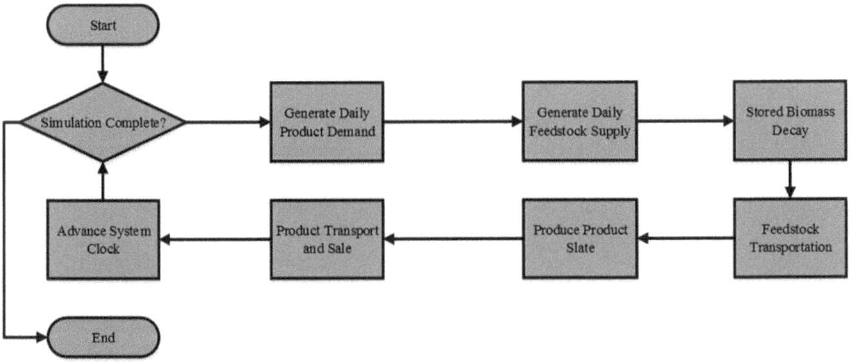

Fig. 5 Sequence of events modeled

model, however, can help assess impacts of seasonal variability in biomass feedstock and other uncertainties. An explicit goal of the DES model is to incorporate uncertainty in the model parameters. This necessitates different data requirements in the form of probability distributions that reasonably represent feedstock supply availability and product demand over time which are used as model inputs. To develop these distributions, aggregate years of production and harvest data must be used to fit distribution functions. Developing these distributions allows capturing the inherent variability and results in reasonable supply and demand values for operational simulation.

When extreme variations in biomass feedstock is experienced, the operation of the DES supply chain will deviate from optimality to source inputs from farther locations (for example, when availability drops) to satisfy daily biomass demand. This results in increased diesel and transportation costs at certain times of the year (typically corresponding to periods of time when biomass is not being harvested).

Applying DES modeling to the optimal supply chain determined via MILP with the same assumptions reveals very insightful details that are relevant to multiple stakeholders and decision makers in the biorefinery supply chain. This approach of modeling allows for a time-varying assessment of the robustness of the optimal supply chain under consideration. During seasonably variable months, for instance, revenues for the biorefinery may plummet to a degree where overcoming the deficit becomes difficult. Accounting for the time value of money via net present value (discounted cash flow method) allows for a more realistic analysis of investment return and payback period. Similar to the MILP model, the profitability of the supply chain in iteration with the chemical process models would potentially alter the technology mix and subsequent supply chain MILP optimization. Including this DES based modeling step allows assessing supply chain long-term performance under uncertain conditions to design more robust supply chains. A detailed discussion of the application of DES for biorefinery supply chain modeling is provided by Amundson (2013).

3.3 *Bioenergy Supply Chain Risk Modeling with BBN*

The nature of agricultural products makes them an inherently risky source of raw materials. Supplies of these materials are dependent on multiple factors such as market demand, weather, and seasonal availability, among others. Each of these influences has a relatively high level of uncertainty; unexpected changes in them could result in sub-par operations at the biorefinery. Most tools currently available for modeling risk and causal relationships in complex supply chains are qualitative (Tang and Musa 2011, Zhao et al. 2010) and descriptive [see Vanany et al. (2009) for a detailed review] making it difficult to fully capture the interdependencies between various risk factors or how they propagate. While stochastic optimization has been considered by some (Gebreslassie et al. 2012; Hytoene and Stuart 2010) to incorporate uncertainty in biomass supply chains, in-depth risk analyses have not been performed.

Bayes Theorem is a fundamental proposition that governs the relationship between conditional events and can be found in many textbooks. The theorem can be presented as follows:

$$P(B|A) = P(A|B) * P(B)/P(A) = P(A|B) * P(B)/\sum_{i=1}^{n} P(A|B_i) * P(B_i) \quad (4)$$

In words, the probability of the occurrence of an event B given the occurrence of event A [P(B|A)]is found to be the product of the probability of event A given the occurrence of event B [P(A|B)] and the probability of event B [P(B)] divided by the sum of *i* products of the probability of event A given the occurrence of event B_i and the probability of event B_i. The chain rule application of Bayes Rule to a causal network yields Bayesian Belief Networks (BBN), a probabilistic modeling technique that permits robust reasoning when there is uncertain knowledge (Medina-Oliva et al. 2009). Better computing power has increased BBN application in a number of domains (Mittal and Kassim 2007): to assess risk level for regulatory noncompliance (Yen and Zeng 2010), fault diagnostics (Fenton and Neil 2007), maintenance (Medina-Oliva et al. 2009), and supplier failure analysis (Lockamy III 2011). BBN has the advantage of being able to perform what-if analyses on the causal risk network to study sensitivity to various events (Fenton and Neil 2007). In supply chain risk assessment, Fernández et al. (2010) describes a supply chain event management model that includes a Bayesian Network. This proposed model monitors actions along the supply chain and has the ability, due to application of Bayes Theorem, to dynamically change after unexpected events occur, allowing for more proactive risk management. Kelepouris et al. (2011) propose a Bayesian approach to utilize serial-level data provided by modern supply chain tracking systems to predict product location and increase visibility into the supply chain. The benefit of such a system for supply chain stakeholders is the ability to quickly optimize operational decisions dependent on product availability. These examples of supply chain application of BBN, and most applications in general, seem to focus on the operational capabilities provided by BBN modeling. Alternatively, a more strategic quantitative method using BBN has been developed by Badurdeen et al. (2014) to assess risks to manufacturing supply chains. This work is composed of the development of a generic risk taxonomy and risk network map, the development of a BBN based software tool for risk modeling, and the application of the tool to an industrial case study.

Using the methodology presented in Badurdeen et al. (2014), Amundson et al. (2012) applied BBN using AgenaRisk© Software specifically to an issue of biomass supply chain risk. Risk events were identified via literature search and brainstorming resulting in a large, generic risk network map consisting of possible risks to biomass supply chains, their causal interconnections, and potential key performance indicators (KPI) for monitoring and risk mitigation policy design.

While not comprehensive, this extensive map served as a template allowing for the selection of a subset of risks, connections, and KPIs. An individualized risk network map can be constructed from the larger, generic map for specific case studies. It should be noted that the probabilistic risk assessment represents only probability of risk occurrence; the scale of the impact experienced by the supply chain stakeholders is not considered (Brown et al. 2012 explores means to address this concern). For this reason, it is important to properly identify the interconnections, risks, and KPIs that are of interest to a client. Conditional probabilities of each risk event in the map can be estimated from historic data or qualitatively through expert opinion. These data can then be used to populate a supply chain risk BBN. An example from Amundson et al. (2012) can be seen as the Base Case in Fig. 6. The probabilities shown for each risk item correspond to the calculated probability of that risk event occurrence, given the provided conditional probabilities.

With this construct, several opportunities are available to leverage the power of BBN modeling. For example, BBN allows for scenario analyses. Zwart et al. (2006) and others have discussed the benefits of biomass pretreatment methods for biorefinery supply chain design. Biomass densification of would result in a decrease in the total number of shipments to supply the biorefinery. In this example, the case where preprocessing is present has been called Scenario 1; the probability of "Lack of Preprocessing" being true is set to 0 %. The case where preprocessing is not present has been called Scenario 2; the probability of the same risk node being true is set to 100 %. Through application of Bayes' Rule (Eq. 1) posterior probabilities for each risk are calculated given the observational evidence. The results of this can be seen in Fig. 6 as Scenario 1 and Scenario 2, respectively. The probabilities for

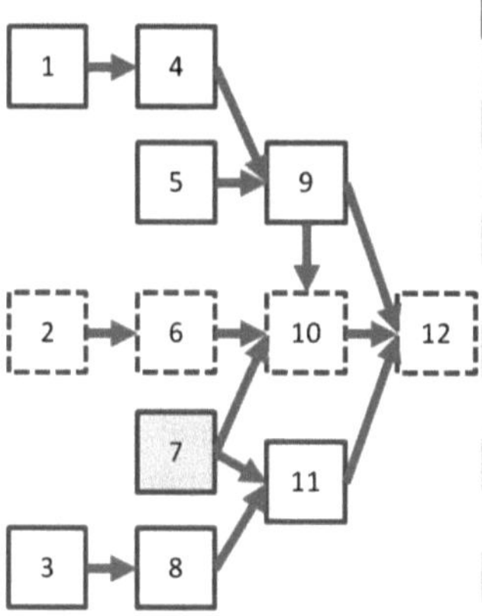

	Scenario		
Description	Base	1	2
Unseasonable Weather (1)	33%	33%	33%
Increased Environmental Regs (2)	70%	70%	70%
Increased Labor Regs (3)	70%	70%	70%
Limited Growing Degree Days (4)	24%	24%	24%
Decreased Rainfall (5)	20%	20%	20%
Increased Diesel Price (6)	54%	54%	54%
Lack of Preprocessing (7)	50%	0%	100%
Decreased Labor Availability (8)	59%	59%	59%
Decrease in Total Biomass Available (9)	24%	24%	24%
Increased Transportation Costs (10)	70%	64%	75%
Increased Labor Costs (11)	51%	45%	57%
Insufficient Feedstock Supply (12)	35%	32%	37%

Fig. 6 Base Case: Boolean probabilities for each risk node, Scenario 1: preprocessing is included, Scenario 2: preprocessing is not included *Note Dashed lines* indicate results from root cause analysis

individual risk items and the KPI (feedstock gate availability) can be compared with the naïve case (Base case in Fig. 6). This simple scenario analysis demonstrates the potential applications of BBN for scenario analysis in supply chain risk.

Another benefit of supply chain risk analysis with BBN is the ability to conduct root cause identification analysis with the model. BBN allow for back propagation to calculate probabilities after observational information is added to the model. For example, if, in Fig. 6 Base Case, observation indicated that the feedstock gate availability was insufficient (meaning the risk "Feedstock Gate Availability" had a 100 % probability for "Insufficient Supply"), applying Eq. 1 would yield probabilities for each parent node that would result in that observation. Selecting the most probable contributing event and repeating the observations and calculation of parent probabilities successively would ultimately result in a probabilistic root cause of the change in KPI. In theory, mitigation strategies along this causal chain would lead to decreases in the probabilities of unfavorable KPI changes. A chain of risk events identified in this way can be seen outlined with dashed lines in Fig. 6 for the Base Case scenario.

Applying these analyses with risks events and KPIs relevant to a region of study for biorefinery supply chain design (following the framework proposed earlier) could yield valuable insights for modeling. Representing feedstock availability probabilistically, for instance, could feed process simulation. With this information, the 'least risky' feedstock could be selected to begin the optimization algorithms. Applying similar BBN results to MILP supply input matrices, for instance, could lead to a more robust optimization reducing the need for excessive iterations after DES modeling. Similarly, probabilistic feedstock supply levels in DES formulation could lead to adaptive supply and demand distribution functions reflecting the most likely scenario as operating conditions change with time.

4 Proof of Concept via Application of a Case Study in Western Kentucky

To show the development of key aspects of the integrated framework, a case study was defined in the Jackson Purchase Region (JPR) of Kentucky (Fig. 7). Located in southwest Kentucky, this region was selected due to the availability of GIS and other county level resource data, the abundance of potential agricultural and forest residue feedstocks and relatively small size. With a population of approximately 200,000 (U.S. Census Bureau 2010) this region also can be assumed to have the necessary supporting infrastructure and demand base for the consumption of liquid fuels, natural gas, and electricity produced by the suggested biorefinery.

This region represents a tractable microcosm for application of the framework. For this region, a chemical process model (Sukumara et al. 2013) for the conversion of biomass feedstock to products via gasification and Fischer-Tropsch synthesis, an MILP supply chain optimization model (Faulkner 2012), and a DES model

Fig. 7 The JPR (Highlighted), Kentucky, USA

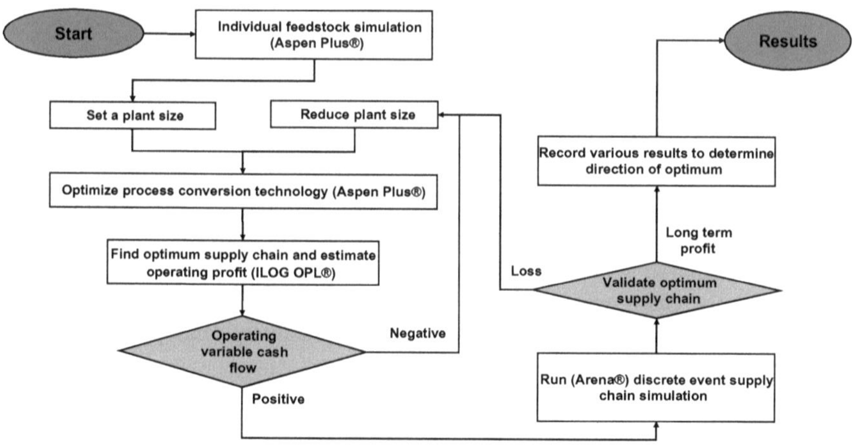

Fig. 8 Methodology to analyze economic viability of biomass gasification in the JPR

(Amundson 2013) were developed. These efforts were pursued in parallel using similar assumptions and data with results from each model informing the development of the others. A BBN based supply chain risk model was also formulated with the case region in mind (Amundson et al. 2012); however, it has not yet been integrated into the framework. Figure 8 shows the flow of data to determine economic performance of biomass gasification in the case study region. Results from the application of each of these models are presented in the sections below.

4.1 Process Simulation Results

The preliminary step to perform the case study is the development of a steady state process simulation model (Aspen Plus®) for biomass gasification using locally available feedstock such as corn stover, forest residue and chicken litter. The process conversion model is developed as described in Sect. 3.1.1. Iteration is initiated by assuming a capacity for a biorefinery that can completely or partially meet the local energy demands. Previous work (Sukumara et al. 2012) shows the detailed assumptions for generating feed distributions in order to start the iterative process. Subsequently, the hydrocarbon products and the power generated are assigned a selling price (EIA 2012). The model is optimized to minimize the variable operating costs. Critical information obtained as an output from this model are optimum feed ratio, utility consumption and product slate. These results are then passed to the MILP model to determine the most profitable biorefinery location and corresponding transportation network. Table 3 shows the various results obtained as an output of the process optimization model. Subsequently, the same information was used to inform the development of the DES model.

4.2 MILP Results

On calculating the amount of feed required at the biorefinery and the products to be sold at the demand locations, respective logistics for transportation needs are determined. Applying the methodology described previously in Sect. 3.1.2, an optimal supply chain was identified for the biomass gasification process. The identified biorefinery location, feedstock supply locations, and product demand locations can be seen in Fig. 9.

The MILP resulted in losses for each month except September, October, November, and December, corresponding to the corn stover harvesting season. As nearby sources of corn stover deplete, the biorefinery must draw from farther sources resulting in higher transportation and fuel expenses. Ideally, at this point, the algorithm would require optimization of a process model utilizing an alternate

Table 3 Outputs from the process simulation model

			Medium biorefinery	Small biorefinery
	Commodities	Units	Operating year	Operating year
Feedstock Requirements	Chicken litter	t/day	239	103
	Corn stover		518	406
	Forest residue		26	17
Operating Cost	Utilities + Chemicals	$/day	20786	14175
Production	Gasoline	gal/month	84969	52555
	Diesel		176988	106578
	Residual fuel oil		481003	286237
	Natural gas	MMcuft/month	10269	5884
	Electricity	MWh/month	3142	1865

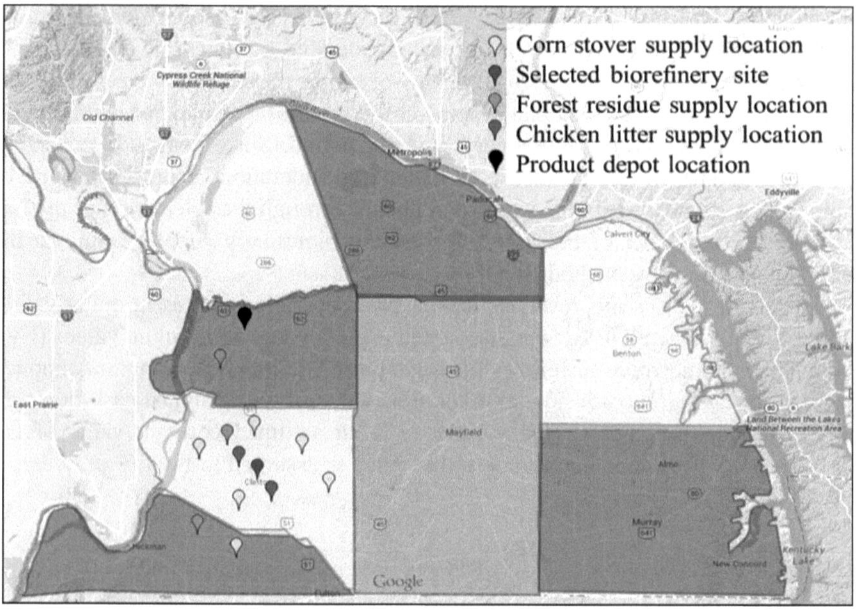

Fig. 9 Optimal supply chain configuration determined via MILP model Faulkner 2012

technology. However, because the purpose of this case study was to produce representative process simulation, MILP, and DES models, and real-time data exchange between models is still not possible, the scenarios were passed to DES to analyze in greater detail the costs incurred during operations.

4.3 Discrete Event Simulation Results

The supply chain activities for the described case study were simulated over a period of 10 years. Costs associated with each transportation, sale, and production aspect of the chain were tracked to evaluate the overall, long-term profitability and economic sustainability of the biorefinery supply chain. A primary goal of this simulation modeling exercise was to learn the characteristics of this type of supply chain and, ultimately draw generalized conclusions from the study. To that end, transportation costs for biomass and biorefinery products, raw material costs, transportation and fuel costs, operating costs, and storage costs were aggregated for each day of operation. Distributions for feedstock supply and product demand were generated using historic data to account for supply chain uncertainty. In Fig. 10, for example, Amundson (2013) compares the historic mean value of gasoline consumption on a county level with the expected value generated through the demand of the determined distributions for each month. It is clear that sampling from these distributions typically provides comparable projections for gasoline demand values of reasonable magnitude. Additionally, income derived from the sale of products in the marketplace was recorded in order to examine the revenues created by the supply chain system. Overall plant capital cost was also taken into consideration for potential evaluations of a biorefinery payback period. Since this study covers an extended period, it was necessary to include consideration of the time-value of money in these analyses. This was accomplished through the calculation of the Net Present Value (NPV) of the investment in the biorefinery supply chain via the discounted cash flow method.

This method discounts revenues from a given operation to present dollar values utilizing Eq. 5 where DPV is the discounted present value of all future values (FV_t) of positive and negative cash flows for a given time period, N, a commonly applied method for taking into consideration the time value of future annual cash flows. In this equation, the value of the variable r is an assumed constant value of the discount rate. For modeling purposes this value is assumed to be 3 % per year.

$$DPV = \sum_{t=0}^{N} \frac{FV_t}{(1+r)^t} \tag{5}$$

As found through the DES, the optimal supply chain (Fig. 10) generates a negative NPV over the period considered. This means that the capital cost of investing in the biorefinery is never recovered. To explain this largely negative cash flow, it is helpful to examine the influencing costs in more detail. For the DES, these can be seen in Fig. 11. It is clear that, throughout the year, diesel costs are the dominant influence on cash flows as biorefineries must deviate from the optimal supply chain to source biomass feedstock demand, thus, driving up costs.

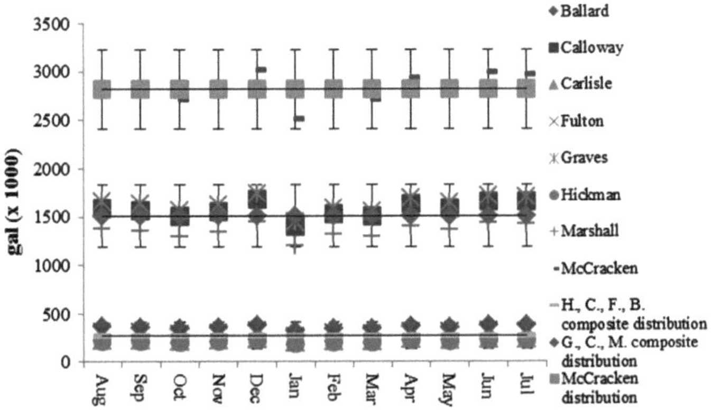

Fig. 10 Gasoline sales historic mean per county compared to distribution expected values, Amundson (2013)

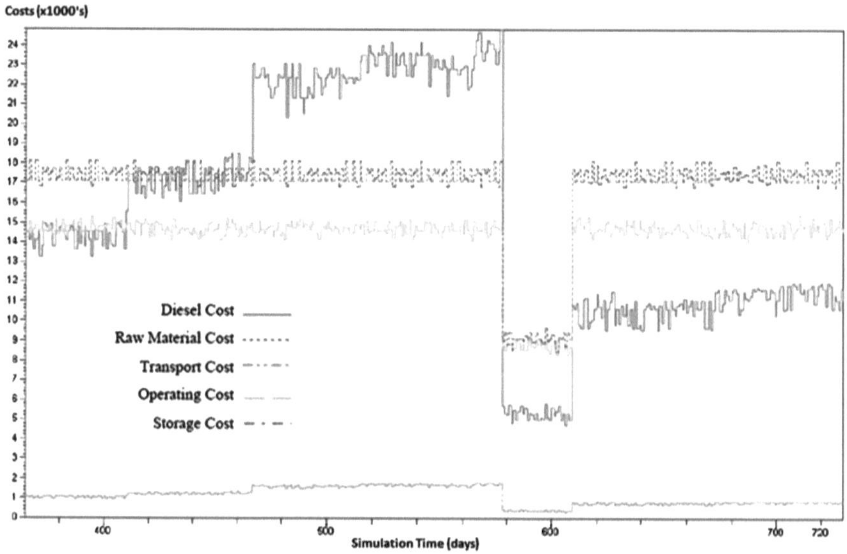

Fig. 11 Costs for a representative year in DES

5 Strategy Development and Policy Analysis with Integrated Decision Support Model

The model developed calculates the detailed impact of a potential biorefinery using locally available feedstock. The framework is run iteratively beginning with the process optimization model to determine optimal feedstock ratio, processing cost, capital investment and corresponding product slates. The results from this model acts as an input to the supply chain optimization model which, thereafter, determines the optimal transportation network and associated costs. Finally, the results from both the models are used to initiate the DES model to study the effect of variability in the supply chain and sensitive parameters over defined period of operation. The results of applying the set of models collectively provide insight towards the long-term economic consequence of practical implementation of biofuel production in the region. Access to such a detailed tool in the beginning of the strategic planning stage will be an informative asset to have. This tool can prove to be helpful to stakeholders in hedging the undesirable impacts of numerous operational uncertainties. The result generated by this comprehensive framework aids various stakeholders to promote investment in the biorefining sector. Specifically, amongst many, the following can be key potential applications of the decision support tool.

- Researchers can use this tool to validate their innovations for practical applications. Currently, on a lab scale, many discoveries for the production of biofuels are proving to be successful. This unique framework can assess economic success of these innovations in reality.
- Investors can use the information obtained from various processes to justify channeling monetary investment to the right direction. Also, this can be used to see impact of any particular risks and their propagation through the supply chain.
- Policy makers can check the outcomes of biofuel subsidies to determine the appropriate supporting laws and credits for future.
- Growers can use this model to check potential benefits in planting or replacing existing crops with alternatives, such as dedicated energy crops.

The formulated model aims to guide investors and policymakers to develop strategies and by providing results regarding multiple operating configurations and corresponding biorefinery locations. Effectiveness of the framework lies in the fact that it can deliver quick impact analysis while being rooted in fundamental details. This feature makes the model not only unique but allows the results to be accurate and credible. With proper interfacing the tool can be used as a yardstick by the investors, producers, growers and policymakers to estimate economic effects of any potential process.

6 Conclusion and Future Work

This chapter presents a novel framework for the design of an economically sustainable biorefinery and its supply chain. Through a discussion of the methodology employed for each aspect of the framework, a generalized picture of a new model-based decision making support method can be seen. By applying the framework to a case study, the results demonstrate the benefits of the integrated models in a specific geographic region. The resulting decision support tool can incorporate fundamental details pertaining to several process and capture broad supply chain configurations, eventually guiding the stakeholders by demonstrating the impact of potential biorefining configurations. With this framework, various scenarios can be explored to inform local supply chain stakeholders and help with the expansion of biofuels.

To expand this work, integration of alternate processing technologies would be beneficial for identifying the most positive option for a regional biorefinery given specific available feedstocks. Faulkner (2012) began this with an exploration of an optimal biorefinery supply chain for a biorefinery producing ethanol from corn stover via biochemical processes. Expansion of process models applied to this scenario would lead to other optimal supply chain structures which could then be examined via DES and BBN based risk modeling. However, to accelerate the progression, existing process and supply chain models must be integrated closely, with potential real time data sharing. For sustainability assessment of a biorefinery supply chain, integration of appropriate environmental and societal metrics is crucial. In addition, rules for deciding precedence with regards to the inherent tradeoffs associated with triple objective optimization with the perspective of multiple stakeholder groups are likely necessary.

References

Aden A, Ruth M, Ibsen K, Jechura J, Neeves K, Sheehan J, Wallace B (2002) Lignocellulosic biomass to ethanol process design and economics utilizing co-current dilute acid prehydrolysis and enzymatic hydrolysis for corn stover. National Renewable Energy Laboratory, Golden, Colorado, USA

Akgul O, Zamboni A, Bezzo F, Shah N, Papageorgiou LG (2011) Optimization-based approaches for bioethanol supply chains. Ind Eng Chem Res 50(9):4927–4938. doi:10.1021/ie101392y

Aksoy B, Cullinan H, Webster D, Gue K, Sukumaran S, Eden M, Sammons N (2011) Woody biomass and mill waste utilization opportunities in Alabama: transportation cost minimization, optimum facility location, economic feasibility, and impact. Environ Prog Sustain Energ 30 (4):720–732. doi:10.1002/ep.10501

Amundson J, Faulkner W, Sukumara S, Seay J, Badurdeen F (2012) A Bayesian Network based approach for risk modeling to aid in development of sustainable biomass supply chains. In: 22nd european symposium on computer aided process engineering ,vol 30, pp 152–156. doi:10.1016/B978-0-444-59519-5.50031-9

Amundson J (2013) Modeling of biorefinery supply chain economic performance with discrete event simulation. Lexington, Manufacturing Systems Engineering, University of Kentucky

An H, Wilhelm WE, Searcy SW (2011) Biofuel and petroleum-based fuel supply chain research: a literature review. Biomass Bioenergy 35(9):3763–3774. doi:10.1016/j.biombioe.2011.06.021

Azapagic A (2014) Sustainability considerations for integrated biorefineries. Trends Biotechnol 32 (1):1–4. doi:10.1016/j.tibtech.2013.10.009

Badurdeen F, Shuaib M, Wijekoon K, Brown A, Faulkner W, Amundson J, Jawahir IS, Goldsby T, Iyengar D, Boden B (2014) Quantitative modeling and analysis of supply chain risks using Bayesian Theory. J Manufact Technol Manage (Accepted for Publication)

Batidzirai B, Smeets EMW, Faaij André PC (2012) New conversion technologies for liquid biofuels production in Africa. In: Janssen Rainer, Rutz Dominik (eds) Bioenergy for sustainable development in Africa. Springer, Netherlands, pp 117–130

Bernardi A, Sara G, Bezzo F (2013) Spatially explicit multiobjective optimization for the strategic design of first and second generation biorefineries including carbon and water footprints. Ind Eng Chem Res 52(22):7170–7180. doi:10.1021/ie302442j

Brown A, Amundson J, Badurdeen F (2012) Bayesian informed simulation for supply chain risk probability and impact assessment. In: 22nd international conference on production research

Corsano G, Vecchietti AR, Montagna JM (2011) Optimal design for sustainable bioethanol supply chain considering detailed plant performance model. Comput Chem Eng 35(8):1384–1398. doi:10.1016/j.compchemeng.2011.01.008

Coyle WT (2010) Next-generation biofuels: near-term challenges and implications for agriculture. Washington

Dal-Mas M, Giarola S, Zamboni A, Bezzo F (2011) Strategic design and investment capacity planning of the ethanol supply chain under price uncertainty. Biomass Bioenergy 35(5):2059–2071. doi:10.1016/j.biombioe.2011.01.060

De Mol RM, Jogems MAH, Beek PV, Gigler JK (1997) Simulation and optimization of the logistics of biomass fuel collection. Neth J Agric Sci 45:219–228

Dunnett A, Adjiman C, Shah N (2007) Biomass to heat supply chains: application of process optimization. Process Saf Environ Prot 85(5):419–429. doi:10.1205/psep07022

Dunnett AJ, Adjiman CS, Shah N (2008) A spatially explicit whole-system model of the lignocellulosic bioethanol supply chain: an assessment of decentralised processing potential. Biotechnol Biofuels 1(1):1–17. doi: 10.1186/1754-6834-1-13

EIA (2012) Refiners'/Gas Plant Operators' Monthly Petroleum Product Sales Report. Energy Information Administration, Washington

EIA (2013a) Annual Energy Outlook 2013. U.S. Energy Information Administration, Washington

EIA (2013b) U.S. Imports by Country of Origin. U.S. Energy Information Administration, Washington

Ekşioğlu SD, Acharya A, Leightley LE, Arora S (2009) Analyzing the design and management of biomass-to-biorefinery supply chain. Comput Ind Eng 57(4):1342–1352. doi:10.1016/j.cie.2009.07.003

EPA (2014) Renewable fuels: regulations & standards. Office of transportation and air quality (OTAQ) 2014 [cited 02/24/2014 2014]. http://www.epa.gov/otaq/fuels/renewablefuels/regulations.htm

Faulkner WH (2012) Economic modeling & optimization of a region specific multi-feedstock biorefinery supply chain. Manufacturing Systems Engineering, University of Kentucky, Lexington

Fenton N, Neil M (2007) Managing risk in the modern world: application of Bayesian Networks

Fernández E, Salomone E, Chiotti O (2010) Model based on Bayesian Networks for monitoring events in a supply chain. In: Vallespir B, Alix T (ed) Advances in production management systems. New Challenges, New Approaches, pp 358–365. Springer, Berlin Heidelberg

Gebreslassie BH, Yao Y, You F (2012) Design under uncertainty of hydrocarbon biorefinery supply chains: Multiobjective stochastic programming models, decomposition algorithm, and a Comparison between CVaR and downside risk. AIChE J 58(7):2155–2179. doi:10.1002/aic.13844

Gold S, Seuring S (2011) Supply chain and logistics issues of bio-energy production. J Clean Prod 19(1):32–42. doi:10.1016/j.jclepro.2010.08.009

Humbird D, Davis R, Tao L, Kinchin C, Hsu D, Aden A (2011) Process design and economics for biochemical conversion of lignocellulosic biomass to ethanol

Hytönen E, Stuart PR (2010) Biofuel production in an integrated forest biorefinery & technology identification under uncertainty. J Biobased Mater Bioenergy 4(1):58–67. doi:10.1166/jbmb.2010.1066

Kelepouris T, Harrison M, McFarlane D (2011) Bayesian supply chain tracking using serial-level information. IEEE Trans Syst Man Cybern Part C Appl Rev 41(5):733–742. doi:10.1109/TSMCC.2010.2093599

Kim J, Realff MJ, Lee JH (2011a) Optimal design and global sensitivity analysis of biomass supply chain networks for biofuels under uncertainty. Comput Chem Eng 35(9):1738–1751

Kim J, Realff MJ, Lee JH (2011b) Optimal design and global sensitivity analysis of biomass supply chain networks for biofuels under uncertainty. Comput Chem Eng 35(9):1738–1751. doi:10.1016/j.compchemeng.2011.02.008

Kumar A, Sokhansanj S (2007) Switchgrass (Panicum vigratum, L.) delivery to a biorefinery using integrated biomass supply analysis and logistics (IBSAL) model. Bioresour Technol 98 (5):1033–1044

Lockamy A III (2011) Benchmarking supplier risks using Bayesian Networks. Int J Benchmarking 18(3):409–427

Mantovani B, Gibson H, Peart RM, Brook RC (1992) A simulation model for analysis of harvesting and transport costs for biomass based on geography, density and plant location. Anal Agric Energy Syst 253–280

Marvin WA, Schmidt LD, Daoutidis P (2012) Biorefinery location and technology selection through supply chain optimization. Ind Eng Chem Res 52(9):3192–3208. doi:10.1021/ie3010463

Medina-Oliva G, Weber P, Simon C, Iung B (2009) Bayesian Networks applications on dependability, risk analysis and maintenance. Paper read at 2nd IFAC workshop on dependable control of discrete system, DCDS'09

Mele FD, Kostin AM, Guillén-Gosálbez G, Jiménez L (2011) Multiobjective model for more sustainable fuel supply chains: a case study of the sugar cane industry in Argentina. Ind Eng Chem Res 50(9):4939–4958. doi:10.1021/ie101400g

Min H, Zhou G (2002) Supply chain modeling: past, present and future. Comput Ind Eng 43(1–2):231–249. doi:10.1016/S0360-8352(02)00066-9

Mittal A, A Kassim (2007). Bayesian network technologies: applications and graphical models: IGI Global

Naik SN, Goud VV, Rout PK, Dalai AK (2010) Production of first and second generation biofuels: a comprehensive review. Renew Sustain Energy Rev 14(2):578–597. doi:10.1016/j.rser.2009.10.003

Nikolopoulou A, Ierapetritou MG (2012) Optimal design of sustainable chemical processes and supply chains: A review. Comput Chem Eng 44:94–103. doi:10.1016/j.compchemeng.2012.05.006

Sammons NE Jr, Yuan W, Eden MR, Aksoy B, Cullinan HT (2008) Optimal biorefinery product allocation by combining process and economic modeling. Chem Eng Res Des 86(7):800–808. doi:10.1016/j.cherd.2008.03.004

Sammons N, Eden M, Yuan W, Cullinan H, Aksoy B (2007) A flexible framework for optimal biorefinery product allocation. Environ Prog 26(4):349–354. doi:10.1002/ep.10227

Santibañez-Aguilar, JE, González-Campos JB, Ponce-Ortega JM, Serna-González M, El-Halwagi MM (2013) Optimal planning and site selection for distributed multiproduct biorefineries involving economic, environmental and social objectives. J Clean Prod (0). doi: 10.1016/j.jclepro.2013.08.004

Sharma P, Sarker BR, Romagnoli JA (2011) A decision support tool for strategic planning of sustainable biorefineries. Comput Chem Eng 35(9):1767–1781

Sokhansanj S, Kumar A, Turhollow AF (2006) Development and implementation of integrated biomass supply analysis and logistics model (IBSAL). Biomass Bioenergy 30(10):838–847. doi:10.1016/j.biombioe.2006.04.004

Sukumara S, Amundson J, Faulkner W, Badurdeen F, Seay J (2012) Multidisciplinary approach in developing region specific optimization tool for sustainable biorefining. In: 22nd European symposium on computer aided process engineering, vol 30, pp 157–161. doi: 10.1016/B978-0-444-59519-5.50032-0

Sukumara S, Faulkner W, Amundson J, Badurdeen F, Seay J (2013) A multidisciplinary decision support tool for evaluating multiple biorefinery conversion technologies and supply chain performance. Clean Technol Environ Policy 1–18. doi: 10.1007/s10098-013-0703-6

Tako AA, Robinson S (2012) The application of discrete event simulation and system dynamics in the logistics and supply chain context. Decis Support Syst 52(4):802–815. doi:10.1016/j.dss.2011.11.015

Tang O, Musa SN (2011) Identifying risk issues and research advancements in supply chain risk management. Int J Prod Econ 133(1):25–34. doi:10.1016/j.ijpe.2010.06.013

US Census Bureau (2010) Intercensal Population Estimates 1970-2010. http://www.census.gov/popest/data/counties/totals/2011/CO-EST2011-01.html. Accessed 12 Jan 2011

Vanany I, Zailani S, Pujawan N (2009) Supply chain risk management: literature review and future research. Int J Inf Syst Supply Chain Manage (IJISSCM) 2(1):16–33. doi: 10.4018/jisscm.2009010102

Yen BPC, Zeng B (2010) A hierarchical assessment method using Bayesian network for material risk detection on green supply chain. In: 2010 IEEE international conference 2010 paper read at industrial engineering and engineering management (IEEM)

You F, Tao L, Graziano DJ, Snyder SW (2012) Optimal design of sustainable cellulosic biofuel supply chains: Multiobjective optimization coupled with life cycle assessment and input–output analysis. AIChE J 58(4):1157–1180. doi:10.1002/aic.12637

Young D, Scharp R, Cabezas H (2000) The waste reduction (WAR) algorithm: environmental impacts, energy consumption, and engineering economics. Waste Manage 20(8):605–615. doi:10.1016/s0956-053x(00)00047-7

Yue D, You F, Snyder SW (2014) Biomass-to-bioenergy and biofuel supply chain optimization: overview, key issues and challenges. Comput Chem Eng (0). doi: http://dx.doi.org/10.1016/j.compchemeng.2013.11.016

Zamboni A, Bezzo F, Shah N (2009a) Spatially explicit static model for the strategic design of future bioethanol production systems. 2. Multi-Objective Environmental Optimization. Energy Fuels 23(10):5134–5143. doi:10.1021/ef9004779

Zamboni A, Shah N, Bezzo F (2009b) Spatially explicit static model for the strategic design of future bioethanol production systems. 1. Cost minimization. Energy Fuels 23(10):5121–5133. doi:10.1021/ef900456w

Zhang F, Johnson DM, Johnson MA (2012) Development of a simulation model of biomass supply chain for biofuel production. Renew Energy 44(0):380–391. doi:10.1016/j.renene.2012.02.006

Zhao B, Li J, Zhang Y (2010) Research on the reward and punishment feedback contract based on asymmetric information under the “Company + Farmer” model. In ICLEM 2010:871–879

Zwart RWR, Boerrigter H, van der Drift A (2006) The impact of biomass pretreatment on the feasibility of overseas biomass conversion to Fischer-Tropsch products. Energy Fuels 20 (5):2192–2197. doi:10.1021/ef060089f

Evaluating Supply Chain Design Models for the Integration of Biomass Co-firing in Existing Coal Plants Under Uncertainty

Didem Cinar, Panos M. Pardalos and Steffen Rebennack

Abstract Co-firing biomass with a primary fuel in existing power plants is a cost effective environmental strategy. Co-firing implementations are carried out considering the impact of emissions reduction, logistic-related costs, plant efficiency and incentive savings. The literature on mathematical programming models for co-firing can be grouped in two categories: supply chain network design and determination of optimal fuel combination. Our literature review reveals that there is a need for models to show the effect of uncertainties in both of these categories. In this chapter, we carry out a traditional sensitivity analysis and propose a stochastic mixed-integer linear programming model within a single-period planning framework for a supply chain design problem which integrates biomass co-firing in existing coal plants. Purchase costs of coal, biomass and the amount of available biomass are considered as the uncertain parameters. The performance of the proposed model is evaluated by computational tests using data from the State of Mississippi. The computational results reveal that the benefit of the two-stage stochastic programming approach is marginal although a considerable uncertainty is taken into account in the model, while our sensitivity analysis shows that strategies making the co-firing option profitable under uncertainties can be revealed by the model.

Keywords Biomass co-firing · Supply chain design · Uncertainty · Two-stage stochastic programming · Optimization

---

D. Cinar
Faculty of Management, Department of Industrial Engineering, Istanbul Technical University, Istanbul, Turkey
e-mail: didemcin@gmail.com

D. Cinar · P.M. Pardalos
Faculty of Engineering, Department of Industrial and Systems Engineering, Center for Applied Optimization, University of Florida, Gainesville, USA
e-mail: pardalos@ufl.edu

S. Rebennack (✉)
Division of Economics and Business, Colorado School of Mines, Golden, CO, USA
e-mail: srebenna@mines.edu

S.D. Eksioglu et al. (eds.), *Handbook of Bioenergy*,
Energy Systems, DOI 10.1007/978-3-319-20092-7_8

1 Introduction

Energy planning has been one of the important issues for governments and companies in terms of environment and cost. In 2012, coal and natural gas provided 36 and 29.5 % of the total electricity generation in the United States (US Energy Information Administration 2014). According to the projections in the U.S. Energy Information Administration's Annual Energy Outlook 2014, these ratios will be 30 and 33 % in 2040. Total U.S. electricity generation by energy source in 2012 and projections for 2040 are given in Fig. 1. These ratios show that coal and natural gas is projected to continue to be the dominant fuels for electricity production in the United States for the next quarter of a century. Thus, air pollution emissions accompanying the coal combustion—such as oxides of sulphur (SO_x) and nitrogen (NO_x), which lead to acid rain, ozone depletion, and greenhouse gas emissions (e.g., CO_2, CH_4)—have become a global concern (Sami et al. 2001). Federal regulations and policies are necessary to deal with the concerns over environmental problems and public health. One example of such government regulations in the USA is the Clean Air Act from the US Environmental Protection Agency from 1963 with several major amendments; most notably the recent Clean Power Plan (Environmental Protection Agency 2014). To reduce carbon emissions, governments focus on utilizing a variety of renewable energy resources which are less environmentally harmful in a sustainable way together with carbon capture and storage technology (Huang et al. 2013). There is a considerable stream of research on environmental policy design and evaluations (Fell and Linn 2013) as well as optimal investment planning under emission caps (Rebennack 2014). According to

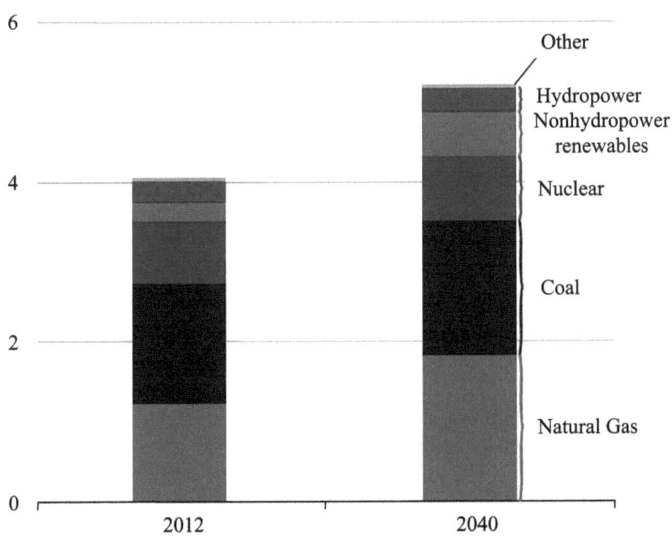

Fig. 1 Total U.S. electricity generation by energy source in billion kilowatt hours (US Energy Information Administration 2014)

the projections given in Fig. 1, total electricity generation of coal increases by 11 % until 2040 while the generation from renewable energy increases by 80 %.

Biomass is one of the non-hydropower renewable energy resources which includes organic substances produced through photosynthesis as well as municipal, industrial and animal wastes. The CO_2 emissions during the combustion of biomass equals the CO_2 absorbed during the growth of the biomass; thus, the net CO_2 emissions cause by biomass combustion is zero. Although biomass is a very clean option for energy generation, shifting significant power generating capacity to biomass is currently not possible, because suppliers and supply chains have not been developed yet (Saidur et al. 2011). However, co-firing biomass with a primary fuel is a cost effective environmental strategy. Co-firing refers to the combustion of biomass with a primary fuel. Unlike other renewable energy sources, biomass can be integrated into existing fossil fuel plants by co-firing. Blending biomass with high quality coal is an attractive energy production strategy rather than combustion of coal as a sole source. Co-firing with natural gas is also expected to have a remarkable potential in the near future (Hu et al. 2011). The reasons that make the co-firing attractive are (Basu et al. 2011; Roni et al. 2014; Tillman 2000):

1. Net greenhouse effect of biomass combustion is zero. Co-firing also reduces emissions of SO_2, NO_2 and heavy metals which cause acid rain and ozone depletion compared to a primary fuel plant alone.
2. Co-firing minimizes the environmental problems related with agricultural waste and its disposal.
3. Unlike other renewable energy production strategies, co-firing does not need major capital investment, because existing coal-fired power plants can be used for biomass co-firing. Coal plants can handle biomass co-firing with their existing boilers.
4. Policies and incentives promoting the use of renewable energy sources for electricity generation and the reduce of greenhouse gas emission are offered by government agencies.
5. Co-firing is a low risk option to produce renewable energy because of the resource availability and no requirement for major capital investment.
6. Since the co-fired plants provide biomass from near surroundings and generate less air pollution, the local communities have economical and environmental benefits from co-firing.

In spite of the various economic, environmental and social advantageous of co-firing, there are also several drawbacks that energy producers have to take into account. With co-firing, overall system efficiency reduces along with the decrease of boilers efficiency (Eksioglu and Karimi 2014). Chlorine-based corrosion and deactivation of selective catalytic reduction system are some possible negative impacts of biomass combustion in coal-fired boilers which should be properly regulated (Touš et al. 2011). Unlike coal reserves, the biomass resources are widely scattered geographically. The physical features of biomass and the location of the resources and energy producing plants cause additional transportation costs (Roni et al. 2014; Touš et al. 2011). Logistics processes—such as loading, unloading, and transportation—can be difficult and expensive. Because of these reasons, co-firing implementations are carried out considering the impact of emissions reduction, logistic-related costs, plant efficiency and incentive savings (Eksioglu and Karimi 2014). Mathematical programming models are commonly used for formulation and planning of energy systems because they can help decision makers to determine the relationship between indicators that affect energy production and to propose optimal energy production strategies.

The majority of supply chain design studies for biomass co-firing assumes that the design parameters of supply chains are deterministic. However, critical parameters—such as demands, prices, raw material availability and resource capacities—are considerably uncertain (Santoso et al. 2005). In other applications, supply chain design problems under uncertainty have gained significant interest and attention (Snyder 2006). Dealing with a huge uncertainty space poses significant challenges to optimization models under uncertainty. Moreover, decision-making under uncertainty is considerably more involved when integer decision variables are present in a multi-period or multi-stage setting (Sahinidis 2004). Stochastic programming has been widely used in the literature to cope with the uncertainties in energy/power system optimization problems (Rebennack 2012). In this study, a stochastic mixed-integer linear programming (MILP) approach is developed within a single-period planning framework for a supply chain design problem which integrates biomass co-firing in existing coal plants. Biomass availability and purchase prices of biomass and coal enter as the uncertain parameters in the problem. The main contributions of this study are threefold:

1. We review and classify the literature on mathematical programming models for biomass co-firing.
2. We carry out a sensitivity analysis and for this purpose, propose also a two-stage stochastic programming model for a supply chain optimization problem taking three sources of uncertainty into account.
3. We evaluate the proposed model through a case study for the State of Mississippi.

The remainder of this chapter is organized as follows. We investigate and discuss mathematical models on supply chain design and on determination of optimal fuel combinations for co-firing in Sect. 2. This motivates our stochastic programming model as presented in Sect. 3. We report and analyze our computational results for the State of Mississippi in Sect. 4. Finally, we conclude in Sect. 5.

2 Mathematical Programming Models for Biomass Co-firing

There are several optimization studies minimizing the environmental impact and costs of biofuels supply chain in the literature (Lam et al. 2010; van Dyken et al. 2010). Among the alternative optimization models, linear programming (LP), multi-objective LP and MILP are most widely used to integrate biomass co-firing with existing primary fuel. We classify the literature on biomass co-firing in two categories: supply chain design and optimal fuel combination.

Supply chain design studies mainly focus on the location of biomass collection centers (hubs) and the amount of biomass transported from resources or centers to plants. The aim is to minimize transportation cost while satisfying energy demand. We discuss supply chain design models in Sect. 2.1.

The studies on the determination of optimal fuel combinations deal with the combustion process. Optimal amount of burned fuels are identified to produce the required energy with minimum cost or maximum profit. We review these mathematical programming models in Sect. 2.2.

2.1 Supply Chain Design

Biomass can be combusted in power generation facilities in two ways: (1) by constructed exclusively for that fuel type in dedicated plants or (2) by co-fired in existing coal power plants.

Smith and Hobbs (2013) develop a mixed-integer, non-linear programming (MINLP) model to obtain an optimum location and type (dedicated or co-fired) of power plants and biomass cultivation regions. Their model minimizes total cost including capital, operations, maintenance, cultivation and transportation costs. Moreover, avoided coal savings and generation expansion credit are considered as savings in the objective function as well. They use an agricultural land-use LP model called POLYSYS to obtain supply curves by region for the agricultural biomass. Then, they use the proposed MINLP model to optimize allocation of co-firing and biomass dedicated plants. Computational experiments are performed on US electric generation on a national base. The results show that the largest cost component is cultivation constituting two-thirds of the total cost. The relatively high cultivation cost is one of the drawbacks of biomass resources. Unlike wind, solar, and geothermal resources, biomass is not that much geographically constrained, but it is available only with significant costs. High costs are also the result of competition in the agricultural sector. Considering the food shortages, millions of acres of traditional crops can be displaced with very high costs. The results also showed that co-firing is preferable option rather than dedicated generation in terms of efficiency. Dedicated biomass plants are smaller and less efficient than the co-fired power plants. This situation can be changed with the improvement of efficient technologies for biomass plants which results in cost reductions.

Roni et al. (2014) propose a MILP model to design supply chain network for biomass co-firing in coal-fired power plants. The problem is modelled as an extension of an uncapacitated network design model and solved with a Benders decomposition-based heuristic approach. The model's goal is to minimize the total transportation cost, hubs location costs and penalty costs which occur from not satisfying demand. Both train and trucks are considered as transportation vehicles. A piecewise linear function is used in objective function to represent the transportation cost of trains. According to the computational results obtained using real data in a realistic case study, co-firing is an economically preferable option when biomass is located within 75 miles from a coal plant. If biomass is located further than 75 miles, co-firing may have higher cost than sole coal firing. In this case, the decision about co-firing will depend on the costs of producers, densifying costs, amount of biomass available, legislations and incentives promoting the renewable energy production.

Akgul et al. (2014) develop a multi-objective MINLP model to optimize the supply chain network of co-firing. Both emission and supply chain cost minimizations are considered as the objective. The problem includes determination of appropriate power stations, decision of load factor and degree of CO_2 capture and co-firing, subject to limited biomass resources. The model is applied towards the electricity generation network of the UK. Both solid waste and conventional biomass resources are taken into account in the model. The level of CO_2 prices is investigated to stimulate the generation of carbon negative electricity. When the availability of biomass resources increases, the price of CO_2 decreases.

Eksioglu and Karimi (2014) use a MILP model to design a supply chain network for biomass co-firing in an existing coal fired plant with maximum profit. Investment, biomass storage and handling, transportation, operating, compressors and dryers costs are taken into consideration with savings due to the displacement of coal and tax credits. A sensitivity analyses is performed with the data from the State of Mississippi. Because of the reduction in profit, power plants would avoid co-firing unless they are subsidized for their efforts. On the other hand, increase of tax credit beyond some threshold does not lead an increase in the production of renewable energy. The authors indicate that would be more efficient to define tax credit as a function of the amount of renewable electricity produced instead of a fixed rate.

Devlin and Talbot (2014) investigate transportation strategies of Ireland under different scenarios for 2015 and 2030. Scenarios are determined based on the priority degree of energy and board sectors. An LP model is used to minimize the transportation cost. The scenario for 2015 is based on the current situation. The cost of optimized problem is less than the actual transportation cost ruled by actual power plants. 30 % co-firing target is predetermined for the 2030 scenario. The amount of subsidy required to make the biomass utilization attractive decreases when the availability of biomass resources increases.

Griffin et al. (2014) develop an LP model to minimize total cost and greenhouse gas emission of biomass co-firing with coal in Peninsular Malaysia. Computational results show that co-firing at the 330 MW level can reduce the annual electricity generation costs by up to $20 million and can reduce the CO_2 emissions by 1.9 million tones (Mt). Furthermore, a significant reduction on import of coal is obtained by the use of biomass co-firing.

2.2 Optimal Combination of Fuels

Although co-firing is an attractive energy generation option for companies and governments due to emission and pollution reduction, its costs and implementation can still be challenging (Nienow et al. 2000). Therefore, fuel choice models are developed to appraise the competitiveness of alternative fuels given different regulations, operating and financial limitations. In the literature, there are studies on modelling the co-firing process to obtain the optimum fuel blending.

Nienow et al. (2000) propose an LP model to obtain an optimal coal and biomass combination that minimizes total fuel cost with disposal and pollution cost. Waste wood and willow trees are considered as the secondary fuel options to co-fire with coal which is taken as the primary fuel. According to the computational results, the most cost effective strategy is blending both wood waste and willow trees.

Ko and Chang (2008) investigate the optimal fuel blending to minimize the net operating cost. They consider refuse-derived fuel (RDF) as an option as well as biomass for co-firing. RDF is a product which is obtained from processing the municipal solid waste by lessening the amount of non-combustible material in waste and producing a fuel convenient for combustion in solid fuel-fired boilers. Along with a deterministic model, the authors develop a grey MINLP model, which represents parameters or variables with possible ranges to provide some flexibility in decision making. Grey optimization model consists of two sub-models composed of upper and lower bounds and can be expressed independently or formulated jointly. Three options considered in the model are co-firing two types of coal with: (1) wood wastes, (2) RDF and (3) both wood wastes and RDF. Stepwise relaxation method is used to cope with the non-linearity of grey optimization model. Computational experiments are performed for the Michigan City power plant. It is shown that co-firing coals with wood wastes is economically and environmentally better option than sole coal-firing. Co-firing RDF with coal can be economically preferable only if reduction on unit cost of RDF is achieved by charging a tipping fee or receiving a subsidy.

Touš et al. (2011) develop an LP model to achieve an optimum biomass integration strategy for the existing large energy producing system with combined heat and power (CHP) production. The problem handled by Ko and Chang (2008) is extended by considering energy demand and subsidies. However, the costs and restrictions caused by emissions and pollution are not taken into account. The objective of the model is maximization of total profit which includes sales, subsidies, allowances and bonuses as income and fuel purchasing, ash disposal and flue gas desulphurization as cost. The proposed model is applied to an existing CHP plant operated in the Czech Republic. Computational results show that feasibility of biomass utilization depends mostly on the amount of the government subsidies for renewable energy production and prices of coal and biomass. Sensitivity analysis is performed to investigate the effect of subsidies and prices on the amount of co-fired biomass fuels. If the amount of subsidies are high enough, then the increasing price may not effect the biomass usage and co-firing remains being profitable. Although biomass is much more expensive than coal, government subsidies and CO_2 allowances make co-firing more profitable than consuming coal only.

Hu et al. (2011) develop a NashCournot competition model for simulating the Taiwanese electricity market with switchgrass and Miscanthus co-firing. The market investigated in the study has a oligopolistic structure where the profit of a firm depends not only on its own sales but also on the sales of other firms in the market. With the proposed model all firms decide their sales at the same time and a Nash equilibrium solution is obtained, in which the best strategy for each firm is the response of the other firms' strategies. The model maximizes the profit while satisfying the demand (total generated power equals to the amount of sales), capacity and biomass combustion rate constraints. The impact of switchgrass and Miscanthus co-firing on CO_2 and SO_2 is investigated. Experimental results indicate that all biomass feedstocks can decrease pollution emissions. Miscanthus decreases production cost and carbon dioxide emission more than switchgrass.

Liu et al. (2014) propose an LP model, which minimizes total cost, to investigate the economic feasibility and emission reduction of integrating biomass co-firing in existing coal plants in Missouri. According to the computational results, biopower production in Missouri is not economically feasible except for some location. Because of the lower prices and transportation costs, corn residues are used more than other feedstocks, such as wheat or sorghum. Also corn is more abundant than others, i.e., it provides the largest source of energy. Because of these reasons, locating the power plants close to corn residue sources is important to make the co-firing option economically more feasible.

Berggren et al. (2008) develop an LP model to examine the potential of biomass co-firing with coal in the electricity market of Poland. Total cost including boilers retrofit, processing, transportation, operation and maintenance costs are minimized. The results show that co-firing increases renewable electricity production by 1.6–4.6 % in total power production. However, it results with an additional implementing cost which amount to 18 % of the electricity price in Poland.

Model structure and the components of the objective function and constraints considered in the studies investigating optimum fuel mixture for co-firing are outlined in Table 1. Next to variable domain restrictions, the constraints can be summarized as follows:

- Heating value requirement constraint: Stability of heat flux in the furnace is a very important issue in a power plant. Heating value of blended fuel should meet the required heat input in boilers which is computed by the multiplication of heat input rate and total operating hours (Ko and Chang 2008).
- Coal blending limitation constraint: The predetermined ratio of coals should be ensured to satisfy the limitation of SO_2 and NO_x emissions.
- Biomass availability limitation constraints: The amount of biomass in fuel mixture cannot exceed the available amount for each source.
- Coal availability limitation constraints: The amount of coal in fuel mixture cannot exceed the available amount.

Table 1 Model structure of the studies on optimal fuel combination

Study	Nienow et al. (2000)	Ko and Chang (2008)	Tous et al. (2011)	Hu et al. (2011)	Liu et al. (2014)	Berrgren et al. (2008)[a]
Model	LP	Nonlinear MIP	LP	Nash-Cournot competition model	LP	LP
Objective	Cost min.	Cost min.	Profit max.	Profit max.	Cost min.	Cost min.
Purchasing cost	X	X	X			
Pollution cost	X					
Ash disposal cost	X		X			
Flue gas desulphurization			X			
Transportation cost					X	X
Harvesting cost					X	
Handling and processing cost				X	X	X
Operation and maintenance cost					X	X
Investment/retrofit cost						X
Subsidies, allowances, bonus			X			
Income from selling heat and electricity			X	X		
Constraints						
Heating value requirement	X	X				
Coal blending limitation	X	X				
Biomass availability limitation	X	X	X	X	X	X
Coal availability limitation			X			X
Share of biomass restriction	X	X	X			X
SO_2 emission and pollution	X	X			X	
Capacity constraints			X	X	X	X
Demand			X	X	X	
Energy flow conservation			X			

[a]The model was not given in the paper explicitly

- Share of biomass restriction constraints: The amount of combusted biomass is limited in terms of heat content and weight to prevent alkali slagging and fouling in the boiler (Ko and Chang 2008).
- SO_2 emission and pollution constraints: Both SO_2 emission and pollution emission resulting with combustion cannot exceed the standards and allowances.
- Capacity constraints: During combustion, the energy input contained in fuels is transferred into water/steam in boilers. Then the steam is transferred to turbines to produce electricity and heat (Touš et al. 2011). The amount of input cannot exceed the capacity of boilers and turbines.
- Demand constraints: Total energy demand includes both demand of customers and self-consumption caused by the production of heat and energy. Heat and electricity production obtained by combustion should meet total demand.
- Energy flow conservation constraints: Total energy transported from boilers to turbines and total energy led on turbines should be equal. This relation can be referred to energy balance equation (Touš et al. 2011).

3 Stochastic Programming Model

As seen in Sect. 2, most of the applications in the biomass co-firing literature have focused on deterministic problems. However, because of the uncertain structure of biomass integration in existing coal plants, such as costs and demands, developing an optimization model considering uncertainties becomes crucial. Stochastic programming (SP) has been widely used to cope with uncertainties in decision making processes. SP integrates the variability of uncertain parameters during modelling and evaluates the solutions in the space of uncertain parameters before optimal solution is detected (Sahinidis 2004).

Two-stage stochastic programming is the most well-known SP model (Birge and Louveaux 2011). We assume that the uncertainty can be characterized by a finite set of scenarios, $s \in \mathbb{S}$. (These are standard assumptions in two-stage stochastic programming.) With each scenario, we associate a specific realization of the uncertain parameter(s), in the second stage. All parameters for the first stage are known. We also assume that we know the probability, p_s, of each scenario $s \in \mathbb{S}$. The decision variables are then categorized as first and second-stage variables. First-stage variables represent the decisions which have to be made before the realization of uncertainties. As such, they hedge against the uncertain outcome. They are deterministic. Decisions for the second stage, the so-called recourse decisions, are made after the realization of the uncertainty. Thus, the second-stage decision variables are scenario dependent. SP optimizes the expected value as given by the distribution of the finite set of scenarios. A very general formulation of the scenario-based, mixed-integer linear SP, minimizing cost, is framed as follows:

$$z^* := \min \mathbf{c}^T \mathbf{x} + \sum_{s \in \mathbb{S}} p_s \mathbf{q}_s^T \mathbf{y}_s \tag{1}$$

$$\text{s.t.} \quad \mathbf{A}\mathbf{x} \leq \mathbf{b} \tag{2}$$

$$\mathbf{T}_s \mathbf{x} + \mathbf{W}_s \mathbf{y}_s \leq \mathbf{h}_s \quad \forall s \in \mathbb{S} \tag{3}$$

$$\mathbf{x} \in \mathbb{R}_+^{n_1 - r_1} \times \mathbb{Z}_+^{r_1} \tag{4}$$

$$\mathbf{y}_s \in \mathbb{R}_+^{n_2 - r_2} \times \mathbb{Z}_+^{r_2} \quad \forall s \in \mathbb{S} \tag{5}$$

Vectors $\mathbf{x}$ and $\mathbf{y_s}$ refer to the first-stage and second-stage variables, respectively. The objective function (1) includes the costs related with the first-stage variables and the weighted sum of the costs associated with all scenarios. Constraints (2) involve only the the first stage decision variables whereas constraints (3) connect the first stage with the second stage decision variables. Constraints (3) have to hold for each scenario $s \in \mathbb{S}$. The decision variables' domains are given in (4) and (5).

SP has been used in several biofuel supply chain planning and management studies. Gebreslassie et al. (2012) develop a multi-objective, mixed-integer and linear SP models for hydrocarbon biorefinery supply chains which minimizes the expected annualized cost and the financial risk. Their proposed model captures uncertainty in costs, levels of production and inventory decisions. The authors employ two decomposition methods to solve their proposed model. Xie and Huang (2013) also propose a multi-objective, mixed-integer and linear SP model in which biomass-to-biofuel conversion rate is treated as uncertain. Awudu and Zhang (2013) use a linear SP model within a single-period planning framework for production planning of a biofuel supply chain. They maximize the expected profit under demand and price uncertainties. Tong et al. (2014) develop a two-stage, mixed integer and linear SP approach to obtain an integrated biofuel and petroleum supply chain system with minimum cost. Biomass availability, product demand, crude oil price and availability, biomass harvesting cost, operation cost, capital cost, and government incentive are considered as uncertain parameters.

We develop a mixed-integer and linear SP model for the biomass co-firing supply chain design problem. Our model extends the work by Eksioglu and Karimi (2014) which we describe next. The aim of their problem is to determine the amount of co-fired biomass from alternative suppliers that maximize total profit. Only the additional costs and savings which result from co-firing are taken into consideration to compute profit. Costs include additional investments to integrate biomass into existing coal plants, purchase cost of biomass and transportation cost of biomass from suppliers to plants. Moreover, savings result from displacement of coal and tax credits obtained from electricity generation from renewable energy. Eksioglu and Karimi (2014) consider a deterministic problem. In this study, we take uncertainties resulting from (coal and biomass) prices and biomass availability into account and investigate the benefit of considering uncertainty.

Our mixed-integer and linear SP model for a supply chain design integrating biomass co-firing in existing coal plants is given as follows:

$$z^* := \max \sum_{s\in\mathbb{S}}\sum_{j\in\mathbb{C}} p_s((\sigma^p_{js}+\sigma^t_j)\sum_{i\in\mathbb{B}} x_{ijs}) - \sum_{s\in\mathbb{S}}\sum_{i\in\mathbb{B}}\sum_{j\in\mathbb{C}} p_s(c^t_{ij}+c^{bm}_{is})x_{ijs} - \sum_{l\in\mathbb{L}}\sum_{j\in\mathbb{C}} I_{lj}y_{lj} \quad (6)$$

$$\text{s.t.} \quad \sum_{j\in\mathbb{C}} x_{ijs} \leq b_{is} \quad \forall i \in \mathbb{B}, \forall s \in \mathbb{S} \quad (7)$$

$$\sum_{i\in S} x_{ijs} \leq \sum_{l\in L} M^{bm}_{lj} y_{lj} \quad \forall j \in \mathbb{C}, \forall s \in \mathbb{S} \quad (8)$$

$$\sum_{l\in L} y_{lj} \leq 1 \quad \forall j \in \mathbb{C} \quad (9)$$

$$x_{ijs} \geq 0 \quad \forall i \in \mathbb{B}, \forall j \in \mathbb{C}, \forall s \in \mathbb{S} \quad (10)$$

$$y_{lj} \in \{0,1\} \quad \forall j \in \mathbb{C}, \forall l \in \mathbb{L} \quad (11)$$

The possible ratios of displaced coal is comprised of set $\mathbb{L}$ where $l = 0, 1, \ldots, |\mathbb{L}|$. For example, $l = 2$ refers to the situation where 2 % coal is displaced by biomass. y_{lj} is the first-stage binary variable which takes 1 if l % of coal is displaced in facility j. x_{ijs} is the second-stage variable which represents the amount of biomass delivered from supplier i to coal plant j for scenario s. In order to meet coherency, same notations with (Eksioglu and Karimi 2014) is used in this study which is given in Table 2.

The objective function (6) maximizes total profit arising from co-firing. Total profit is computed using expected values of total savings minus costs. Resource availability and prices are considered as uncertain parameters. σ^p_{js} is the saving obtained by displacement of one ton coal from plant j in scenario s and σ^t_j is the tax credits for plant j. Unit transportation cost from supplier i to plant j is given by c^t_{ij} while unit purchasing cost for biomass from supplier i for scenario s is represented by c^{bm}_{is}. I_{lj} is the investment cost for plant j when l % of coal is displaced during co-firing.

The biomass supply limit constraints are stated in (7) which ensure that the total amount of biomass received from supplier i cannot exceed the available amount of biomass. Scenario dependent parameter b_{is} refers to the amount of available biomass of supplier i under the scenario s. Constraints (8) guarantee that the amount of supplied biomass satisfies required coal displacement for plant j. M^{bm}_{lj} is the amount of delivered biomass when plant j displace l % of coal which is a function of plant capacity, efficiency and the percentage of biomass co-fired. Constraints (9) ensure that at most one co-firing strategy is selected for every plant j. Constraints (10) and (11) are sign restrictions for decision variables. Derivations of the parameters

Table 2 Notation

Sets:	
$\mathbb{B}$	Set of biomass suppliers with index i
$\mathbb{C}$	Set of coal plants with index j
$\mathbb{L}$	Set of displacement ratio with index l
$\mathbb{S}$	Set of scenarios with index s
Decision variables:	
x_{ijs}	Amount of biomass delivered from supplier i to coal plant j in scenario s [tons]
y_{lj}	Binary variable for displacement of l % of coal in facility j
Parameters:	
σ^p_{js}	Saving from the displacement of coal for plant j [$/ton]
σ^t_j	Saving from tax credits for plant j [$/ton]
c^t_{ij}	Unit transportation cost from supplier i to plant j [$/ton]
c^{bm}_{is}	Unit biomass purchasing cost from supplier i [$/ton]
I_{lj}	Investment cost for plant j when l % displacement of coal [$]
b_{is}	Amount of available biomass in supplier i [tons]
M^{bm}_{lj}	Amount of delivered biomass when plant j displace l % of coal [tons]

σ^p_{js}, σ^t_j, c^{bm}_{is}, c^t_{ij}, I_{lj} and M^{bm}_{lj}, are given in Appendix A. For detailed explanation, readers are referred to Eksioglu and Karimi (2014).

The SP model (6)–(11) has a very special structure. (1) The recourse matrix **W** is scenario independent, i.e., the SP model has a fixed recourse. (2) The technology matrix **T** is deterministic as well. (3) The binary decision variables are first stage decision variables only; the second stage decision variables are continuous. The latter allows for efficient solution via a Benders decomposition approach (Benders 1962). Because current off-the-shelf MILP software can solve our SP model for the Mississippi case study very fast, there is no need to implement a decomposition approach. We leave this as future research directions and solve the model as monolith instead.

4 Experimental Analysis

We perform a sensitivity analysis and as part of that study the proposed two-stage SP model through computational experiments for the case of the State of Mississippi, USA. We use the data from Eksioglu and Karimi's deterministic MIP model (Eksioglu and Karimi 2014). All models are implemented in GAMS 23.7 and are solved with CPLEX 12.1. SCENRED is also used to reduce the two-stage scenario tree, modelling the random data processes (Heitsch and Römisch 2009).

Computational experiments are performed on an Amd Ompteron 6128 computer with 2.00 GHz 8 processors and 12.0 GB RAM.

4.1 Configuration of the Case Study

The investigated supply chain consists of 171 biomass supply sources and 5 coal firing power plants. Forest products and residues are considered as the available biomass at the county level, because they are low in sulphur and chemicals. Only the power plants with capacity larger than 1 MW are taken into account. Feedstock is transported by trucks to the power plants. Transportation cost of a truck includes distance variable cost and distance fixed cost. Distance fixed cost comprises loading and unloading costs while distance variable cost contains fuel and labor costs. Total profit increases when the size of set $\mathbb{L}$ increases. On the other hand, benefit from increment of $|\mathbb{L}|$ decreases while the computational effort increases (Eksioglu and Karimi 2014). This is the reason that the size of set $\mathbb{L}$ is taken as 51 in this study and its elements are represented with integer values, $\mathbb{L} = \{0, 1, 2, \ldots, 50\}$, as in (Eksioglu and Karimi 2014). Deterministic values for all parameters are taken from (Eksioglu and Karimi 2014) and given in Appendix B.

4.2 Uncertain Parameters

In the proposed model, biomass availability, coal price and biomass purchase price are considered as uncertain parameters.

For coal price, three scenarios are taken into account: (1) the base case which is provided from the data, (2) increment in coal price by 7 %, and (3) increment in coal price by 10 %. Second and third scenarios are based on (Tong et al. 2014). Coal price is not directly used in our model. It effects parameter σ_{js}^{p} which refers to saving obtained from the displacement of coal. The equation for this parameter is given as follows (Eksioglu and Karimi 2014):

$$\sigma_{js}^{p} = c_j^{coal} \cdot \frac{LHV_j^{bm}}{LHV_j^{coal}} \tag{12}$$

where c_j^{coal} is the unit purchasing cost for biomass for plant j, and LHV_j^{coal} and LHV_j^{bm} are the lower heating values of coal and biomass fired at plant j, respectively.

Since log-normal distributions provide non-negativity and are convenient to represent economic stochastic parameters (Santoso et al. 2005), they are used for the amount of available biomass and purchase price of biomass. Deterministic data is taken as mean and 2.5 % of mean value is received as standard deviation for both of these parameters. 60 scenarios are generated for each of the parameter. On the basis of these assumptions, a total of 10,800 (3 × 60 × 60) scenarios are generated.

4.3 Computational Results

Firstly, the performances of deterministic and stochastic models are compared. In the deterministic model, the mean values are taken into account for the amount of available biomass and biomass price and the base case is conducted for the coal price. The reports for models are shown in Table 3. The deterministic model does not use expected values for the uncertain parameters. The solution obtained by expected values are given later (in Table 5). The size of the resulting stochastic programming model is very large compared to the size of the deterministic model. To obtain a computationally tractable model, we reduce the 10,800 scenarios to 300 scenarios (we also test 108, 147, 192 and 243 scenarios) to compute a first stage solution. We then compute optimal recourse solutions, for the first stage solution computed for the reduced scenarios, for the 10,800 scenarios case. Profits from the stochastic model exceed the ones achieved by the deterministic model, because the stochastic model incorporates increased coal prices (thus, the coal price in the deterministic model is much smaller than the average coal price of the three scenarios in the stochastic approach). Most significantly, the stochastic solutions provide more biomass utilization than the deterministic solution.

In two-stage stochastic programming, the decisions on first stage variables are made before the realization of uncertainties which can be called as "here-and-now" (HN) solution. On the other hand, a "wait-and-see" (WS) solution is obtained if the decisions are performed after getting the information on uncertain parameters. The expected value of perfect information (EVPI) refers to the price that the decision maker would be ready to pay to obtain perfect information and is shown as the following equation:

Table 3 Comparison of deterministic and stochastic models

	Deterministic model	Number of scenarios	
		300	10,800
Total profit ($)	137,475,956.02	154,948,948.33	152,172,314.32[a]
Number of binary variables	255	255	255
Number of continuous variables	856	256,501	8,550,001
Number of constraints	431	53,055	1,760,255
CPU time	1.193	35.111	2,203.307[a]
Relative gap (%)	–	12.71	10.69[a]

[a]10,800 scenarios are reduced to 300 by scenred. The objective value and CPU time is obtained by the stochastic model including the selected 300 scenarios

$$EVPI = HN - WS \quad (13)$$

Another way to solve a stochastic programming is using the expected value of uncertain parameters in deterministic model in which the obtained solution is called as expected value (EV) solution. If EV is applied to the first stage decision and the second stage decision variables are adjusted for all scenarios, the expected result when using the EV (EEV) solution is obtained. The value of the stochastic solution (VSS) is then defined as the following:

$$VSS = EEV - HN \quad (14)$$

The EVPI values for different number of scenarios are given in Table 4. Since the case with 300 scenarios has a relatively small EVPI value, 10,800 scenarios can be reduced to 300 to find the stochastic solution. The stochastic measures for 300 and 10,800 scenarios case are given in Table 5. We compare the results for 300 and 10,800 scenario cases. The solution for 10,800 scenario case is obtained as explained for Table 3. Since our problem is a maximization problem, EVPI and VSS values are obtained by considering the solutions as negative. The wait-and-see solution is computed as the expected value of optimum solution for all selected scenarios. It can be inferred that the cost of using stochastic model is less than 0.05 % of total profit of co-firing. As seen in Table 3, VSS values are significantly small especially for the case with 10,800 scenarios. This shows that the value of knowing and using distributions on future outcomes is very small.

Table 4 Stochastic solutions for different number of scenarios

# scenarios	Uncertain parameters	Wait-and-see solution	Stochastic solution	EVPI
108	$(3 \times 6 \times 6)$	154,269,318.82	154,201,564.56	67,754.26
147	$(3 \times 7 \times 7)$	152,872,279.29	152,787,855.99	84,423.30
192	$(3 \times 8 \times 8)$	153,689,762.61	152,916,847.31	772,915.30
243	$(3 \times 9 \times 9)$	153,935,575.29	153,825,637.32	109,937.97
300	$(3 \times 10 \times 10)$	155,024,870.50	154,948,948.33	75,922.17

Table 5 Performance analysis for stochastic programming

	Number of scenarios	
	300	10,800
Perfect information (wait-and-see) ($)	155,024,870.50	152,294,384.40
Stochastic solution (here-and-now) ($)	154,948,948.33	152,172,314.32
EEV ($)	154,774,516.74	152,115,493.39
Expected value of perfect information (EVPI)	75,922.17	122,070.08
Value of the stochastic solution (VSS)	174,431.59	56,820.93

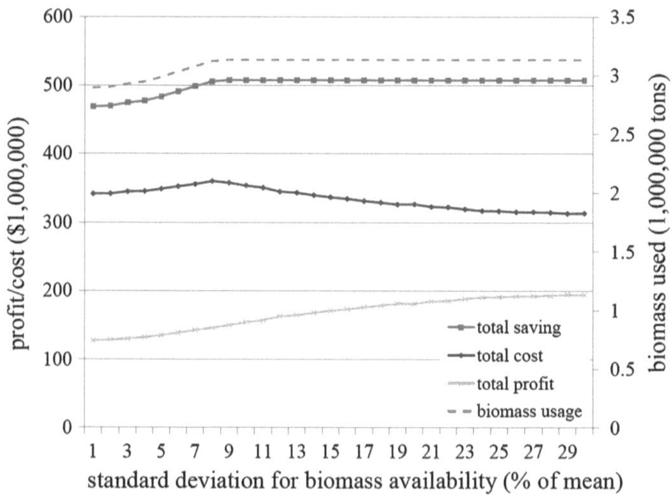

Fig. 2 Effects of uncertainty of biomass availability on total profit due to co-firing

Finally, the effect of uncertainties on total profit arising co-firing is investigated in this study. In order to investigate the effect of an uncertain parameter other uncertain parameters are fixed with their mean values. The amount of available biomass for co-firing can be unsteady because of several external reasons, such as weather or soil conditions. That is why, it is important to foresee the changes on cost and/or profit caused by uncertainty. Figure 2 shows the total profit, cost and biomass usage by changing the variance of biomass availability. The standard deviation of biomass availability is represented by the percentage of mean and changes from 1 to 30 %. When uncertainty of biomass availability increases, biomass usage increases until the standard deviation of 9 % of mean. From this point, the amount of biomass usage is not changed significantly. While total saving shows the same behaviour with biomass usage, it can be easily observed that the reduction in total cost results with profit increase. Although the biomass usage remains same, providing biomass with less purchasing cost causes an increment in total profit. This can be inferred that under high uncertainty of biomass availability, decision makers tend to provide biomass from suppliers having less purchasing cost.

The effect of changes in variance of biomass price on total profit and total biomass utilization is shown in Fig. 3. It is interesting to see that although the amount of used biomass decreases, total profit due to co-firing increases with the increase of uncertainty in biomass price. Increasing the variation of biomass price causes a decrease in biomass utilization. However, the amount of savings obtained from coal displacement and tax credits is the reason of the profit increase under price uncertainty. This result shows us that, under the situations with uncertain biomass price, government incentives have a very significant effect on co-firing investment strategies.

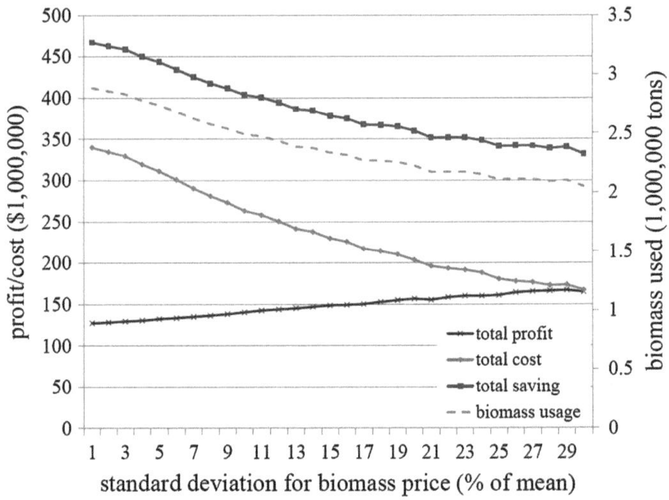

Fig. 3 Effects of uncertainty of biomass price on total profit and total biomass utilization

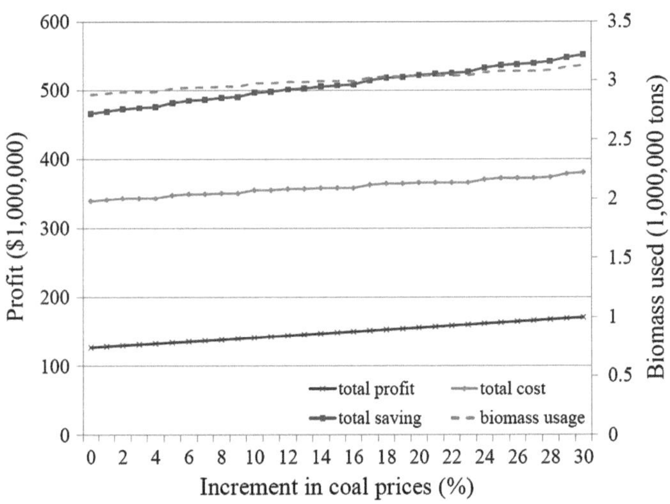

Fig. 4 Relationship of biomass utilization and total profit for changes of coal price

The influence of coal prices is seen in Fig. 4. The coal price is increased by a percentage of base value which is derived from the deterministic data. The investigated increment on coal price varies from 1 to 30 %. When the coal price increases, total profit due to co-firing increases with the amount of biomass as well. Although the total cost of biomass increases with the amount of used biomass, savings from coal displacement and tax credits make the co-firing option profitable under price uncertainty.

5 Concluding Remarks

Co-firing of biomass and fossil fuels has the potential to lead to the generation of clean energy without major capital investments. Efficiency of co-firing processes is the main focus of optimization studies in the co-firing literature. In this study, mathematical programming models developed for biomass co-firing with a primary fuel are investigated and the literature on biomass co-firing is classified into two categories: supply chain design and optimal fuel combination. The motivation of this study is to investigate the effect of uncertainties on supply chain or operational level decision processes for biomass co-firing as this has attracted very little attention in the literature. We proposed a stochastic programming model to develop an optimum co-firing strategy under uncertainties.

Our model treats biomass availability and prices of coal as well as biomass as uncertain parameters for a supply chain design problem which maximizes the total profit obtained from co-firing. The performance of the proposed model is evaluated on the data from the State of Mississippi. The computational experiments show that total profit due to co-firing found by the proposed model exceeds the one achieved by deterministic model. The value of the stochastic solution is exceptionally small which means that including stochasticity has not much benefit for our problem. As well as the stochastic approach, an extensive sensitivity analysis is also performed to demonstrate the effect of uncertain parameters on total profit arising co-firing. When the uncertainty of the available amount of biomass is over a certain limit, the amount of co-fired biomass remains unchanged. This relation is an important point to investigate in further study. Government incentives also play a crucial role for co-firing investment under uncertainty. Specifically, the effect of uncertainty can also be investigated to compute optimal fuel combination for biomass co-firing.

Acknowledgements The authors thank Sandra Eksioglu (Clemson University) for providing the Mississippi data for the case study. D. Cinar is supported by TUBITAK- BIDEB 2219-International Postdoctoral Research Scholarship Program during this study at University of Florida.

Appendix A: Equations (Eksioglu and Karimi 2014)

See Table 6

Efficiency loss caused by displacing $\%l$ of coal

$$EL_j = 0.0044l^2 + 0.0055 \tag{15}$$

Table 6 Notation

I_j^S	Investment for biomass storage [$]
I_j^S	Investment for biomass handling [$]
I_j^S	Investment for compressors and dryers [$]
Q_j^0	The initial annual heating input of plant j [MW]
OH_j	The number of operating hours plant j [Hours]
LHV_j^{bm}	The lower heating value of biomass fired at plant j [BTU/Ton]
LHV_j^{coal}	The lower heating value of coal fired at plant j [BTU/Ton]
ρ_j^b	The boiler efficiency rate of plant j [%]
ρ_j^{rp}	The efficiency rate of the rest of plant j [%]
TC_j	The nameplate capacity of plant j [MV]
f_j	The capacity factor of plant j [%]
EL_j	Efficiency loss of plant j [%]
α_j	A constant for each plant j

Amount of biomass required for co-firing:

$$M_{lj}^{bm} = \frac{1}{\frac{1}{l} - \alpha_j} \cdot \frac{Q_j^0 \cdot OH_j \cdot C^{wb}}{LHV_j^{coal}} \cdot \frac{\rho_j^b}{\rho_j^b - EL_j} \tag{16}$$

$$\alpha_j = 1 - \frac{LHV_j^{coal}}{LHV_j^{bm}} \tag{17}$$

Investment cost when $l \leq 4\ \%$:

$$I_{lj} = I^{cap} \times \frac{l}{1-l} \tag{18}$$

$$I^{cap} = 50,000 \times TC_j \times f_i \times \frac{LHV_j^{bm}}{LHV_j^{coal}} \tag{19}$$

Investment cost (I_{lj}) when $l > 4\ \%$:

$$I_{lj} = I_j^S + I_j^H + I_j^{CD} \tag{20}$$

$$I_j^S = I_j^s \times \left(\frac{l}{1-l}\right)^{0.5575}, \quad I_j^H = I_j^h \times \left(\frac{l}{1-l}\right)^{0.9554}, \quad I_j^{CD} = I_j^{cd} \times \left(\frac{l}{1-l}\right)^{0.5575} \tag{21}$$

$$I_j^s = 136{,}578 \times \left(TC_j \times f_j \times \frac{LHV_j^{bm}}{LHV_j^{coal}}\right)^{0.5575} \tag{22}$$

$$I_j^h = 55{,}780 \times \left(TC_j \times f_j \times \frac{LHV_j^{bm}}{LHV_j^{coal}}\right)^{0.9554} \tag{23}$$

$$I_j^{cd} = 13{,}646 \times \left(TC_j \times f_j \times \frac{LHV_j^{bm}}{LHV_j^{coal}}\right)^{0.5575} \tag{24}$$

Appendix B: Data of the State of Mississippi (Eksioglu and Karimi 2014)

Ses Tables 7, 8 and 9

Table 7 Parameters of case study

	Coal plants (j)				
	1	2	3	4	5
σ^t (\$/Ton)	113.32	113.32	113.32	113.32	113.32
Q_j^0 (MW)	926.56	15.90	728.28	2,779.14	1,077.20
OH_j (Hours)	8,760	8,761	8,763	8,764	8,765
LHV_j^{bm} (BTU/Ton)	16,811,000	16,811,000	16,811,000	16,811,000	16,811,000
LHV_j^{coal} (BTU/Ton)	22,460,600	22,460,600	22,460,600	22,460,600	22,460,600
ρ_j^b (%)	0.93	0.93	0.93	0.93	0.93
ρ_j^{rp} (%)	0.43	0.43	0.43	0.43	0.43
TC_j (MW)	513.7	59	400	2,229	1,215.5
f_j (%)	0.7213	0.1078	0.7281	0.4986	0.3544

Table 8 Unit transportation cost for biomass from supplier i to plant j

	Coal plants (j)						Coal plants (j)				
i	1	2	3	4	5	i	1	2	3	4	5
1	19.89	21.68	3.73	11.17	9.73	86	8.22	10.75	14.92	21.34	21.21
2	14.04	18.64	10.86	14.59	15.51	87	9.94	3.07	21.75	28.75	28.24
3	20.15	23.89	8.04	8.32	9.56	88	19.51	18.53	10.66	17.98	15.89
4	19.89	21.68	3.73	11.17	9.73	89	8.02	14.73	21.80	26.10	27.10
5	7.76	14.59	18.60	22.92	23.87	90	20.22	21.14	6.01	13.08	11.12
6	20.13	23.11	5.70	8.73	8.67	91	9.20	14.99	24.40	28.84	29.80
7	7.44	9.92	24.08	29.86	30.17	92	5.51	7.50	20.56	26.80	26.83
8	14.04	18.64	10.86	14.59	15.51	93	7.87	12.64	15.02	20.59	20.92
9	20.07	22.36	3.76	9.97	8.94	94	10.59	14.55	12.33	17.92	18.19
10	22.79	26.22	8.65	5.66	7.15	95	7.76	14.59	18.60	22.92	23.87
11	20.15	23.89	8.04	8.32	9.56	96	5.23	12.07	20.41	25.49	26.13
12	8.04	6.55	23.08	29.52	29.43	97	23.59	25.09	6.98	10.10	7.44
13	25.85	28.29	9.50	6.53	3.32	98	20.13	23.11	5.70	8.73	8.67
14	11.93	16.66	27.62	32.00	33.02	99	21.73	21.26	9.71	16.24	13.78
15	25.87	29.08	10.73	3.77	5.62	100	9.70	12.64	26.40	31.76	32.31
16	16.53	19.40	6.50	12.42	12.26	101	13.53	16.97	29.84	34.56	35.44
17	19.89	21.68	3.73	11.17	9.73	102	12.90	8.90	27.49	34.22	33.94
18	7.76	14.59	18.60	22.92	23.87	103	12.60	12.56	12.40	19.77	18.89
19	5.23	12.07	20.41	25.49	26.13	104	15.38	15.80	9.44	17.05	15.86
20	20.13	23.11	5.70	8.73	8.67	105	13.83	15.81	9.13	16.05	15.53
21	21.73	21.26	9.71	16.24	13.78	106	23.12	25.63	6.89	7.17	5.69
22	12.60	12.56	12.40	19.77	18.89	107	14.18	16.18	30.87	36.07	36.72
23	13.83	15.81	9.13	16.05	15.53	108	15.24	19.07	31.20	35.55	36.60
24	14.18	16.18	30.87	36.07	36.72	109	20.88	21.29	7.36	14.09	11.87
25	16.69	20.76	8.75	11.80	12.72	110	14.28	7.85	21.89	29.40	28.31
26	13.94	15.00	30.66	36.19	36.67	111	16.69	20.76	8.75	11.80	12.72
27	7.44	9.92	24.08	29.86	30.17	112	24.95	22.45	15.95	22.22	19.48
28	14.04	18.64	10.86	14.59	15.51	113	5.34	11.51	17.74	23.17	23.61
29	6.72	12.88	22.35	27.23	27.98	114	23.52	20.56	16.43	23.22	20.67
30	17.07	15.91	11.45	19.10	17.41	115	23.02	21.59	12.47	18.85	16.22
31	20.07	22.36	3.76	9.97	8.94	116	6.57	8.39	17.52	23.99	23.85
32	21.41	19.49	13.12	20.11	17.73	117	13.94	15.00	30.66	36.19	36.67
33	22.79	26.22	8.65	5.66	7.15	118	7.44	9.92	24.08	29.86	30.17
34	20.15	23.89	8.04	8.32	9.56	119	3.56	9.71	19.29	25.08	25.36
35	8.04	6.55	23.08	29.52	29.43	120	18.63	15.66	15.11	22.67	20.72
36	25.85	28.29	9.50	6.53	3.32	121	14.04	18.64	10.86	14.59	15.51

(continued)

Table 8 (continued)

	Coal plants (*j*)						Coal plants (*j*)				
i	1	2	3	4	5	*i*	1	2	3	4	5
37	14.26	12.48	13.74	21.37	20.09	122	6.72	12.88	22.35	27.23	27.98
38	11.93	16.66	27.62	32.00	33.02	123	17.07	15.91	11.45	19.10	17.41
39	25.87	29.08	10.73	3.77	5.62	124	16.84	18.52	6.31	13.77	12.81
40	16.53	19.40	6.50	12.42	12.26	125	20.07	22.36	3.76	9.97	8.94
41	19.89	21.68	3.73	11.17	9.73	126	21.41	19.49	13.12	20.11	17.73
42	8.22	10.75	14.92	21.34	21.21	127	22.79	26.22	8.65	5.66	7.15
43	19.51	18.53	10.66	17.98	15.89	128	20.15	23.89	8.04	8.32	9.56
44	20.22	21.14	6.01	13.08	11.12	129	8.04	6.55	23.08	29.52	29.43
45	9.20	14.99	24.40	28.84	29.80	130	25.85	28.29	9.50	6.53	3.32
46	7.87	12.64	15.02	20.59	20.92	131	14.26	12.48	13.74	21.37	20.09
47	7.76	14.59	18.60	22.92	23.87	132	8.97	6.36	18.46	25.47	24.95
48	5.23	12.07	20.41	25.49	26.13	133	11.93	16.66	27.62	32.00	33.02
49	23.59	25.09	6.98	10.10	7.44	134	25.87	29.08	10.73	3.77	5.62
50	20.13	23.11	5.70	8.73	8.67	135	13.58	16.97	9.42	15.04	15.19
51	21.73	21.26	9.71	16.24	13.78	136	17.37	18.43	6.81	14.49	13.16
52	12.60	12.56	12.40	19.77	18.89	137	16.53	19.40	6.50	12.42	12.26
53	15.38	15.80	9.44	17.05	15.86	138	19.89	21.68	3.73	11.17	9.73
54	13.83	15.81	9.13	16.05	15.53	139	11.40	16.71	13.37	17.48	18.38
55	23.12	25.63	6.89	7.17	5.69	140	18.31	18.42	8.33	15.90	14.14
56	14.18	16.18	30.87	36.07	36.72	141	8.22	10.75	14.92	21.34	21.21
57	20.88	21.29	7.36	14.09	11.87	142	9.94	3.07	21.75	28.75	28.24
58	16.69	20.76	8.75	11.80	12.72	143	19.51	18.53	10.66	17.98	15.89
59	24.95	22.45	15.95	22.22	19.48	144	8.02	14.73	21.80	26.10	27.10
60	23.52	20.56	16.43	23.22	20.67	145	11.03	9.99	14.94	22.21	21.44
61	23.02	21.59	12.47	18.85	16.22	146	20.22	21.14	6.01	13.08	11.12
62	6.57	8.39	17.52	23.99	23.85	147	9.20	14.99	24.40	28.84	29.80
63	13.94	15.00	30.66	36.19	36.67	148	5.51	7.50	20.56	26.80	26.83
64	7.44	9.92	24.08	29.86	30.17	149	7.87	12.64	15.02	20.59	20.92
65	3.56	9.71	19.29	25.08	25.36	150	10.59	14.55	12.33	17.92	18.19
66	14.04	18.64	10.86	14.59	15.51	151	7.76	14.59	18.60	22.92	23.87
67	6.72	12.88	22.35	27.23	27.98	152	5.23	12.07	20.41	25.49	26.13
68	17.07	15.91	11.45	19.10	17.41	153	23.59	25.09	6.98	10.10	7.44
69	16.84	18.52	6.31	13.77	12.81	154	20.13	23.11	5.70	8.73	8.67
70	20.07	22.36	3.76	9.97	8.94	155	21.73	21.26	9.71	16.24	13.78
71	21.41	19.49	13.12	20.11	17.73	156	9.70	12.64	26.40	31.76	32.31

(continued)

Table 8 (continued)

	Coal plants (j)						Coal plants (j)				
i	1	2	3	4	5	i	1	2	3	4	5
72	22.79	26.22	8.65	5.66	7.15	157	13.53	16.97	29.84	34.56	35.44
73	20.15	23.89	8.04	8.32	9.56	158	12.90	8.90	27.49	34.22	33.94
74	8.04	6.55	23.08	29.52	29.43	159	12.60	12.56	12.40	19.77	18.89
75	25.85	28.29	9.50	6.53	3.32	160	15.38	15.80	9.44	17.05	15.86
76	14.26	12.48	13.74	21.37	20.09	161	13.83	15.81	9.13	16.05	15.53
77	8.97	6.36	18.46	25.47	24.95	162	23.12	25.63	6.89	7.17	5.69
78	11.93	16.66	27.62	32.00	33.02	163	14.18	16.18	30.87	36.07	36.72
79	25.87	29.08	10.73	3.77	5.62	164	15.24	19.07	31.20	35.55	36.60
80	13.58	16.97	9.42	15.04	15.19	165	20.88	21.29	7.36	14.09	11.87
81	17.37	18.43	6.81	14.49	13.16	166	14.28	7.85	21.89	29.40	28.31
82	16.53	19.40	6.50	12.42	12.26	167	16.69	20.76	8.75	11.80	12.72
83	19.89	21.68	3.73	11.17	9.73	168	4.76	9.57	21.45	27.18	27.50
84	11.40	16.71	13.37	17.48	18.38	169	24.95	22.45	15.95	22.22	19.48
85	18.31	18.42	8.33	15.90	14.14	170	5.34	11.51	17.74	23.17	23.61
						171	9.01	8.28	24.86	31.11	31.15

Table 9 Unit purchase cost for biomass $\left(c_i^{bm}\right)$ and the available amount of biomass (b_i) for each supplier i

i	c_i^{bm} ($/Ton)	b_i (Ton)	i	c_i^{bm} ($/Ton)	b_i (Ton)	i	c_i^{bm} ($/Ton)	b_i (Ton)
1	50	25,900	58	80	32,900	115	100	46,500
2	60	95,100	59	80	3,600	116	100	43,500
3	60	37,200	60	90	6,000	117	100	0
4	60	17,300	61	90	9,700	118	100	100
5	60	41,300	62	90	1,000	119	100	16,500
6	60	42,000	63	90	26,000	120	100	100
7	70	34,800	64	90	0	121	100	29,400
8	70	30,100	65	90	500	122	100	100
9	70	36,400	66	90	49,800	123	100	31,600
10	70	65,700	67	90	8,500	124	100	3,800
11	70	11,800	68	90	21,400	125	100	12,500
12	70	21,200	69	90	7,300	126	100	0
13	70	23,500	70	90	4,700	127	100	18,500
14	70	18,500	71	90	12,100	128	100	3,900
15	70	101,300	72	90	37,700	129	100	1,600
16	70	9,700	73	90	15,000	130	100	78,900
17	70	0	74	90	3,100	131	100	3,200
18	70	13,100	75	90	139,700	132	100	11,900
19	70	7,700	76	90	32,600	133	100	44,500

(continued)

Table 9 (continued)

i	c_i^{bm} ($/Ton)	b_i (Ton)	i	c_i^{bm} ($/Ton)	b_i (Ton)	i	c_i^{bm} ($/Ton)	b_i (Ton)
20	70	13,300	77	90	22,300	134	100	13,800
21	70	9,800	78	90	1,000	135	100	2,700
22	70	7,200	79	90	55,100	136	100	15,900
23	70	19,300	80	90	5,100	137	100	39,300
24	70	83,800	81	90	300	138	100	0
25	70	27,000	82	90	36,000	139	100	17,800
26	80	26,300	83	90	0	140	100	56,500
27	80	7,700	84	90	300	141	100	11,000
28	80	0	85	90	2,200	142	100	8,100
29	80	64,700	86	90	400	143	100	18,500
30	80	46,900	87	90	31,400	144	100	12,700
31	80	34,900	88	90	24,200	145	100	400
32	80	11,200	89	90	600	146	100	11,600
33	80	7,600	90	90	58,200	147	100	0
34	80	8,100	91	90	41,900	148	100	56,400
35	80	14,100	92	90	1,000	149	100	7,600
36	80	0	93	90	200	150	100	10,000
37	80	33,800	94	90	20,500	151	100	53,800
38	80	19,500	95	90	12,400	152	100	9,200
39	80	109,500	96	90	17,500	153	100	9,700
40	80	0	97	90	57,300	154	100	0
41	80	18,300	98	90	21,500	155	100	0
42	80	33,600	99	90	0	156	100	5,300
43	80	18,200	100	90	200	157	100	20,100
44	80	36,100	101	90	300	158	100	2,500
45	80	46,700	102	90	4,600	159	100	33,600
46	80	65,300	103	90	1,500	160	100	3,900
47	80	7,800	104	90	20,000	161	100	0
48	80	6,900	105	90	0	162	100	0
49	80	75,000	106	90	0	163	100	100
50	80	21,500	107	90	0	164	100	26,900
51	80	6,500	108	90	1,200	165	100	0
52	80	4,900	109	90	4,500	166	100	9,000
53	80	18,800	110	90	16,900	167	100	1,000
54	80	0	111	90	14,400	168	100	100
55	80	23,900	112	90	35,800	169	100	31,300
56	80	0	113	90	37,800	170	100	72,400
57	80	4,400	114	100	3,100	171	100	200

References

Akgul O, Mac Dowell N, Papageorgiou LG, Shah N (2014) A mixed integer nonlinear programming (minlp) supply chain optimisation framework for carbon negative electricity generation using biomass to energy with CCS (BECCS) in the UK. Int J Greenhouse Gas Control 28:189–202

Awudu I, Zhang J (2013) Stochastic production planning for a biofuel supply chain under demand and price uncertainties. Appl Energy 103:189–196

Basu P, Butler J, Leon MA (2011) Biomass co-firing options on the emission reduction and electricity generation costs in coal-fired power plants. Renewable Energy 36(1):282–288

Benders JF (1962) Partitioning procedures for solving mixed-variables programming problems. Numer Math 4(1):238–252

Berggren M, Ljunggren E, Johnsson F (2008) Biomass co-firing potentials for electricity generation in Poland—matching supply and co-firing opportunities. Biomass Bioenergy 32 (9):865–879

Birge JR, Louveaux F (2011) Introduction to stochastic programming. Springer series in operations research and financial engineering, Springer, New York

Devlin G, Talbot B (2014) Deriving cooperative biomass resource transport supply strategies in meeting co-firing energy regulations: a case for peat and wood fibre in Ireland. Appl Energy 113:1700–1709

Eksioglu SD, Karimi H (2014) An optimization model in support of biomass co-firing decisions in coal fired power plants. In: Pawlewski P, Greenwood A (eds) Process simulation and optimization in sustainable logistics and manufacturing, EcoProduction, Springer International Publishing, Switzerland, pp 111–123

Environmental Protection Agency (2014) Carbon pollution emission guidelines for existing stationary sources: electric utility generating units; proposed rules. Fed Reg 79(117):34830–34958

Fell H, Linn J (2013) Renewable electricity policies, heterogeneity, and cost effectiveness. J Environ Econo Manage 66(3):688–707

Gebreslassie BH, Yao Y, You F (2012) Design under uncertainty of hydrocarbon biorefinery supply chains: multiobjective stochastic programming models, decomposition algorithm, and a comparison between cvar and downside risk. AIChE J 58(7):2155–2179

Griffin WM, Michalek J, Matthews HS, Hassan MNA (2014) Availability of biomass residues for co-firing in Peninsular Malaysia: implications for cost and ghg emissions in the electricity sector. Energies 7(2):804–823

Heitsch H, Römisch W (2009) Scenario tree modeling for multistage stochastic programs. Math Program 118(2):371–406

Hu M-C, Lin C-H, Chou C-A, Hsu S-Y, Wen T-H (2011) Analysis of biomass co-firing systems in Taiwan power markets using linear complementarity models. Energy Policy 39(8):4594–4600

Huang Y, Rebennack S, Zheng QP (2013) Techno-economic analysis and optimization models for carbon capture and storage: a survey. Energy Syst 4(4):315–353

Ko AS, Chang N-B (2008) Optimal planning of co-firing alternative fuels with coal in a power plant by grey nonlinear mixed integer programming model. J Environ Manage 88(1):11–27

Lam HL, Varbanov PS, Klemeš JJ (2010) Optimisation of regional energy supply chains utilising renewables: P-graph approach. Comput Chem Eng 34(5):782–792

Liu Z, Altman I, Johnson TG (2014) The feasibility of co-firing biomass for electricity in Missouri. Biomass Bioenergy 69:12–20

Nienow S, McNamara KT, Gillespie AR (2000) Assessing plantation biomass for co-firing with coal in Northern Indiana: a linear programming approach. Biomass Bioenergy 18(2):125–135

Rebennack S (2012) Combining sampling-based and scenario-based nested Benders decomposition methods: application to stochastic dual dynamic programming. Submitted, May 2012

Rebennack S (2014) Generation expansion planning under uncertainty with emissions quotas. Electr Power Syst Res 114:78–85

Roni MS, Eksioglu SD, Searcy E, Jha K (2014) A supply chain network design model for biomass co-firing in coal-fired power plants. Transp Res Part E Logistics Transp Rev 61:115–134

Sahinidis NV (2004) Optimization under uncertainty: state-of-the-art and opportunities. Comput Chem Eng 28(67):971–983

Saidur R, Abdelaziz EA, Demirbas A, Hossain MS, Mekhilef S (2011) A review on biomass as a fuel for boilers. Renew Sustain Energy Rev 15(5):2262–2289

Sami M, Annamalai K, Wooldridge M (2001) Co-firing of coal and biomass fuel blends. Prog Energy Combust Sci 27(2):171–214

Santoso T, Ahmed S, Goetschalckx M, Shapiro A (2005) A stochastic programming approach for supply chain network design under uncertainty. Eur J Oper Res 167(1):96–115

Smith RK, Hobbs BF (2013) Biomass combustion for electric power: allocation and plant siting using non-linear modeling and mixed integer optimization. J Renew Sustain Energy 5(5)

Snyder LV (2006) Facility location under uncertainty: a review. IIE Trans 38(7):547–564

Tillman DA (2000) Biomass cofiring: the technology, the experience, the combustion consequences. Biomass Bioenergy 19(6):365–384

Tong K, Gong J, Yue D, You F (2014) Stochastic programming approach to optimal design and operations of integrated hydrocarbon biofuel and petroleum supply chains. ACS Sustain Chem Eng 2(1):49–61

Touš M, Pavlas M, Stehlík P, Popela P (2011) Effective biomass integration into existing combustion plant. Energy 36(8):4654–4662

US Energy Information Administration (2014) US Department of Energy. Annual energy outlook, April 2014. http://www.eia.gov/forecasts/aeo/

van Dyken S, Bakken BH, Skjelbred HI (2010) Linear mixed-integer models for biomass supply chains with transport, storage and processing. Energy 35(3):1338–1350

Xie F, Huang Y (2013) Sustainable biofuel supply chain planning and management under uncertainty. Transp Res Rec J Transp Res Board 2385(1):19–27

Economic and Land-Use Optimization of Lignocellulosic-Based Bioethanol Supply Chains Under Stochastic Environment

Jun Zhang and Atif Osmani

Abstract The ever increasing concerns such as energy security and climate change calls for a wide range of alternate renewable sources of energy. Bioethanol produced from lignocellulosic feedstock show enormous potential as an economically and environmentally sustainable renewable energy source. In recent years considerable research has focused on the economic feasibility of lignocellulosic-based biofuel supply chains while analytical understanding of land-usage for biomass cultivation has remained limited. Switchgrass is considered as one of the best lignocellulosic feedstock for bioethanol production that can be cultivated on both marginal land with arid soil and crop land. Switchgrass cultivated on crop land normally gives twice the yield when compared with marginal land: however, the higher yield is obtained due to higher input costs. Crop lands are a finite resource and their widespread use for growing energy crops rather than food crops like corn and wheat has resulted in land-use issues such as food versus fuel debate. Therefore, cultivation of switchgrass on marginal land is being studied intensively to minimize the use of crop land for biomass cultivation. This work proposes a novel dual-objective stochastic optimization model to maximize the expected profit and simultaneously minimize usage of crop land to cultivate switchgrass for a lignocellulosic-based bioethanol supply chain (LBSC) under uncertainties of biomass supply, bioethanol demand and bioethanol sale price. The model determines the optimum allocation of marginal land and crop land for switchgrass cultivation, biorefinery locations, and biomass processing capacity of biorefineries. The *e*–constraint method is applied to trade-off among the competing objectives of profit maximization and land-use minimization. In order to solve the proposed stochastic model efficiently and effectively, the Sample Average Approximation method is

J. Zhang (✉)
Gordon and Jill Bourns College of Engineering, California Baptist University,
8432 Magnolia Avenue, Riverside, CA 92504, USA
e-mail: jzhang@calbaptist.edu

A. Osmani
College of Business and Innovation, Minnesota State University Moorhead,
Center for Business 100, Moorhead, MN 56563, USA
e-mail: atif.osmani@mnstate.edu

S.D. Eksioglu et al. (eds.), *Handbook of Bioenergy*,
Energy Systems, DOI 10.1007/978-3-319-20092-7_9

utilized. A case study based in the state of Alabama in the U.S. illustrates the application of the proposed stochastic model. In addition, sensitivity analyses are conducted to provide insights on the important factors that impact on the profitability and land usage in the LBSC.

1 Introduction

Due to the energy crisis (i.e. depletion of fossil-fuels) and environmental issues (i.e. increase in carbon emissions), researchers have been attracted to develop sources of renewable energies to secure the energy consumption and protect the environment. Biofuel is one type of the renewable energies that can be used to substitute fossil-fuel energy. Bioethanol is one type of biofuel that is currently widely used in transportation sector as a gasoline substitute (Schnepf 2011). Currently, 1st generation bioethanol production is commercialized around the world. While 1st generation has provided economic benefits, wide use of 1st generation bioethanol has resulted in new social issues such as food versus fuel debate (i.e. use of finite amount of crop land for energy purposes rather than for food) and higher corn price, since 1st generation bioethanol is produced from edible biomass such as corn and sugarcane. In addition, bioethanol produced from 1st generation biomass feedstocks is not carbon neutral (Charles et al. 2007).

The need for new types of biomass that can improve environmental and social aspects of sustainability has resulted in the emergence of 2nd generation biomass/bioethanol. 2nd generation bioethanol is produced from lignocellulosic biomass feedstocks which is mainly composed of cellulose, hemicellulose, and lignin. Major sources 2nd generation lignocellulosic-based biomass includes, but is not limited to, crop residues and dedicated energy crops (like native grasses) that can be cultivated on both crop land and marginal land that consumes less water and chemical fertilizers. Plenty of research has shown that lignocellulosic-based bioethanol substantially reduces the greenhouse gas (GHG) emissions and relieves the social concerns. In order to improve various aspects of sustainability in a lignocellulosic-based bioethanol supply chain (LBSC), many nations have enacted legislation to promote 2nd generation bioethanol production. For instance, the U.S. Renewable Fuel Standard (RFS) requires that biofuels satisfy at least 20 % of the demand for liquid transportation fuels by 2022. The RFS mandates that at least two-thirds of the bioethanol demand should be met from 2nd generation by the year 2022. Although 2nd generation bioethanol has gained great attraction from researchers, policy makers, and investors, the production of lignocellulosic-based bioethanol has not been commercialized due to lack of comprehensive understanding of the challenges in the entire LBSC. Much work has been done in developing financially viable LBSCs. However, these efforts do no incorporate land usage analysis, which is an important social and environmental concern (Carriquiry et al. (2011). Therefore, research is needed to design optimum LBSCs that are both financially viable as well as considering the concerns on agricultural land usage.

In North America, switchgrass and crop residue are two types of the promising primary sources of lignocellulosic biomass (Cherubini and Ulgiati 2010; Wu et al. 2010). Switchgrass, a perennial grass native to U.S. and Canada, is suitable for cultivation on marginal land (with arid soil) without competing for cropland with other agriculture products (Zhang et al. 2012). It is considered as one of the best second generation bioethanol feedstock due to the following economic, environmental and social benefits: (1) easy to grow; (2) low cost of production; (3) low soil nutrient requirement; (4) not consuming too much water; (5) high net energy yield per unit of cultivated land; (6) adapted to a wide range of environments including marginal soils and arid climates; (7) improved soil conservation; and (8) reduction of greenhouse gas emissions (Sokhansanj et al. 2009). However, switchgrass cultivated on crop land will lead to higher profit due to higher yield, at lower production cost, when compared with marginal land. The higher yield is obtained due to intensive agricultural practices including but not limited to the use of pesticides, chemical fertilizers, and irrigation. Work by Carriquiry et al. (2011) unfavorably evaluates the environmental impact of using crop land (for biomass cultivation) on GHG emissions. Trade-off needs to be made between use of crop land that results in higher profit versus marginal land usage which reduces the practice of using finite resources of crop land for biomass cultivation, but at the cost of reduced profit.

Another challenge in the LBSCs is that various uncertainties relating to biomass supply, bioethanol demand, and bioethanol sales price exist (Awudu and Zhang 2012). These uncertainties introduce significant risk in the decision making process, making it imperative that robust decisions are made concerning the key logistics variables in a stochastic environment. Therefore, decision making should consider the uncertainties. This increases the complexity of the problem as "deterministic" parameters cannot be exclusively used to obtain the optimal values of the decision variables.

The decisions made regarding the key logistics variables are likely to greatly impact on the financial and land-use performances of the LBSC (Mabee et al. 2011). In order to sustainably meet the RFS target for lignocellulosic-bioethanol production, a comprehensive optimization of the various logistical components along the entire supply chain is essential by taking into account the: (1) dual optimization criteria; and (2) stochastic nature of the LBSC. Reasons are elaborated below.

This work is the first research effort in optimization of a LBSC that simultaneously maximizes the expected profit while minimizing crop land usage by considering multiple uncertainties in supply, demand, and prices. Stochastic dual-objective mixed integer linear programming (MILP) model is proposed to determine the key logistic decisions such as land allocation for biomass cultivation, location and processing capacity of biorefineries, etc. Dual-objective optimization generates a set of designs (i.e. Pareto optimum sets) of optimal LBSC configuration in the solution space, which are superior to all other feasible designs (Chen et al. 1999). However, within this set, no design is superior to another in all criteria, and no improvement can be made with respective to one objective without worsening the other objective.

The work incorporates the following specific LBSC characteristics: (1) land-use impact is quantified through crop land used for switchgrass cultivation; (2) land-use impact is traded-off against the financial objective (i.e. profit maximization) using the *e*–constraint method; (3) uncertainties in lignocellulosic-biomass supply, bioethanol demand, and bioethanol selling price are considered jointly; (4) The Sample Average Approximation (SAA) method is used to make the stochastic optimization problem computationally tractable in order to obtain optimal solutions within reasonable time; and (5) to demonstrate the effectiveness of the research, the proposed stochastic mathematical model is used to determine the infrastructure and operational requirements for sustainable lignocellulosic-bioethanol production from switchgrass (cultivated on marginal and/or crop land) and crop residue in the Southern state of Alabama.

The rest of the paper is structured as follows. Section 2 summarizes the literature review. Section 3 gives a summary of the problem statement and evaluates the uncertain parameters. A stochastic dual-objective MILP (SDOMILP) optimization model is proposed in Sect. 4. Section 5 presents a solution methodology for SDOMILP. The case study is described in Sect. 6. The case study results are discussed in Sect. 7. Final conclusions and further research are outlined in Sect. 8.

2 Literature Review

In recent years considerable research has been done on the economic feasibility of 2nd generation lignocellulosic biomass-based supply chains (Papapostolou et al. 2011; Van Dyken et al. 2010). Huang et al. (2010) develop a MILP to design an optimal biowaste based bioethanol supply chain. They suggest that 2nd generation bioethanol is feasible when the bioethanol production is below $1.7 per gallon. Eksioglu et al. (2009) develop an MILP model to design a cost effective forest residue based bioethanol supply chain. The study determines the optimal number, size and location of bioethanol plants. However most work done so far regarding optimization of integrated LBSC has been confined to using the "deterministic" parameters to obtain the optimal production capacity, and locations of biorefineries (Leduc et al. 2010), site selection and allocation of land for biomass cultivation (Zhang et al. 2012). Most work on the deterministic optimization of LBSC only considers the financial objective. Recently, a number of authors (Cucek et al. 2012; Frombo et al. 2009; Giarola et al. 2011; Kocoloski et al. 2011; Zamboni et al. 2009) have presented research that also takes into account the environmental impact.

Literature review (Osmani and Zhang 2013) has highlighted some of the key uncertainties inherent in the life cycle of a LBSC. However, the preliminary work on stochastic optimization of LBSCs only considers one type of uncertainty such as uncertain bioethanol demand, or bioethanol sale price uncertainty (Dal-Mas et al. 2011; Kim et al. 2011; Kostin et al. 2012). Work by Chen and Fan (2012) considers an LBSC where the supply and demand uncertainties are considered separately but not jointly. Only a few recent works (Osmani and Zhang 2013) jointly consider the multiple uncertainties in biomass supply and purchase price, bioethanol demand and sales price. The multiple uncertainties if not considered in the decision making process will result in non-optimal (or even infeasible) solutions which generate lower profits.

Although stochastic modeling approaches provide more reliable results when compared to deterministic models (Osmani and Zhang 2013), the resulting heavier computational burden means that stochastic MILP models cannot be accurately solved using traditional algorithms (Awudu and Zhang 2012). In solving stochastic MILP optimization problems the main computational burden is imposed by the binary/integer decision variables (Osmani and Zhang 2013) which need to be optimized over all the scenarios. The application of heuristics and/or decomposition techniques is required to obtain optimal solutions within reasonable time frame (Lam et al. 2011). The SAA method is a common sampling based approach used to make the optimization problem computationally tractable (Escudero et al. 2010).

The literature on dual-objective stochastic optimization of LBSC is sparse with most work only considering the financial objective (An et al. 2011; Dal-Mas et al. 2011; Kim et al. 2011). Only a few recent works have incorporated the environmental impact (Abdallah et al. 2012; Leduc et al. 2010). Giarola et al. (2011) design a 2nd generation based bioethanol supply chain under uncertainties that aims to reduce cost under carbon emission trading schemes. Gebreslassie et al. (2012) develop a stochastic MILP model to design an economically viable 2nd generation bioethanol supply chain. The objective is to simultaneously maximize profit and minimize risk. Work by You et al. (2012) presents a MILP model along with Pareto optimal curves that trade-off among the economic and environmental performances while developing sustainable biofuel supply chains. However parametric uncertainty is not considered in the research.

To the best of our knowledge, no comprehensive work has been carried out on the stochastic optimization of multi-feedstock biomass-to-bioethanol supply chains under multiple uncertainties where the financial objective is optimized by also taking into account the impact of land-use change. To fill the identified gap in current literature and to advance the state-of-art in LBSC optimization, this work proposes a two-stage stochastic MILP formulation to maximize the expected profit of a multi-feedstock LBSC while simultaneously minimizing use of crop land for switchgrass cultivation.

3 Problem Statement

This research studies a comprehensive multi-feedstock (i.e. switchgrass and crop residue) dual-objective LBSC under multiple uncertainties. The objectives are to maximize the expected profit while at the same time minimizing the land usage (especially crop land) for switchgrass cultivation. A list of indices, parameters, and decision variables is given in Appendix A. This work assumes that: (1) switchgrass and crop residues are harvested once a year in late fall; (2) only road haulage for the transportation of lignocellulosic biomass and bioethanol is considered; and (3) the total bioethanol requirement is proportional to the population in each demand zone.

Due to the complex tradeoffs involved, various competing supply chain and logistics decisions that affect the LBSC cannot be made independently. Therefore, comprehensive management of all the individual logistical components along the entire supply chain is essential to optimize the objectives (e.g. minimize land usage and/or maximize the profit). The production of lignocellulosic-based bioethanol has not yet been commercialized (Osmani and Zhang 2013), and this research aims to demonstrate whether it is sustainable to meet the RFS mandate under the best-case scenario (i.e. centralized decision making) where there is no competition among the different players (i.e. biomass producers, biomass/bioethanol transporters, bioethanol producers, and bioethanol retailers).The proposed decision-making framework can be used as a reference guide by the key LBSC stakeholders (such as renewable energy policy makers, and bioethanol producers) to evaluate their decisions.

The major logistics activities in a LBSC are shown in Fig. 1. Switchgrass can be cultivated on two different types of agricultural land (i.e. marginal land and/or crop land) and harvested from biomass supply zone i. Crop residue can also be procured from biomass supply zone i. The biomass feedstock m is then transported from supply zone i to biorefinery r. The supplied biomass feedstock is converted into bioethanol and co-product (i.e. bioelectricity) by biorefinery r using the biochemical pathway. The volume of bioethanol produced is driven by the ethanol demand and

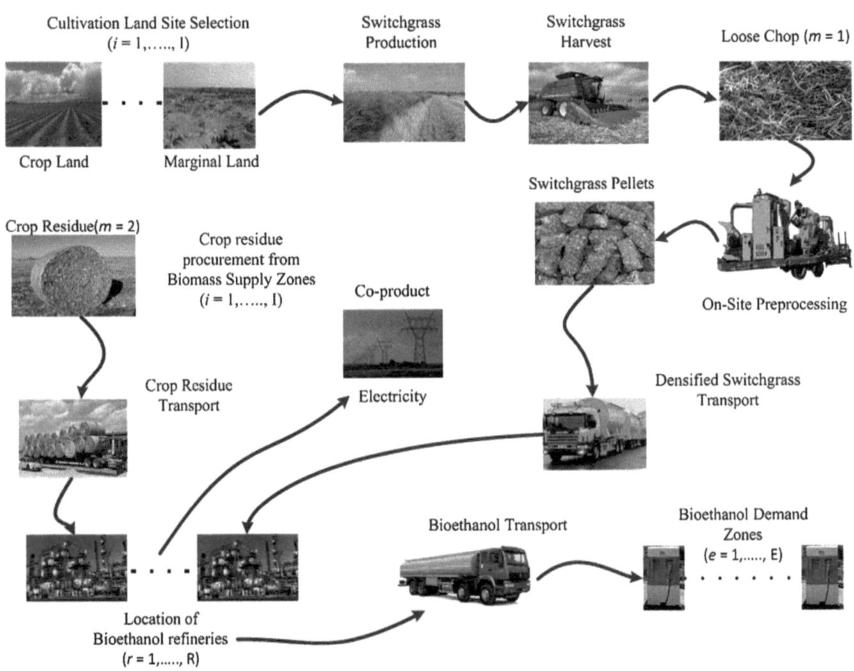

Fig. 1 Major logistics activities in a LBSC

limited by the maximum biomass processing capacity of biorefinery r. The bioethanol is transported from biorefinery r to biofuel demand zone e. After satisfying the total bioethanol requirement, any excess volume is directly sold from biorefinery r. If the volume of bioethanol produced is not sufficient to meet the demand, shortfall in bioethanol requirement incurs a high penalty cost.

This study jointly considers three of the major sources of uncertainties, namely: (1) switchgrass yield due to unpredictable weather conditions; (2) demand for bioethanol; and (3) sale price for bioenergy products.

(1) Switchgrass yields (cultivated on marginal and/or crop land) exhibit a great range of variations on an annual basis. 90 % of the yield variations are caused by the variation in rainfall level at the cultivation site during a given year (Lee and Boe 2005). This randomness in switchgrass yield has a major impact in deciding how much of the available marginal and/or crop land to allocate for switchgrass cultivation in each supply zone. The challenge is to determine the amount of land to be cultivated that is the best possible combination under all possible switchgrass yield scenarios. $o(\omega)$ represents the switchgrass supply level for each stochastic scenario ω and is given by Eq. 1a.

$$o(\omega) = [\text{Switchgrass yield}(\omega)/\text{Average Switchgrass yield}] \tag{1a}$$

(2) The RFS requires biofuels to satisfy at least 20 % of the demand for liquid transportation fuels by 2022. However, demand for gasoline (and hence ethanol) is not deterministic and fluctuates on an annual basis (Carriquiry et al. 2011). $\pi(\omega)$ represents the bioethanol demand level and is given by Eq. 1b.

$$\pi(\omega) = [\text{Bioethanol demand}(\omega)/\text{Average bioethanol demand}] \tag{1b}$$

(3) The sale price of bioenergy products depends on the inherent energy content and is largely influenced by gasoline price (Osmani and Zhang 2013). $\sigma(\omega)$ represents the bioenergy sale price level and is given by Eq. 1c.

$$\sigma(\omega) = [\text{Price of gasoline}(\omega)/\text{Average price of gasoline}] \tag{1c}$$

In this work, all stochastic scenarios are governed by the three above mentioned independent random variables (IRVs) which are not correlated. The reasons are elaborated below. Energy prices are usually governed by the available supply and the market demand (Parker et al. 2010). However in the case of bioethanol, the relationship between sale price and supply/demand is distorted by government action via the RFS mandate. The sale price of bioethanol is only partly influenced by the production cost and is mainly dictated by the price of gasoline and government incentives for cellulosic bioethanol production. Uncertainty in energy prices and their level of supply/demand is commonly modeled using known probability distributions which are based on statistical analysis of historical data (see Sect. 6 for further details).

In order to maximize the expected LBSC profit, and minimize the use of cropland for switchgrass cultivation, the following strategic (i.e. first-stage) decisions need to be optimized across all the stochastic scenarios:

- Site selection from *i* supply zones and allocation of available marginal and/or crop land for switchgrass cultivation.
- Bioethanol refinery sites selection from *r* potential locations.
- Annual biomass processing capacity of each selected biorefinery.

The following tactical (i.e. second stage) decisions also need to be optimized for each stochastic scenario:

- Amount of switchgrass to be harvested and densified from biomass supply zones.
- Amount of crop residues to be procured from biomass supply zones.
- Material flow of procured lignocellulosic feedstock *m* from supply zone *i* to biorefinery *r*.
- Amount of biomass type *m* to be processed by each biorefinery *r*.
- Volume of bioethanol to be produced by the *r* biorefineries.
- Material flow of bioethanol from the *r* refineries to the *e* biofuel demand zones.
- Volume of unmet bioethanol requirement for the *e* biofuel demand zones.

4 Model Formulation

The goal of the study is to determine the optimal configuration of the LBSC (i.e. first-stage decisions) along with the associated operational decisions (i.e. second-stage) under uncertainties. A SDOMILP model with dual objectives is proposed to respectively: (1) maximize economic performance; and (2) minimize land-usage for switchgrass cultivation. The formulation including the objective functions and constraints of the model is explained in the following sections. All continuous decision variables are non-negative, while all integer variables have 0–1 (i.e. binary) restriction.

4.1 *Objective Functions of the LBSC*

The proposed model has two distinct objective functions, namely: (1) maximization of expected profit; and (2) minimization of land usage (especially crop land) for the cultivation of switchgrass.

4.1.1 Financial Objective of the LBSC

The financial objective of the LBSC is profit maximization (i.e. revenue—cost) subject to meeting the RFS mandates relating to bioethanol production. Equation 2a refers to the expected profit (θ) which needs to be maximized.

$$\begin{aligned} Max \quad \theta = & -\sum_{r=1}^{R} G_r Y_r - \sum_{r=1}^{R} H_r K_r - \sum_{i=1}^{I} \Theta 1_i CP_i - \sum_{i=1}^{I} \Theta 2_i MG_i \\ & + E_\omega [\sum_{r=1}^{R} \sigma(\omega) \iota_r Z_r(\omega) + \sum_{r=1}^{R} \sigma(\omega) \chi_r S_r(\omega) + \sum_{r=1}^{R} \sum_{e=1}^{E} \tau_{re} P_{re}(\omega) - \sum_{i=1}^{I} lc_i [X1_i(\omega) + X2_i(\omega)] - \sum_{i=1}^{I} \sum_{r=1}^{R} pp_i F_{1ir}(\omega) \\ & - \sum_{r=1}^{R} \sum_{m=2}^{M} \sum_{i=1}^{I} \lambda_{mi} F_{mir}(\omega) - \sum_{r=1}^{R} \sum_{m=1}^{M} \sum_{i=1}^{I} \eta_{mir} D_{ir} F_{mir}(\omega) - \sum_{r=1}^{R} U_r Z_r(\omega) - \sum_{r=1}^{R} \sum_{e=1}^{E} \psi_{re} D_{re} P_{re}(\omega) - \sum_{e=1}^{E} \phi_e O_e(\omega)] \end{aligned} \tag{2a}$$

The individual cost/revenue components of Eq. 2a are explained below:

- $\sum_{r=1}^{R} G_r Y_r$ calculates the fixed cost of biorefineries, and is the sum-product of annualized biorefinery fixed cost parameter and total number of biorefineries.
- $\sum_{r=1}^{R} H_r K_r$ calculates the variable capacity cost of biorefineries, and is the sum-product of annualized biorefinery variable cost parameter and biomass processing capacity of all biorefineries.
- $\sum_{i=1}^{I} \Theta 1_i CP_i$ calculates the switchgrass cultivation cost using crop land, and is the sum-product of cultivation cost parameter using crop land and total crop land area used for switchgrass cultivation.
- $\sum_{i=1}^{I} \Theta 2_i MG_i$ calculates the switchgrass cultivation cost using marginal land, and is the sum-product of cultivation cost parameter using marginal land and total marginal land area used for switchgrass cultivation.
- $E_\omega[\sum_{r=1}^{R} \sigma(\omega) \iota_r Z_r(\omega)]$ calculates the expected revenue from the sale of bioethanol, and is the sum-product of bioethanol sale price and total volume of bioethanol produced.
- $E_\omega[\sum_{r=1}^{R} \sigma(\omega) \chi_r S_r(\omega)]$ calculates the expected revenue from the sale of bioelectricity, and is the sum-product of bioelectricity sale price and total amount of bioelectricity produced.
- $E_\omega[\sum_{r=1}^{R} \sum_{e=1}^{E} \tau_{re} P_{re}(\omega)]$ calculates the expected tax credits accrued from bioethanol production, and is the sum-product of tax credit parameter for bioethanol production and total volume of subsidized bioethanol produced.

- $E_\omega\{\sum_{i=1}^{I} lc_i[X1_i(\omega) + X2_i(\omega)]\}$ calculates the expected harvest cost of switchgrass, and is the sum-product of harvest cost parameter and total land area (i.e. crop land + marginal land) used for switchgrass cultivation.
- $E_\omega[\sum_{i=1}^{I}\sum_{r=1}^{R} pp_i F_{1ir}(\omega)]$ calculates the expected densification cost of switchgrass, and is the sum-product of densification cost parameter and total amount of densified switchgrass sent from supply zones to biorefineries.
- $E_\omega[\sum_{r=1}^{R}\sum_{m=2}^{M}\sum_{i=1}^{I} \lambda_{mi} F_{mir}(\omega)]$ calculates the expected procurement cost of crop residue, and is the sum-product of crop residue purchase cost parameter and total amount of crop residue sent from supply zones to biorefineries.
- $E_\omega[\sum_{r=1}^{R}\sum_{m=1}^{M}\sum_{i=1}^{I} \eta_{mir} D_{ir} F_{mir}(\omega)]$ calculates the expected biomass transportation cost, and is the sum-product of transport cost parameter for biomass type m and total amount of biomass type m sent to biorefineries.
- $E_\omega[\sum_{r=1}^{R} U_r Z_r(\omega)]$ calculates the expected bioethanol production cost and is the sum-product of production cost parameter for bioethanol and total volume of bioethanol produced by biorefineries.
- $E_\omega[\sum_{r=1}^{R}\sum_{e=1}^{E} \psi_{re} D_{re} P_{re}(\omega)]$ calculates the expected bioethanol transportation cost, and is the sum-product of ethanol transport cost parameter and total volume of subsidized bioethanol sent from biorefineries to demand zones.
- $E_\omega[\sum_{e=1}^{E} \phi_e O_e(\omega)]$ calculates the expected penalty cost of unmet bioethanol demand, and is the sum-product of penalty cost parameter and total volume of unmet demand for subsidized bioethanol.

4.1.2 Land-Usage Objective of the LBSC

The land-use objective of the LBSC is minimization of land used for switchgrass cultivation subject to meeting the RFS mandates relating to bioethanol production. Equation 2b refers to the "weighted" amount of land used for switchgrass cultivation (*JB*) which needs to be minimized. The different components of *JB* (for each scenario ω) respectively refer to: weighted amount of marginal land used for switchgrass cultivation; and weighted amount of crop land used for switchgrass cultivation. w_1 and w_2 refer to the assigned weightage for marginal land usage and crop land usage respectively.

$$Min \quad JB = w_1 \sum_{i=1}^{I} MG_i + w_2 \sum_{i=1}^{I} CP_i \tag{2b}$$

Equation 2b can be converted into a maximization problem by changing the signs of the RHS terms (see Eq. 2c)

$$Max \quad JB = -w_1 \sum_{i=1}^{I} MG_i - w_2 \sum_{i=1}^{I} CP_i \tag{2c}$$

4.2 Capacity Constraints

The capacity constraints are given by Eq. 3–11. Equations 3 and 4 ensures that in biomass supply zone i, the allocated lands for switchgrass cultivation do not exceed the land availability. Equations 5 and 6 ensures that in biomass supply zone i, the harvested lands do not exceed the allocated lands for switchgrass cultivation. Equation 7 ensures that in biomass supply zone i, the amount of biomass type $m = 1$ sent to all biorefineries is not more than the amount of available densified switchgrass during scenario ω. Equation 8 ensures that in biomass supply zone i, the amount of biomass type $m \neq 1$ sent to all biorefineries during scenario ω is not more than the amount of available biomass type m. Equation 9 ensures that maximum of one biorefinery is situated at location r. Equation 10 ensures that a biorefinery (if built at location r) must annually process more biomass than minimum design capacity (ρ_j^{min}) and cannot process more biomass than maximum design capacity (ρ_j^{max}). Equation 11 ensures that during scenario ω a biorefinery does not process more biomass than its processing capacity.

$$CP_i \leq B1_i \quad \forall i \tag{3}$$

$$MG_i \leq B2_i \quad \forall i \tag{4}$$

$$X1_i(\omega) \leq CP_i \quad \forall i, \omega \tag{5}$$

$$X2_i(\omega) \leq MG_i \quad \forall i, \omega \tag{6}$$

$$\sum_{r=1}^{R} F_{1ir}(\omega) \leq o(\omega)\upsilon_i[X1_i(\omega) + X2_i(\omega)A] \quad \forall i, \omega, m = 1 \tag{7}$$

$$\sum_{r=1}^{R} F_{mir}(\omega) \leq \zeta_{mi} \quad \forall i, \omega, m \neq 1 \tag{8}$$

$$Y_r \leq 1 \quad \forall r \tag{9}$$

$$\rho_{\min} Y_r \leq K_r \leq \rho_{\max} Y_r \quad \forall r \tag{10}$$

$$\sum_{m=1}^{M} \sum_{i=1}^{I} F_{mir}(\omega) \leq K_r \quad \forall r \tag{11}$$

4.3 Material Balance Constraints

The material balance constraints are given by Eqs. 12–15. Equation 12 ensures that the cumulative amount of biomass (from all type m feedstocks) used by biorefinery r is converted into bioethanol (i.e. primary product) during scenario ω. Equation 13 ensures that the volume of bioethanol produced by biorefinery r is either sold as unsubsidized bioethanol from the refinery-gate, or sent as subsidized biofuel to demand zones. Equation 14 ensures that the cumulative amount of biomass (from all type m feedstocks) used by biorefinery r is also converted into electricity (i.e. co-product) during scenario ω. Equation 16 ensures that during scenario ω, the volume of unmet bioethanol requirement (in each demand zone) plus the volume of biofuel (from all biorefineries) transported to demand zone e, is equal to the bioethanol requirement in biofuel demand zone.

$$\sum_{m=1}^{M} \sum_{i=1}^{I} \kappa_m F_{mir}(\omega) = Z_r(\omega) \quad \forall r, \omega \tag{12}$$

$$Z_r(\omega) = L_r(\omega) + \sum_{e=1}^{E} P_{re}(\omega) \quad \forall r, \omega \tag{13}$$

$$\sum_{m=1}^{M} \sum_{i=1}^{I} \mu_m F_{mir}(\omega) = S_r(\omega) \quad \forall r, \omega \tag{14}$$

$$O_e(\omega) + \sum_{r=1}^{R} P_{re}(\omega) = v_e \pi(\omega) \quad \forall e, \omega \tag{15}$$

5 Solution Methodology for the Proposed Stochastic Dual-Objective MILP Model

In solving SDOMILP optimization problems the main computational burden is imposed by the binary/integer decision variables which need to be optimized over all the uncertain scenarios. The SAA method, a sampling based approach, is used to make the optimization problem computationally tractable. For dual-objective optimization the e –constraint method is applied for providing a representative subset of the Pareto optimal set (obtained by calling the SAA subroutine).

5.1 Sample Average Approximation Method

For stochastic models with a non-trivial number of scenarios, sampling based approaches have been proposed to estimate objective function values. In the SAA algorithm, a sample set is randomly generated from the total number of N scenarios, and then an optimization problem specified by the generated sample set is solved. The use of the SAA algorithm to solve stochastic optimization problems is depicted in Fig. 2 and described as follows (Kleywegt et al. 2002).

Start: Input the stochastic optimization problem Max $f(\omega)$ with large number of scenarios N

Step 1: Select initial A sample sets of size B (randomly drawn without replacement from the total N scenarios) such that A*B = N. Choose B to be sufficiently small, such that A is sufficiently large number of replications. If say N = 1000, then B can be: B = {1, 2, 4, 5, 8, 10, 20, 25, 40, 50,100, 125, 200, 250, 500, 1000}.

Step 2: For a = 1, 2, 3…A, do steps 2.1, 2.2, 2.3 and 2.4

2.1 Start with A sample sets of size B = 1, and A = N/B

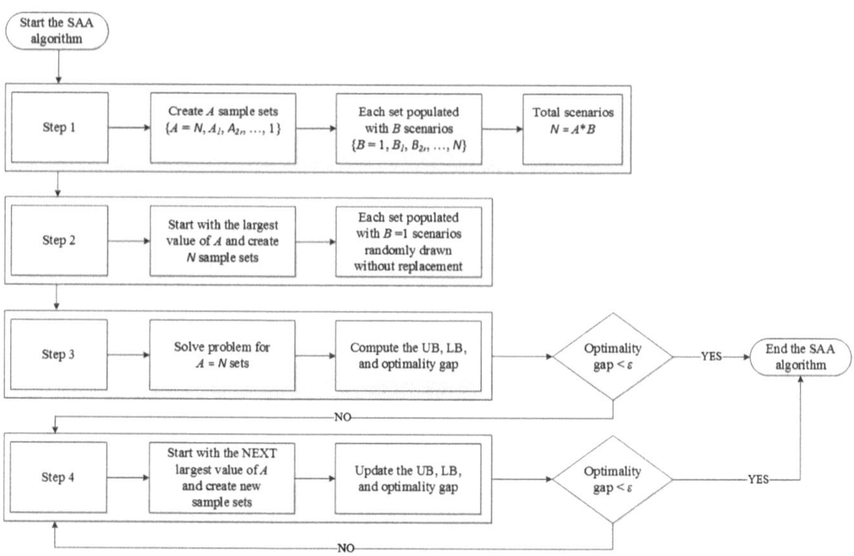

Fig. 2 Flow chart of the SAA method

2.2 Solve the problem with a sample size of B in order to obtain the objective function value $\hat{Z}_B^a$. Any feasible solution represents the lower bound (LB)

2.3 Estimate the upper bound (UB) $Z^* = \sum_{a=1}^{A} \hat{Z}_B^a / A$ and the optimality gap = $[(Z^* - \hat{Z}_B^a)/Z^*]$

2.4 If optimality gap is sufficiently small, go to step 4. In this work optimality gap ≤1 % is used

Step 3: If the optimality gap is too large, increase the sample size to the next possible level of B and create A = (N/B) sample sets and return to step 2

Step 4: Choose the best solution among all the candidate solution.

Stop: Terminate the SAA algorithm.

5.2 Dual-Objective Optimization Using the E*–Constraint Method*

In developing sustainable LBSC, the challenge is to simultaneously maintain financial viability and reduce usage of crop land for biomass cultivation. In the face of these competing objectives, no improvement can be made with respective to one objective without worsening the other objective(s). In this work the *e*–constraint method (You et al. 2012) is used for conducting trade-off analysis among different performance criteria.

A generic dual-objective optimization (DOO) problem is depicted in Eq. 16 where x is the vector of decision variables, $f_1(x)$, $f_2(x)$ are the objective functions and S is the feasible region.

$$\text{Maximize}\, \{f_1(x), f_2(x)\} \tag{16}$$

In the *e*–constraint method we optimize one of the objective functions using the other objective function as constraint, incorporating them in the constraint part of the model as shown in Eq. 17. By parametrical variation in the RHS of the constrained objective functions (*e*) the efficient solutions of the problem are obtained.

$$\begin{aligned} &\text{Maximize}\, f_1(x) \\ \text{st} \quad & f_2(x) \geq e \end{aligned} \tag{17}$$

The stochastic model is optimized for economic performance with Eq. 2a representing the first objective function and the value of land-use performance (represented by Eq. 2c) is calculated based on the decisions obtained from economic optimization. The steps of the algorithm (where both objective functions are for maximization) are explained below:

Step 1: Optimize the first objective function (i.e. maximize profit) by calling the SAA subroutine, obtaining max $f_1 = z_1^*$. Then optimize the second objective function by adding the constraint $f_1 = z_1^*$ in order to keep the optimal solution of the first optimization. Assume that max $f_2 = z_2^*$ is obtained by calling the SAA subroutine.

Step 2: Optimize the second objective function (i.e. maximize reduction in weighted land-usage for switchgrass cultivation) by calling the SAA subroutine, obtaining max $f_2 = z_2^*$. Then optimize the first objective function by adding the constraint $f_2 = z_2^*$ in order to keep the optimal solution of the second optimization. Assume that max $f_1 = z_1^*$ is obtained by calling the SAA subroutine.

Step 3: Set upper and lower bound for each objective function. The best value is the maximum of the individual optimization results obtained from Steps 1 and 2. The worst value over the efficient set (nadir value) is the minimum of the individual optimization results obtained from Steps 1 and 2.

Step 4: Calculate the objective function range R (= max f_2 – min f_2) of the second objective function that is going to be used as constraint.

Step 5: Set number of grid points by dividing the range of the second objective function to q equal intervals using $(q - 1)$ intermediate equidistant grid points. In total $(q + 1)$ grid points are used to vary parametrically the RHS (e) of the second objective function. Total number of runs becomes $(q + 1)$.

Step 7: Solve each run by calling the SAA subroutine. Exit from the algorithm when the problem becomes infeasible.

Step 8: Construct the Pareto efficient frontier using results from Step 7.

6 Case Study Set-up

A case study is presented to demonstrate the effectiveness of the proposed stochastic dual-objective MILP model. The case study is set in the Southern state of Alabama (AL) which has substantial amount of crop land and marginal land that are suitable for switchgrass cultivation. It is aimed to show if 20 % of the annual demand for gasoline (by the year 2022) can be met by the production of bioethanol from multiple lignocellulosic biomass feedstocks (i.e. switchgrass and crop residue) while minimizing the land usage for switchgrass cultivation.

In this work the values of stochastic parameters are assumed to follow Normal probability distributions (see Table B1 in Appendix B and are based on statistical analysis of historical data from sources such as U.S. Department of Agriculture

(USDA), U.S. Energy Information Administration (EIA), etc. Readers with interests can refer to Osmani and Zhang (2013) for the detail method to derive stochastic parameters. The key deterministic parameters used in this case study are displayed in Table B2 (see Appendix B) and obtained from current literature. The values of the following deterministic parameters $\Theta1_i$, $\Theta2_i$, $B1_i$, $B2_i$, υ_i, ζ_{mi}, η_{mir}, ν_e for each of the 67 counties (*i*, *e*, *r*) in AL are obtained from USDA (www.usda.gov) and EIA (www.eia.gov) sources and are available upon request as a data file. Similarly, the distances between the 67 county seats (D_{ir}, D_{re}) are obtained from the U.S. Geological survey (www.usgs.gov).

6.1 Model Assumptions

The various assumptions and indices used in the stochastic MILP model are displayed in Table 1 and also explained below.

(1) All 67 counties of AL are potential biorefinery locations, biomass supply zones, and bioethanol demand zones. Biomass availability and bioethanol demand are centered at the county seat. For example, Birmingham is the seat of Jefferson county.
(2) Demand for co-products i.e. electricity, is assumed to be always greater than the supply.
(3) Biorefineries with a biomass processing capacity of less than 1 million tons per year (MTPY) are not economically viable while those with a production capacity of more than 2 MTPY have not yet been commercialized. Therefore in this work an installed biorefinery has a processing capacity (K_r) between 1 MTPY and 2 MTPY.
(4) Revenues and costs are considered on an annual basis. Equation 18 is used to annualize the initial investment of a biorefinery with life *n* years and interest rate of *q* %. For example, a 1 MTPY biochemical refinery requires an initial investment of $835 Million. With biorefinery life (*n*) of 20 years and interest rate (*q*) of 5 %, the annualized cost is $67 Million obtained from Eq. 18.

$$Annualized\ Cost = [q(Initial\ Investment)]/[1 - (1+q)^{-n}] \tag{18}$$

(5) In Eq. (1a)–(1c), annual fixed cost of biorefinery (G_r) and annual variable cost of bioethanol refinery (H_r) are obtained by the following approximation.

Table 1 Indices used in the case study

e	Bioethanol demand zones (*e* = 1, 2, ……, 67)
i	Lignocellulosic biomass supply zones (*i* = 1, 2, ……, 67)
m	Lignocellulosic feedstocks *m* = 1 (Switchgrass); *m* = 2 (Crop Residue)
r	Biorefinery locations (*r* = 1, 2, ……, 67)
ω	Stochastic scenarios (*ω* = 1, 2, ……, *N*)

Equation 18 is used to calculate the total annual cost (Q_r) of a bioethanol refinery with capacity K_r, where α is a scaling factor, K_r0 is a reference capacity, and Q_r0 is the annualized cost of a bioethanol refinery with capacity of K_r0. In this work, $\alpha = 0.96$ (Osmani and Zhang, 2013). Using Eq. 19, Q_r of a 2 MTPY biochemical refinery is calculated as \$113.3 M for a reference capacity of 1 MTPY with annualized cost of \$67 M.

$$Q_r = Q_r0(K_r/K_r0)^{\alpha} \quad \forall r \tag{19}$$

In this work, K_r is set in the interval of (1, 2) MTPY. For this interval Eq. 19 is linearized and approximated by Eq. 19 in order to avoid non-linear term in the proposed stochastic model. For a biochemical refinery the best value of G_r is \$21.3 Million and H_{rj} is \$46/ton. Using Eq. 20 the annualized cost (Q_r) of a 2 MTPY biochemical refinery is estimated at \$113 M which compares favorably to the value obtained using Eq. 18.

$$Q_r = G_r + H_r K_r \quad \forall r \tag{20}$$

(6) In any scenario if there is excess bioethanol volume leftover after meeting 20 % of AL's bioethanol requirements, it can be sold outside the state. However, out-of-state sale of bioethanol will not qualify for the tax credit.
(7) Not all the crop residue is available for bioethanol production, since significant portion of the residue should be kept on the field to prevent soil erosion and higher operating cost. Most researchers advise that the percentage of agriculture residue that can be sustainably removed from the field be less than 30 % (Osmani and Zhang 2013).
(8) Not all the crop land is available for switchgrass cultivation. USDA's guidelines recommend that at most 25 % of crop land to be used for non-food production.
(9) At each biomass supply zone, the obtained switchgrass yield cultivated on crop land is twice that obtained from cultivation on marginal land. The obtained switchgrass yield cultivated on marginal land is different for each i biomass supply zones, however the ratio A "switchgrass yield ratio using crop land for cultivation" is always equal to 2. The ratio A is given by Eq. 21.

$$A = [\text{Switchgrass yieldusing crop land}/\text{Switchgrass yieldusing marginal land}] \tag{21}$$

6.2 *Modeling the Uncertainties in a LBSC*

In this work, all stochastic scenarios are governed by 3 independent random variables (IRVs) which are not correlated. The first IRV, $o_t(\omega)$ is used to model

Table 2 Discretized levels of independent random variables (IRVs)

Supply level of switchgrass		Demand level of bioethanol		Price level of bioenergy	
Level	Mean value	Level	Mean value	Level	Mean value
L_o_{01}	0.68	L_π_{01}	0.84	L_σ_{01}	0.75
L_o_{02}	0.80	L_π_{02}	0.90	L_σ_{02}	0.86
L_o_{03}	0.87	L_π_{03}	0.94	L_σ_{03}	0.91
L_o_{04}	0.93	L_π_{04}	0.96	L_σ_{04}	0.95
L_o_{05}	0.97	L_π_{05}	0.99	L_σ_{05}	0.98
L_o_{06}	1.03	L_π_{06}	1.01	L_σ_{06}	1.02
L_o_{07}	1.07	L_π_{07}	1.04	L_σ_{07}	1.05
L_o_{08}	1.13	L_π_{08}	1.06	L_σ_{08}	1.09
L_o_{09}	1.20	L_π_{09}	1.10	L_σ_{09}	1.14
L_o_{10}	1.32	L_π_{10}	1.16	L_σ_{10}	1.25

switchgrass supply level. The second IRV, $\pi_t(\omega)$ is used to model bioethanol demand level. The third IRV, $\sigma_t(\omega)$ is used to model gasoline price level, which acts as a surrogate for price level of bioethanol and electricity. The three IRVs are assumed to follow Normal probability distributions.

6.3 *Discretization of Continuous Stochastic Parameters*

A set of possible scenarios with a given probability of occurrence are used to describe the random events. The use of continuous probability distributions to model the uncertainty is likely to result in an infinite number of stochastic scenarios (Osmani and Zhang 2013). In order to make the problem computationally tractable, each IRV is discretized into 10 levels (see Table 2). The number of discretized levels used is sufficient to ensure that the entire range of the probability distribution is captured. The resulting total number of stochastic scenarios $N = 10^3 = 1000$.

7 Case Study Results

In the "base-case" the values of w_1 "weight for marginal land usage" and w_2 "weight for crop land usage" are set at 0 and 1 respectively. This means that the dual-objective is to maximize the financial profit while minimizing the use of crop land for switchgrass cultivation. There is no limit on the amount of available marginal land that can be used for switchgrass cultivation. Work by Osmani and Zhang (2013) has shown that stochastic models outperform deterministic models

Table 3 LBSC performances under different optimization objectives

	Profit	Land usage ('000 acres)		
Optimization objective	($ Million)	Marginal	Crop	Weighted
Maximize financial performance	497	0	765	765
Minimize weighted land usage	359	1,008	0	0

under uncertainties. As such this work will only focus on the decision results and analysis of the stochastic model.

7.1 *Strategic Decisions*

The strategic decisions (i.e. allocation of marginal and/or crop land for switchgrass cultivation, site selection and biomass processing capacity of bioethanol refineries) reached under different optimization objectives are discussed below.

7.1.1 Optimization of Financial Performance

The financial performance is optimized by maximizing θ (see Eq. 2a) while subject to constraints represented by Eqs. 3–15. The value for the land-use performance (see Eq. 2b) is computed based on the decisions obtained from financial optimization. The optimal first-stage decisions taken are summarized in Table 3 and also displayed in Fig. 3. The depicted size of the biorefinery icon is correlated to its biomass processing capacity.

Across the state of Alabama, the production cost of switchgrass cultivated on marginal land is generally higher than that from crop land. As a result, marginal land is not allocated for switchgrass cultivation under objective of profit maximization. Switchgrass cultivated on crop land is preferred as the main source of biomass feedstock as it has the lowest production cost. Due to high purchase price of crop residue it only comprises on average 7 % of the total amount of lignocellulosic biomass feedstock consumed for the production of bioethanol. The central and north-eastern part of Alabama has large amount of crop land available for switchgrass cultivation at the lowest production cost ($/ton of harvested biomass) state-wide. A total of 765,000 acres of crop land are allocated for switchgrass cultivation. This represents 25 % of the total crop land available in the state of AL.

7.1.2 Optimization of Land-Use Performance

The land usage is optimized by minimizing *JB* (see Eq. 2b) while subject to constraints represented by Eqs. 3–15. The value for the financial performance (see

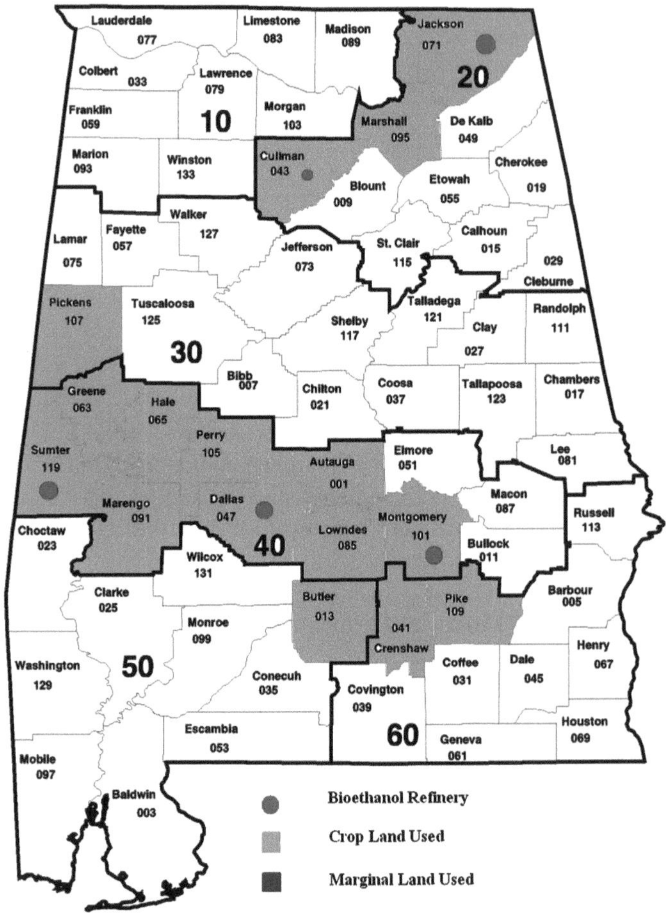

Fig. 3 Optimization of financial performance (*base-case*)

Eq. 2a) is computed based on the decisions obtained from land-use optimization. The optimal first-stage decisions taken are summarized in Table 3 and also displayed in Fig. 4. The depicted size of the biorefinery icon is correlated to its biomass processing capacity. Under both objectives, a total of 5 biorefineries with a combined biomass processing capacity of 9 MTPY are selected for installation. However, the locations of the biorefineries are different under both objectives (see Figs. 3 and 4).

The state of Alabama also has large amount of marginal land available for switchgrass cultivation. In addition, large quantities of crop residues are also available as biomass feedstock. Sufficient amount of biomass (to meet the required bioethanol demand) can be procured from the above mentioned sources albeit at higher cost when compared to biomass production from crop land. As a result, crop

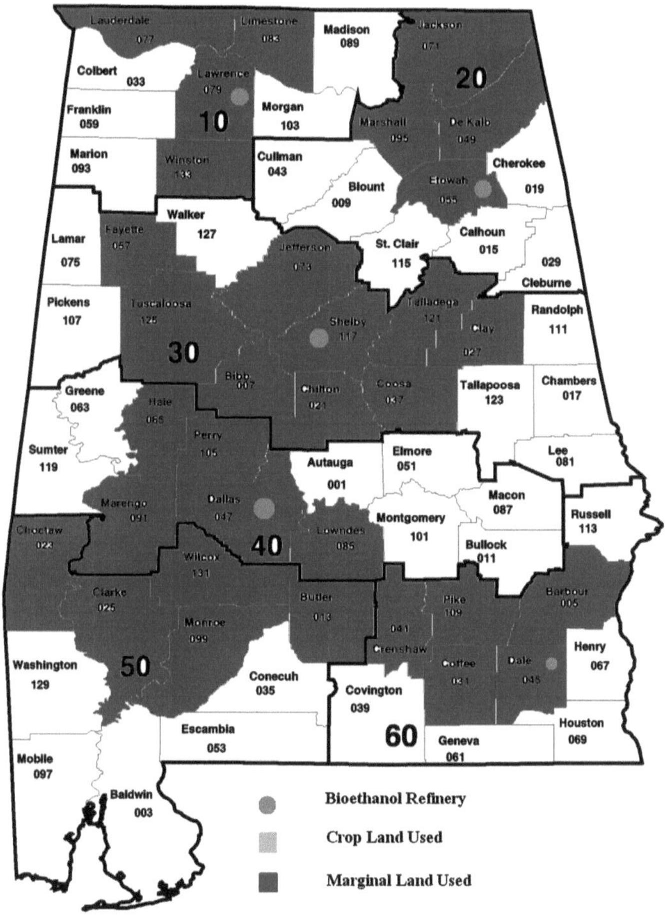

Fig. 4 Optimization of land-use performance (*base-case*)

land is not allocated for switchgrass cultivation under objective of minimizing crop land usage. Switchgrass cultivated on marginal land is preferred as the main source of biomass feedstock as it has lower procurement cost when compared to crop residue. The south-central and northern part of Alabama has large amount of marginal land available for switchgrass cultivation at the lowest production cost ($/ton of harvested biomass) state-wide. A total of 1 million acres of marginal land are allocated for switchgrass cultivation. This represents almost all of the total marginal land available in the state of AL. This necessitates the purchase of expensive crop residue to meet the shortfall in biomass feedstock requirement. Crop residue comprises on average 30 % of the total amount of procured biomass feedstock.

7.2 *Trade-off Between Economic Performance and Land-Usage*

LBSC performances under different optimization objectives are summarized in Table 4 and also displayed in Fig. 5a. Results show that the objective of minimization of crop land usage is contradictory to the objective of maximizing financial profit (and vice versa). Profit maximization (i.e. economic performance) favors switchgrass cultivation on crop land while minimization of weighted land-usage favors switchgrass cultivation on marginal land (under the base-case).

Figure 5b displays the trade-off between economic performance and crop land usage. Compared to the base-case, 18 % decrease in crop land usage results in only a 10 % reduction in expected profit. A further 7 % decrease in crop land usage results in a staggering 20 % reduction in expected profit. The Pareto curve shown in Fig. 5b provides insights for decision makers to make informed choices. For example, if the decision makers decide to allocate no more than 10 % of crop land (for switchgrass cultivation) then reading from the Pareto curve the profit they can expect is $460 M. This is a decrease of 20 % compared to the maximum profit of $497 M obtained from using 25 % of crop land for biomass cultivation.

7.3 *Sensitivity Analysis*

The impact of the uncertain parameters has already been incorporated into the stochastic model and sensitivity analysis is conducted using the "base-case" to measure the impact (on the expected profit and the key decision variables) of the following deterministic parameters: (1) crop residue purchase price; (2) switchgrass

Table 4 Profit versus cultivated land and crop residue usage

Profit ($ M)	Land usage ('000 acres)			Crop residue ('000 tons)
	Crop	Marginal	Total	
497	765	0	765	584
493	688	150	838	584
489	612	303	915	584
483	535	455	990	584
477	459	608	1,067	584
469	382	719	1,101	584
459	306	838	1,144	610
445	229	900	1,129	800
424	153	966	1,119	1,233
401	76	999	1,075	1,756
359	0	999	999	2,772

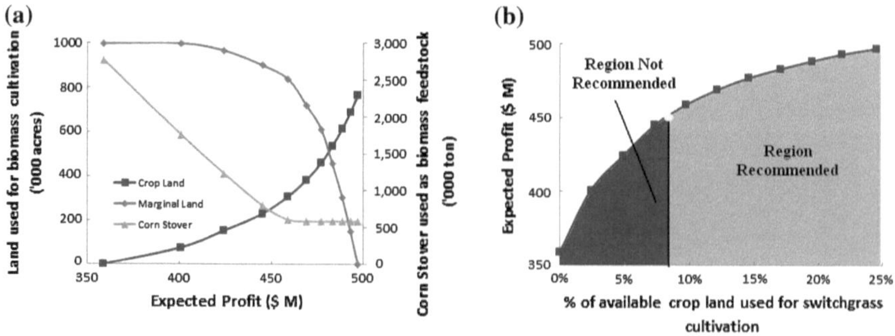

Fig. 5 **a** Profit versus cultivated land and crop residue usage (*base-case*). **b** Economic performance versus crop land usage (*base-case*)

yield cultivated on crop land; (3) limit on crop land availability for switchgrass cultivation; and (4) weightage of w_1 "weight for marginal land usage" and w_2 "weight for crop land usage" on the land usage for switchgrass cultivation.

In the "base-case", crop residue price is \$85/ton, the switchgrass yield ratio (using crop land for cultivation) is 2, the limit on crop land availability (for switchgrass cultivation) is 25 %, and the value of w_1 and w_2 is 0 and 1 respectively.

7.3.1 Impact of Crop Residue Purchase Price

Crop residue purchase price largely determines the amount of crop residue to be procured (i.e. second-stage decision) as the secondary source of biomass feedstock. Crop residue purchase price is varied from \$35/ton to \$95/ton to analyze the impact on the expected profit and the amount of crop land to be used for switchgrass cultivation (i.e. first-stage decision). Figure 6a indicates that the crop residue purchase price negatively impacts on the expected profit in almost a linear manner. As the crop residue purchase price increases, the amount of crop land (used for switchgrass cultivation) also increases. The biggest jump in crop land usage occurs when the crop residue purchase price is between \$45/ton and \$55/ton. Figure 6b indicates that as the crop residue purchase price increases, the amount of crop residue (used as biomass feedstock) decreases. The biggest drop in crop residue usage occurs when the crop residue price increases from \$45/ton to \$55/ton.

7.3.2 Impact of Switchgrass Yield Cultivated on Crop Land

The switchgrass yield cultivated on crop land largely determines the allocation of agricultural land for switchgrass cultivation. The value of A is varied from 1 to 3 to analyze the impact on land used for switchgrass cultivation (i.e. first-stage

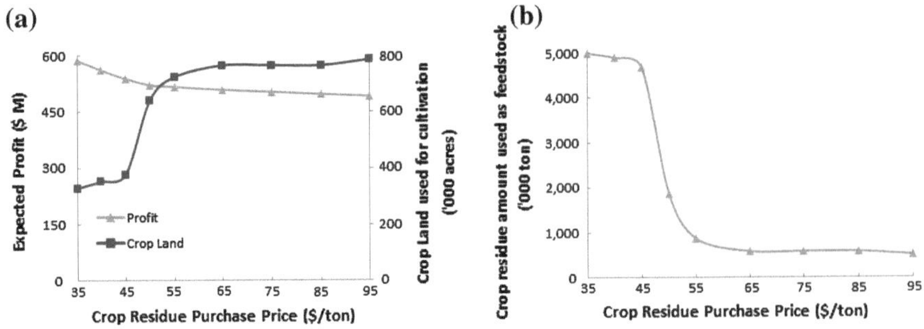

Fig. 6 **a** Impact of crop residue price on profit and land usage. **b** Impact of crop residue price on crop residue usage

decision), expected profit, and average amount of crop residue used as biomass feedstock (i.e. second-stage decision).

The results are presented in Figs. 7 and 8 and show that when $A = 1$ the maximum amount of marginal land and crop residue is used. As the value of A increases the amount of crop land used increases while the amount of marginal land and crop residue used decreases. The maximum amount of crop land is used when $2.00 < A < 2.25$, as A exceeds 2.25 the amount of crop land used begins to decrease as the required amount of switchgrass is obtainable from progressively smaller amount of crop land. Marginal land and crop residue usage is negligible when $A > 2$ and $A \gg 3$ respectively. Figure 8 also indicates that the value of A positively impacts on the expected profit in almost a linear manner.

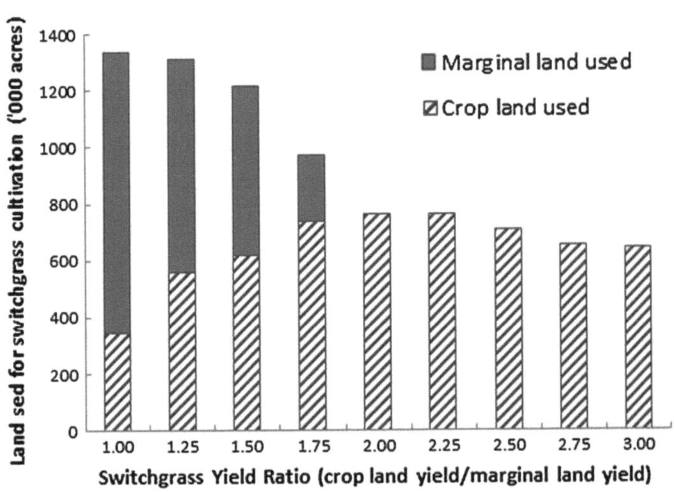

Fig. 7 Economic performance versus weighted land usage

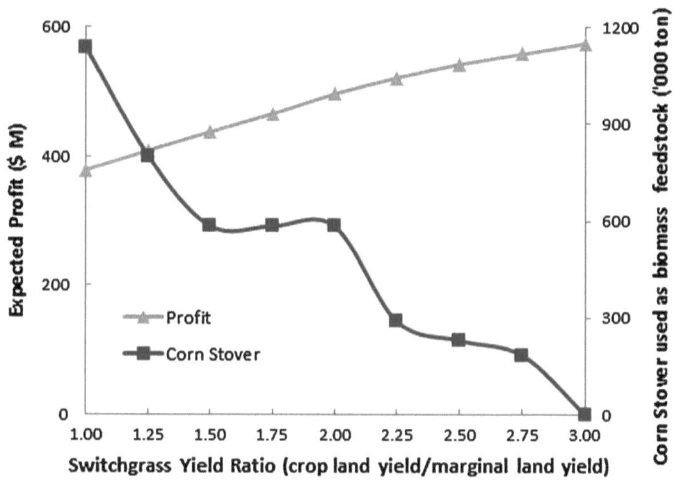

Fig. 8 Economic performance versus crop land usage

7.3.3 Impact of Limit on Crop Land Availability for Switchgrass Cultivation

The limit on crop land availability largely determines the allocation of agricultural land for switchgrass cultivation. The limit on crop land availability (for switchgrass cultivation) is varied from 0 % (i.e. crop land not available for switchgrass cultivation) to 25 % (i.e. maximum allowable use of crop land by the USDA for biomass cultivation) to analyze the impact on amount of crop and marginal land to be used for switchgrass cultivation (i.e. first-stage decision), expected profit, and average amount of crop residue used as biomass feedstock (i.e. second-stage decision). The results are presented in Figs. 9 and 10 and show that when crop land availability is 0 % the maximum amount of marginal land (for switchgrass cultivation) and crop residue (as biomass feedstock) is used. As the crop land availability limit increases the amount of crop land used increases while the amount of marginal land and crop residue used decreases. Marginal land and crop residue usage is lowest when crop land availability is 25 %. Figure 10 also indicates that crop land availability positively impacts on the expected profit in almost a linear manner.

7.3.4 Impact of Weightage of W_1 and W_2 for Switchgrass Cultivation

Sensitivity analysis is carried out to determine the impact of w_1 "weight for marginal land usage" and w_2 "weight for crop land usage" on the land usage for switchgrass cultivation. The value of w_2 is varied from 1.0 to 0.6 while the value of w_2 is varied from 0.4 to 0. It is ensured that $w_1 + w_2$ is always equal to 1.

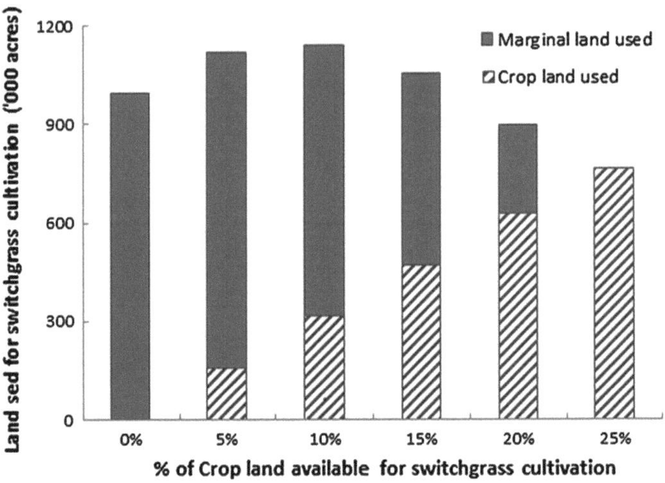

Fig. 9 Economic performance versus weighted land usage

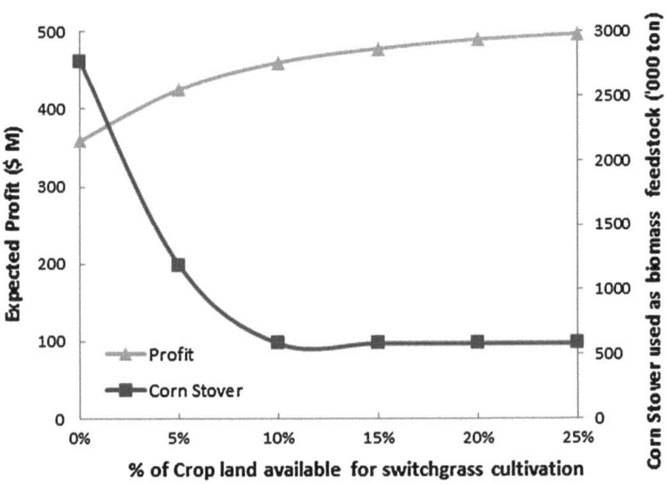

Fig. 10 Economic performance versus crop land usage

The results are presented in Figs. 11 and 12 and show that for a given value of crop land allocated for switchgrass cultivation, the expected profit rises as w_2 increases from 0.6 to 1 (and w_1 decreases from 0.4 to 0). This is due to the fact that when $w_1 = 0$ (i.e. no restriction on usage of marginal land for switchgrass cultivation), use of crop land can be avoided altogether and the required biomass can be obtained from switchgrass cultivated on marginal lands and procured from the available crop residues.

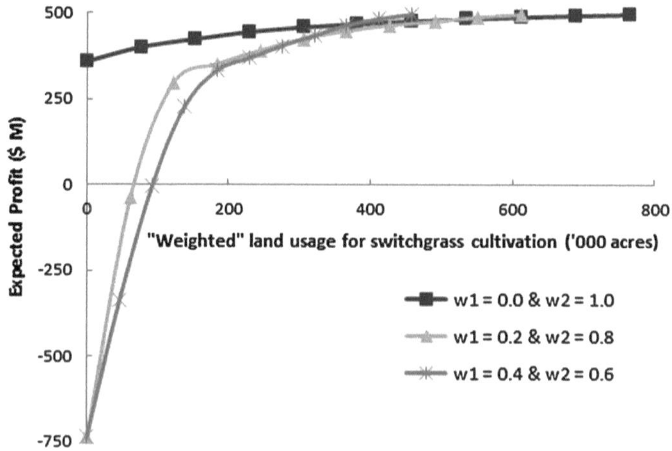

Fig. 11 Economic performance versus weighted land usage

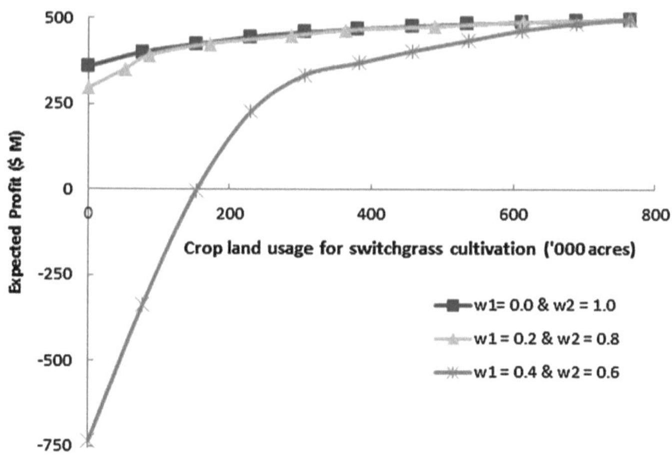

Fig. 12 Economic performance versus crop land usage

8 Conclusions and Future Works

This work studies a novel multi-feedstock dual-objective LBSC under multiple uncertainties in biomass supply, bioethanol demand and price. The unique contribution of this work is analyzing the new concept of land usage (especially crop land) for biomass cultivation. A SDOMILP model is proposed to determine the Pareto optimal sets of the strategic logistic decisions for maximizing the expected profit while at the same time minimizing the land usage (especially crop land) for switchgrass cultivation. The first-stage decisions include both continuous and integer decision variables.

The main computational burden of the stochastic model is due to the first-stage integer variables. In order to solve the proposed SDOMILP model with greater accuracy and within reasonable time, this work utilizes a sampling based approach (i.e. SAA method) to make the optimization problem computationally tractable. The *e*–constraint method is used for conducting trade-off analysis among different performance criteria in order to obtain a set of Pareto optimal solutions.

A case study based in Alabama demonstrates the effectiveness of the proposed stochastic model by determining the strategic requirements (i.e. first-stage decisions) and operational requirements (i.e. second stage decisions) for bioethanol production. Results show that in a stochastic environment it is cost effective to meet up to 20 % of Alabama's annual demand of gasoline energy equivalent requirement from locally produced bioethanol by using switchgrass (cultivated on both crop land and marginal land) as the primary source of biomass feedstock. Crop residue is used as secondary sources of lignocellulosic feedstock. Result also shows that there is no single solution which simultaneously optimizes both performance criteria as the objective of minimization of crop land usage is contradictory to the objective of maximizing financial profit (and vice versa). Profit maximization (i.e. economic performance) favors switchgrass cultivation on crop land while minimization of weighted land-usage favors switchgrass cultivation on marginal land (under the base-case).

Sensitivity analyses are conducted to evaluate the impact of different levels of the following four key deterministic parameters: crop residue purchase price; switchgrass yield cultivated on crop land; limit on crop land availability for switchgrass cultivation; and weightage of w_1 "weight for marginal land usage" and w_2 "weight for crop land usage" on the land usage for switchgrass cultivation. The results show that: (1) the crop residue purchase price negatively impacts on the expected profit in almost a linear manner. As the crop residue purchase price increases, the amount of crop land (used for switchgrass cultivation) also increases; (2) when switchgrass yield cultivated on crop land is at the lowest level the maximum amount of marginal land and crop residue is used. As the switchgrass yield cultivated on crop land increases the amount of crop land used increases while the amount of marginal land and crop residue used decreases. The maximum amount of crop land is used when switchgrass yield cultivated on crop land is at the median level. Marginal land and crop residue usage is negligible when switchgrass yield cultivated on crop land exceeds the median level; (3) when crop land availability is 0 % the maximum amount of marginal land (for switchgrass cultivation) and crop residue (as biomass feedstock) are used. As the crop land availability limit increases the amount of crop land used increases while the amount of marginal land and crop residue used decreases. Marginal land and crop residue usage is lowest when crop land availability is 25 %; and (4) for a given value of crop land allocated for switchgrass cultivation, the expected profit falls as the value of w_1 increases and the value of w_2 decreases. When there are no restriction on usage of marginal land for switchgrass cultivation, use of crop land can be avoided altogether and the required biomass can be obtained from switchgrass cultivated on marginal lands and procured from the available crop residues (including corn stover and wheat straw).

This work demonstrates the optimization of the financial and land-use performance for a limited geographic region. Future work can include a larger geographic area that has uncorrelated uncertainties and a wider land portfolio. Perspective of different decision makers can also be incorporated in the decision model. More efficient stochastic solution techniques will also need to be developed to cope with the much larger scale optimization problem.

Appendix A: Nomenclature

Indices	
e	Bioethanol demand zones (e = 1, …, E)
i	Lignocellulosic biomass supply zones (i = 1, …, I)
m	Lignocellulosic biomass feedstocks (m = 1, …, M)
r	Biorefinery locations (r = 1, …, R)
ω	Stochastic scenarios (ω = 1, …, N)
First stage decision variables	
CP_i	Crop land used for switchgrass cultivation at supply zone i (acres)
MG_i	Marginal land used for switchgrass cultivation at supply zone i (acres)
K_r	Biomass processing capacity of biorefinery at location r (tons/year)
Y_r	{1, if biorefinery setup in location r; Else 0}
Second stage decision variables	
$F_{mir}(\omega)$	Amount of biomass m sent from zone i to biorefinery r during scenario ω (tons)
$X1_i(\omega)$	Crop land in supply zone i harvested for switchgrass (acres)
$X2_i(\omega)$	Marginal land in supply zone i harvested for switchgrass (acres)
$L_r(\omega)$	Volume of unsubsidized ethanol sold from refinery r during scenario ω (gallons)
$O_e(\omega)$	Volume of unmet bioethanol requirement in demand zone e during scenario ω (gallons)
$P_{re}(\omega)$	Volume of subsidized ethanol from refinery r sent to zone e during scenario ω (gallons)
$S_r(\omega)$	Amount of bioelectricity produced by biorefinery r during scenario ω (MWh)
$Z_r(\omega)$	Volume of ethanol produced by biorefinery r during scenario ω (gallons)
Deterministic parameters	
$\Theta1_i$	Switchgrass cultivation cost parameter in supply zone i using crop land ($/acre)
$\Theta2_i$	Switchgrass cultivation cost parameter in supply zone i using marginal land ($/acre)
A	Switchgrass yield ratio using crop land for cultivation
$B1_i$	Crop land area available for switchgrass cultivation in biomass supply zone i (acres)
$B2_i$	Marginal land area available for switchgrass cultivation in biomass supply zone i (acres)

(continued)

(continued)

υ_i	Average switchgrass yield in biomass supply zone *i* using marginal land (tons/acre)
D_{ir}	Distance between biomass supply zone *i* and biorefinery *r* (mile)
D_{re}	Distance between biorefinery *r* and bioethanol demand zone *e* (mile)
G_r	Annualized fixed cost of biorefinery at location *r* ($)
H_r	Variable cost parameter of biorefinery at location *r* ($/ton)
lc_i	Switchgrass harvest cost parameter in supply zone *i* ($/acre)
pp_i	Switchgrass densification cost parameter in supply zone *i* ($/ton)
U_r	Ethanol production cost parameter of biorefinery at location *r* ($/gallon)
ζ_{mi}	Amount of biomass type $m \neq 1$ available in supply zone *i* (tons)
η_{mir}	Transport cost parameter of biomass *m* from supply zone *i* to biorefinery *r* ($/ton x mile)
ι_r	Average sale price of unsubsidized bioethanol at location *r* ($/gallon)
κ_m	Bioethanol yield parameter for biomass type *m* (gallons/ton)
λ_{mi}	Purchase price of biomass type $m \neq 1$ at supply zone *i* ($/ton)
μ_m	Electricity yield parameter for biomass type *m* (MWh/ton)
ν_e	Average bioethanol demand (energy equivalent to 20 % of gasoline requirement) in zone *e* (gallons)
ρ_{max}	Maximum amount of biomass that can be processed by biorefinery *r* (tons/year)
ρ_{min}	Minimum amount of biomass that must be processed by biorefinery *r* (tons/year)
τ_{re}	Tax credit for bioethanol production in location *r* for consumption in demand zone *e* ($/gallon)
φ_e	Penalty cost parameter for unmet bioethanol requirement at biofuel demand zone *e* ($/gallon)
χ_r	Sale price of electricity at location *r* ($/MWh)
ψ_{re}	Transport cost parameter of bioethanol from biorefinery *r* to biofuel demand zone *e* ($/gallon x mile)
w_1	Weight for marginal land usage
w_2	Weight for crop land usage
Stochastic parameters	
$o(\omega)$	Supply level of switchgrass (cultivated on marginal and/or crop land) during scenario ω
$\pi(\omega)$	Demand level of bioethanol during scenario ω
$\sigma(\omega)$	Price level of bioenergy during scenario ω

Appendix B: Input Parameters

See Tables B1 and B2

Table B1 Values of stochastic parameters (Osmani and Zhang 2013)

Input parameter	Value
Switchgrass supply level	$o(\omega)$ ~ N(1, 0.0342) and truncated on the interval (0.45, 1.55)
Bioethanol demand level	$\pi(\omega)$ ~ N(1, 0.0086) and truncated on the interval (0.72, 1.28)
Energy price level	$\sigma(\omega)$ ~ N(1, 0.0193) and truncated on the interval (0.58, 1.42)

Table B2 Values of key deterministic parameters (Zhang et al. 2012)

Input parameter	Value
Switchgrass yield ratio using crop land for cultivation	$A = 2$
Annualized fixed cost of biorefinery at location r ($ M)	$G_r = 21.3$ (for all r)
Variable cost of biorefinery at location r ($/ton)	$H_r = 46$ (for all r)
Switchgrass harvest cost parameter in supply zone i ($/acre)	$lc_i = 9.2$ (for all i)
Switchgrass densification cost parameter in supply zone i ($/ton)	$pp_i = 22$ (for all i)
Ethanol production cost of biorefinery at location r ($/gallon)	$U_r = 0.907$ (for all r)
Transport cost of biomass m from supply zone i to biorefinery r ($/ton x mile)	$\eta_{mir} = 0.21$ (for all m, i, r)
Sale price of unsubsidized bioethanol at location r ($/gallon)	$\iota_r = 2$ (for all r)
Bioethanol yield for biomass type m from (gallons/ton)	$\kappa_m = 90$ (for all m)
Purchase price of biomass type $m \neq 1$ at supply zone i ($/ton)	$\lambda_{2i} = 85$ (for all i)
Electricity generation from biomass type m (MWh/ton)	$\mu_m = 0.2$ (for all m)
Maximum biomass amount that can be processed by refinery r (MTPY)	$\rho^{max} = 2$
Minimum biomass amount that can be processed by refinery r (MTPY)	$\rho^{min} = 1$
Penalty cost for unmet bioethanol requirement at biofuel demand zone e ($/gallon)	$\varphi_e = 4$ (for all e)
Transport cost of bioethanol from refinery r to biofuel demand zone e ($/gallon x mile)	$\psi_{re} = 0.00017$ (for all r, e)
Sale price of renewable electricity generated at location r ($/MWh)	$\chi_r = 88.1$ (for all r)
Weight for marginal land usage	$w_1 = 0$
Weight for crop land usage	$w_2 = 1$
Tax credit for bioethanol production in location r for satisfying demand in zone e ($/gallon)	$\tau_{re} = 0.5$ (for all r, e)

References

Abdallah T, Farhat A, Diabat A, Kennedy S (2012) Green supply chains with carbon trading and environmental sourcing: formulation and life cycle assessment. Appl Math Model 36:4271–4285

An H, Wilhelm WE, Searcy SW (2011) A mathematical model to design a lignocellulosic biofuel supply chain system with a case study based on a region in Central Texas. Bioresour Technol 102:7860–7870

Awudu I, Zhang J (2012) Stochastic production planning for a biofuel supply chain under demand and price uncertainties. Appl Energy 103:189–196

Carriquiry MA, Du X, Timilsina GR (2011) Second generation biofuels: economics and policies. Energy Policy 39:4222–4234
Chen CW, Fan Y (2012) Bioethanol supply chain system planning under supply and demand uncertainties. Transp Res Part E Logistics Transp Rev 48:150–164
Chen W, Wiecek MM, Zhang J (1999) Quality utility—a compromise programming approach to robust design. J Mech Des 121:179–187
Charles MB, Ryan R, Ryan N, Oloruntoba R (2007) Public policy and biofuels: the way forward? Energy Policy 35:5737–5746
Cherubini F, Ulgiati S (2010) Crop residues as raw materials for biorefinery systems–A LCA case study. Appl Energy 87:47–57
Cucek L, Varbanov PS, Klemeš JJ, Kravanja Z (2012) Total footprints-based multi-criteria optimisation of regional biomass energy supply chains. Energy 44:135–145
Dal-Mas M, Giarola S, Zamboni A, Bezzo F (2011) Strategic design and investment capacity planning of the ethanol supply chain under price uncertainty. Biomass Bioenergy 35:2059–2071
Eksioglu SD, Acharya A, Leightley LE, Arora S (2009) Analyzing the design and management of biomass-to-biorefinery supply chain. Comput Ind Eng 57:1342–1352
Escudero LF, Garín MA, Merino M, Pérez G (2010) An exact algorithm for solving large-scale two-stage stochastic mixed-integer problems: some theoretical and experimental aspects. Eur J Oper Res 204:105–116
Frombo F, Minciardi R, Robba M, Rosso F, Sacile R (2009) Planning woody biomass logistics for energy production: a strategic decision model. Biomass Bioenergy 33:372–383
Gebreslassie BH, Yao Y, You F (2012) Design under uncertainty of hydrocarbon biorefinery supply chains: multiobjective stochastic programming models, decomposition algorithm, and a comparison between CVaR and downside risk. AIChE J 58:2155–2179
Giarola S, Zamboni A, Bezzo F (2011) Spatially explicit multi-objective optimisation for design and planning of hybrid first and second generation biorefineries. Comput Chem Eng 35:1782–1797
Huang Y, Chen CW, Fan Y (2010) Multistage optimization of the supply chains of biofuels. Transp Res Part E Logistics TranspRev 46:820–830
Kim J, Realff MJ, Lee JH (2011) Optimal design and global sensitivity analysis of biomass supply chain networks for biofuels under uncertainty. Comput Chem Eng 35:1738–1751
Kleywegt AJ, Shapiro A, Homem-de-Mello T (2002) The sample average approximation method for stochastic discrete optimization. SIAM J Optim 12:479–502
Kocoloski M, Michael Griffin W, Scott Matthews H (2011) Estimating national costs, benefits, and potential for cellulosic ethanol production from forest thinnings. Biomass Bioenergy 35:2133–2142
Kostin AM, Guillén-Gosálbez G, Mele FD, Bagajewicz MJ, Jiménez L (2012) Design and planning of infrastructures for bioethanol and sugar production under demand uncertainty. Chem Eng Res Des 90:359–376
Lam HL, Klemes JJ, Kravanja Z (2011) Model-size reduction techniques for large-scale biomass production and supply networks. Energy 36:4599–4608
Leduc S, Starfelt F, Dotzauer E, Kindermann G, McCallum I, Obersteiner M, Lundgren J (2010) Optimal location of lignocellulosic ethanol refineries with polygeneration in Sweden. Energy 35:2709–2716
Lee DK, Boe A (2005) Biomass production of switchgrass in central South Dakota. Crop Sci 45 (2583):2590
Mabee WE, McFarlane PN, Saddler JN (2011) Biomass availability for lignocellulosic ethanol production. Biomass Bioenergy 35:4519–4529
Osmani A, Zhang J (2013) Stochastic optimization of a multi-feedstock lignocellulosic-based bioethanol supply chain under multiple uncertainties. Energy 59:157–172
Papapostolou C, Kondili E, Kaldellis JK (2011) Development and implementation of an optimisation model for biofuels supply chain. Energy 36:6019–6026

Parker N, Tittmann P, Hart Q, Nelson R, Skog K, Schmidt A, et al (2010) Development of a biorefinery optimized biofuel supply curve for the Western United States. Biomass Bioenergy 34:1597–1607

Schnepf R. (2011). Renewable Fuel Standard (RFS): overview and issues. DIANE Publishing, Darby

Sokhansanj S, Mani S, Turhollow A, Kumar A, Bransby D, Lynd L, Laser M (2009) Large-scale production, harvest and logistics of switchgrass (Panicum virgatum L.)–current technology and envisioning a mature technology. Biofuels Bioprod Biorefin 3:124–141

Van Dyken S, Bakken BH, Skjelbred HI (2010) Linear mixed-integer models for biomass supply chains with transport, storage and processing. Energy 35:1338–1350

Wu J, Sperow M, Wang J (2010) Economic feasibility of a woody biomass-based ethanol plant in Central Appalachia. J Agric Resour Econ 35:522–544

You F, Tao L, Graziano DJ, Snyder SW (2012) Optimal design of sustainable cellulosic biofuel supply chains: multiobjective optimization coupled with life cycle assessment and input–output analysis. AIChE J 58:1157–1180

Zamboni A, Bezzo F, Shah N (2009) Spatially explicit static model for the strategic design of future bioethanol production systems. 2 Multi-objective environmental optimization. Energy & Fuels 23:5134–5143

Zhang J, Osmani A, Awudu I, Gonela V (2012) An integrated optimization model for switchgrass-based bioethanol supply chain. Appl Energy 102:1205–1217

Biofuel Supply Chain Design and the Impacts on Transportation Systems and Infrastructure

Yun Bai and Yanfeng Ouyang

Abstract The biofuel production and distribution system is supported by a number of infrastructure subsystems, including agriculture, transportation, water supply, etc., that are interdependent on one another. Transportation of the bulky, low energy density biomass feedstock (and bioethanol) incurs one of the major operational costs in the biofuel supply chain. The large volume of shipping trucks imposes additional pressure on the already congested and aging transportation infrastructure, causing traffic congestion and pavement damage especially in local areas with biorefineries. The resulting transportation impacts and environmental issues may raise community resistance, and could in turn influence refinery location choice and supply chain efficiency. Therefore, the planning of biorefinery locations and biofuel logistics should be made cautiously with a long term objective of establishing a sustainable bioenergy production and distribution system. This chapter extensively discusses the interdependencies between biofuel supply chain and its supporting transportation infrastructure. We will review the methodologies that quantify the bidirectional impacts and incorporate the transportation externalities into integrated system optimization. Insights are drawn upon the studies of sustainable system design.

Corn starch based ethanol has become a significant energy source in the United States (Brown et al. 2007), while new technologies are being developed for new generations of cellulose-based ethanol. The 2007 Renewable Fuel Standard requires 36 billion gallon of biofuel to be blended into gasoline at existing blending facilities by 2022 (US EPA, 2007). In response to this mandate, many new biofuel refineries (particularly the cellulose-based ones) will be needed in the coming decades. Development of a reliable, efficient and sustainable biofuel supply chain is essential for the national economy, environment and society.

Y. Bai (✉) · Y. Ouyang
Department of Civil and Environmental Engineering, University of Illinois at Urbana-Champaign, Urbana-Champaign, IL, USA
e-mail: yunbai1@illinois.edu

Y. Ouyang
e-mail: yfouyang@illinois.edu

S.D. Eksioglu et al. (eds.), *Handbook of Bioenergy*,
Energy Systems, DOI 10.1007/978-3-319-20092-7_10

The biofuel production and distribution system is supported by a number of infrastructure subsystems, including transportation, agriculture, water supply, etc. It is such a complex system that biofuel supply chain and other critical infrastructure subsystems are interdependent on one another. Due to the low energy density of feedstock biomass (especially cellulosic biomasses such as miscanthus grass, switch grass, and corn stover), transportation of the bulky feedstock (and bioethanol) incurs one of the major operational costs in the biofuel supply chain, that imposes additional pressure on the already congested and aging transportation infrastructure. The large number of heavy trucks to ship biomass and biofuel has been concerned for causing traffic congestion and pavement damage in local areas with biorefineries. The resulting transportation impacts (e.g., excessive transportation cost for both the public and the biofuel industry) and environmental issues may raise community resistance. On the other hand, the social impacts in turn influences refinery location decisions and supply chain efficiency. Therefore, the planning of biorefinery locations and biofuel logistics should be made cautiously with a long term objective of establishing a sustainable bioenergy production and distribution system.

This chapter extensively discusses the interdependencies between biofuel supply chain and its supporting transportation infrastructure. We will review the methodologies that quantify the bidirectional impacts and incorporate the transportation externalities into integrated system optimization. Insights are drawn upon the studies of sustainable system design.

1 Interdependencies Between Biofuel Supply Chain and Transportation Subsystems

This section discusses the interdependencies between the biofuel supply chain and the transportation subsystem in terms of highway traffic and pavement. We will also show that in order to minimize social costs, it is essential to account for these potential negative impacts in system design and planning for long term sustainability of both the biofuel industry and the society.

1.1 Facility Location, Biofuel Shipment Routing and Traffic Congestion

The biofuel supply chain design has several levels of decisions, including strategic (e.g., location and size of refineries), tactical (e.g., feedstock procurement contract), and operational (e.g., routing of transportation) decisions. Biorefinery location is one of the most critical strategic decisions. In general, a large capital investment is required to build a biorefinery plant. For example, a corn-based biorefinery with a 100 MGY (million gallons per year) capacity costs roughly $200 million for construction (Swenson 2008). As large biorefineries become more favorable, the investment in refinery construction constitutes a major portion of the total supply chain cost. More importantly, the location of refineries directly affects the operations of a biofuel supply chain by spatially defining the destination of biomass feedstock and the origin of ethanol product flows.

While a refinery investment incurs one-time fixed cost, the transportation of bulky biomass materials (and ethanol) makes up one major operational cost, because of the low energy density of biomass, especially cellulosic biomasses (Jacobson et al. 2009). To fulfill the ambitious ethanol production mandate, large amount of materials and products need to be transported regularly, inducing substantial truck traffic to ship sufficient low-energy-density biomass to supply daily production and customer demand. It is thus concerned to worsen the traffic congestion in the U.S. highway system which is already burdened with dramatically increasing freight flows (Vedenov et al. 2010). Therefore, development of biofuel industry may raise significant social concerns and objections from the public due to the potential negative social and environmental impacts of congestions. More importantly, severe congestion could also greatly increase the transportation cost for the movement of large volume of biomass and biofuel by heavy trucks.

Strategic biofuel supply chain design and planning should endogenously and simultaneously account for refinery location and biomass/biofuel transportation decisions. On one hand, the construction of biorefineries directly induces or diverts daily traffic demand and alters the congestion pattern (and hence transportation costs) in the network. On the other hand, the decision on refinery locations directly depends on the spatial distribution of biomass supply, ethanol consumption, and the associated shipment costs. The congestion caused by biomass and ethanol shipment may result in transportation cost increase and community resistance, which in turn are likely to influence refinery location decisions. Therefore, without systematic and strategic supply chain planning, it would add to the severe congestion in the part of transportation network where there is already heavy freight flow. Separating the decisions of biorefinery location and shipment routing and omitting the road congestion impact (especially in areas with high existing traffic) may cause not only unnecessary high transportation cost, but also negative socio-economic impacts.

1.2 Pavement Damage and the Impact on Road Users

As discussed in Sect. 1.1, industrial land-use development and expansion of biofuel production facilities, especially those in congested urban/suburban areas, often results in increased traffic (e.g., heavy trucks) for shipping biomass feedstocks and biofuel products in the neighboring transportation network. Many existing highway and local roads are reaching their capacities to accommodate the existing traffic demand. Besides increasing network congestion and causing delay to the existing traffic, biofuel supply chain development could also accelerate the deterioration of pavement and undermine ride quality.

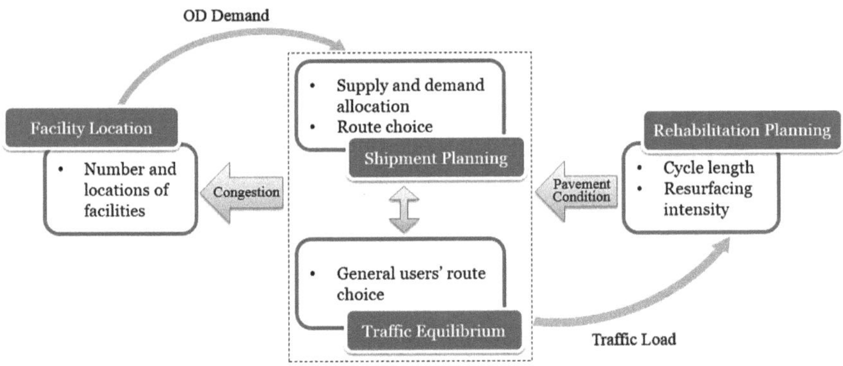

Fig. 1 Interactions among biofuel facility location, biomass/biofuel freight shipment, other road users and pavement rehabilitation (Hajibabai et al. 2014)

Heavy trucks in traffic flow are the major cause of highway pavement damages, as distresses from heavy truck loads are generally higher than those from passenger cars by orders of magnitude (Huang 1993). A typical fully loaded 5-axle tractor semi-trailer is equal to approximately 10,000 passenger cars in terms of pavement damage. The rapid loss of pavement serviceability (e.g., surface roughness) often requires frequent rehabilitation activities and incurs high agency costs. For instance, Iowa's renewable energy industries have brought about significant impacts on the quality of its transportation infrastructure. It was reported that pavement repairs and maintenance costs in multiple Iowa counties had increased significantly during and after the construction of biofuel production plants (Gkritza 2010). The growth of the biofuel industry imposes pressure on both highway operations and pavement quality, and may lead to damaged pavement conditions, higher user and agency costs. It may further cause thousands or millions of the existing individual road users to switch routes. As such, beyond just the direct congestion caused by the new biofuel freight shipments as discussed in Sect. 1.1, there could be additional congestion from travelers' route changes due to highway maintenance and reduced speed from adverse pavement conditions.

In summary, these unintended consequences of biofuel development on transportation infrastructure not only increase social cost (e.g., due to traffic delay and highway maintenance), but also impede the efficiency of the biofuel supply chain itself. Hence, the design of biofuel supply system should holistically account for the major impacts of biofuel production and logistics on the existing highway infrastructure. Figure 1 in Hajibabai et al. (2014) illustrates the interdependencies among industry development, transportation infrastructure, and human activities. These important impacts have long been overlooked in most biofuel logistics literature, but could seriously hinders the sustainable development of the industry. While biofuel companies, pavement maintenance agencies, and public road users may be non-cooperative decision makers and stakeholders, proper mechanisms could be developed to provide the biofuel industry with incentives to evaluate potential impacts of freight-oriented land-use development plans toward social cost minimum.

1.3 Integration of Transportation Infrastructure Impacts in Biofuel Supply Chain Design—A Challenge

As mentioned in the previous Sect. 1.1, the biofuel supply chain design problem has multiple decision levels: number and location of refineries, size of refineries, supply and demand allocation, routing of biomass and ethanol transportation, and so on. As a key strategic network design problem, biorefinery location decision making has been modeled as a facility location problem, such as Kang et al. (2010), Bai et al. (2011), Bowling (2010) and Housh et al. (2014), etc. For example, Kang et al. (2010) presented a multi-year dynamic multi-product supply chain design model for production of ethanol and by-products. Besides the extensive research on strategic location models, there are also many studies on operational biofuel transportation and logistics. For example, Sokhansanj et al. (2006) proposed an integrated analysis on different stages in a biofuel supply chain including collection, storage, and shipment; while Tembo et al. (2003) and Graves and Willems (2005) developed multi-region, multi-period models to focus on technical alternatives for production, shipment, and processing. There is another school of literature estimating shipment costs under numerous factors (e.g., collection days, frequency, and capacity of facilities) and various possible shipment modes (Mapemba 2005; Tatineni and Demetsky 2005; Graves and Willems 2005; Searcy et al. 2007; Brown et al. 2007; Mahmudi and Flynn 2006; Gonzales et al. 2013). Most efforts on supply chain design consider distance as the measure of transportation cost (e.g., Graves and Willems 2005).

The biorefinery location problem is often addressed by discrete fixed-charge facility location models that are essentially NP-hard due to the complexity of integer decision variables (especially in large scale instances); see Drezner (1995) and Daskin (1995) for systematic reviews of classic discrete facility location models. On the other hand, traffic congestion is generally considered through nonlinear link travel time functions with respect to traffic flow. As such, the biomass and ethanol shipment routing under traffic congestion essentially can be modeled as a traffic assignment problem, which determines traffic flow pattern on a highway network to achieve certain optimal criteria (e.g., user equilibrium (UE) for decentralized system with non-cooperative drivers or system optimal (SO) for centralized systems with cooperative drivers). The traffic assignment problems alone can be solved efficiently by polynomial time algoirithms, such as the convex combination method (Frank and Wolfe 1956; Sheffi 1985), the disaggregated simplicial decomposition method (Larsson and Patriksson 1992), the gradient projection method (Jayakrishnan et al. 1994), and the origin-based assignment method (Bar-Gera 2002). However, integrating traffic congestion impact into biofuel supply chain design is not trivial. To solve the biorefinery location design and traffic assignment with nonlinear link travel time simultaneously, customized solution algorithms were developed (Bai et al. 2011).

Furthermore, incorporating pavement deterioration and rehabilitation costs and optimization attributes to one more layer of complexity to the supply chain design. Aside from refinery location and routing decisions, planning of pavement maintenance and rehabilitation activities has been studied for decades, relying on either applied optimal control models (Friesz and Fernandez 1979; Fernandez and Friesz 1981) or discrete optimization approaches such as mixed-integer programming (Ouyang and Madanat 2004) and dynamic programming (Ouyang 2007). Among them, exact analytical formulas are derived by Ouyang and Madanat (2006) for single pavement optimal rehabilitation timing and intensity, which provides a useful building block for more complex system design models.

Finally, the background travelers' individual route choices (i.e., non-biofuel traffic) considering congestion and pavement condition can be modeled in the form of traffic equilibrium (i.e., UE). The integrated supply chain design problem under UE falls into a category of transportation network design problem (NDP), which has a bi-level structure of hierarchical decision making: the upper level problem determines the biofuel manufacturer's the supply chain design (e.g., biorefinery location, biomass/biofuel shipment routing); the lower level problem determines equilibrium network flow via numerous travelers' individual route choices based on traffic and pavement conditions in response to the upper level decisions.

In summery, modeling the interdependencies among freight facility location, truck shipment routing, pavement deterioration and rehabilitation, and traffic equilibrium (as shown in Fig. 1) in an integrated system manner is a major challenge, both conceptually and computationally, although each of these problems/decisions has been extensively studied individually. Facility location problems are NP hard, while the traffic congestion and pavement rehabilitation models are highly non-linear. The bi-level NDP has non-convex nature, and if discrete location variable is included, it is a variant of the discretely constrained mathematical programs with equilibrium constraints (DC-MPEC). It is regarded as of the most difficult mathematical programs, and there is hardly an effective global optimal solution approach.

The remainder of this chapter reviews two mathematical models that integrate the transportation infrastructure impacts to address the challenging integrated problems. In Sect. 2.1, we will first briefly review the model in Bai et al. (2011) that explicitly incorporates shipment routing decisions and traffic congestion impact into the refinery facility location problem. It is a three echelon SO supply chain design model that minimizes the total system cost including the transportation costs (for biomass/biofuel shipment and the background travelers) and refinery investment cost. Section 2.2 reviews an extended model in Hajibabai et al. (2014) that simultaneously optimizes facility location, shipment routing under traffic equilibrium, and pavement rehabilitation in a bi-level biofuel supply chain network design framework. Key findings from the case studies are summarized in Sect. 3, followed by a discussion on managerial insights in Sect. 4.

2 Methodology Review

2.1 SO Formulation Under Traffic Congestion

This subsection introduces the three echelon SO supply chain design model under traffic congestion in our earlier work Bai et al. (2011). It is formulated into a mixed integer nonlinear program (MINLP) that integrates a fixed-charge facility location model and the traffic assignment model to incorporate the traffic congestion impact.

This model embodies a node-link network representation of the biofuel logistics system. The nodes include discrete sets of biomass supply regions (i.e., farms), ethanol demand regions (i.e., blending facilities), candidate biorefinery locations, as well as intermediate nodes (i.e., transportation intersections). The nodes are connected by a set of directed highway links. At each supply and demand node, quantities of biomass supplies and biofuel demands (that need to be shipped out of farms and into blending facilities) are estimated. Each candidate biorefinery location with a given production capacity has its site-specific fixed cost depending on social-economical factors such as land price and availability. The optimal biorefinery locations are to be selected as a subset of candidate nodes, represented by binary decision variables. Given the location of existing and candidate biorefinery locations, biomass and ethanol need to be shipped to or from refineries through the transportation network.

On each roadway link, the link flow consists of background traffic flow (i.e., non-biofuel traffic), biomass truck flow and ethanol truck flow. The background traffic flow is assumed fixed and independent of biomass or ethanol shipments. The total flow can not exceed the link capacity. This assumption may result in an overestimation of the congestion impact of biomass/ethanol shipments since travelers' route switching behavior is not captured. So this model mainly focus on biomass and bioethanol traffic routing only, and the relaxation of this assumption will be discussed in the second model in the next subsection. As a convention for traffic analysis, the biomass supplies and biofuel demands need to be converted to equivalent hourly passenger car flows. The readers are referred to HCM (2000) for details. Traffic congestion is modeled by the flow dependent link travel time. Empirical models such as the BPR function reveals the nonlinear relationship (e.g., quartic polynomial) between delay and total link flow (BPR 1970). This is one of the major differences with most of the existing biofuel supply chain models which in general assume constant link travel times.

The MINLP model optimizes refinery location and biomass and bioethanol shipment routing. It integrates a SO traffic assignment model with objective of minimizing the costs for building refineries, shipping biomass and ethanol, and travel cost for the background traffic. The constraints mainly define different types of flows, flow conservation at nodes, refinery capacity cap, as well as the relationship between flow and location variables. To solve this optimization model, a Lagrangian relaxation (LR) algorithm and a branch-and-bound solution approach are developed to reduce the optimality gap. Please see Bai et al. (2011) for more details of the model formulations.

2.2 Integrated Model with Pavement Infrastructure Rehabilitation and Traffic Equilibrium

Building on the first model considering congestion, Hajibabai et al. (2014) later extended to an integrated joint optimization model, that is readily applicable to the biofuel context. The enhancement includes: (1) network pavement deterioration and rehabilitation are incorporated; (2) background travelers are allowed to switch routes; (3) a more advanced bi-level decision making process is considered, with the biofuel supply chain planning in the upper level and traffic UE for public road users in the lower level.

In a similar highway network, each link (corresponding to a pavement facility) has a given traffic capacity, which carries both biofuel related freight flow and background flow. Here the passenger car and (non-biofuel) truck traffic in the background flow are differentiated since they have different impacts on pavement. The pavement performance and rehabilitation model in Ouyang and Madanat (2006) was adopted that describes the evolution of pavement roughness over time under any traffic load and the analytical optimal rehabilitation strategy for each individual highway link. It is important to note that the performance function in Ouyang and Madanat (2006) is actually based on empirical data from Paterson (1990), and thus is much simplified with only a single time variable such that the roughness increases exponentially over time between any two consecutive rehabilitation activities. To account for traffic impact, Hajibabai et al. (2014) adjusted it to add traffic load variables. The detailed effects of structural and functional distresses such as fatigue and environmental cracking, however, are not captured in this model. Assuming an infinite time horizon, the pavement rehabilitation decisions reduce to rehabilitation cycle length and the rehabilitation intensity (i.e., thickness of the overlay).

The life-cycle user and agency costs under optimal rehabilitation policies are then derived in closed form. The agency cost for performing a pavement rehabilitation activity is assumed to be proportional to rehabilitation intensity. In addition to delay cost under congestion, both freight vehicles and the public travelers also experience vehicle operating cost due to pavement roughness caused extra fuel consumption, vehicle repair and maintenance needs, tire wear and lubrication, and driver discomfort (Chatti and Zaabar 2012; Tan et al. 2012). Both congestion and pavement roughness directly influence the user cost, and consequently affect travelers' route choices. By assuming that travelers perceive time-average condition for making routing decision, the authors showed that the users' route choices would form a steady traffic flow pattern because in the long run their perceived travel cost is equivalent to the actual cost.

The integrated model can be used to optimize biorefinery location and biomass and bioethanol shipment routing, and endogenously consider its impacts on travel delay and pavement rehabilitation costs under UE background traffic. The objective is to minimize the total costs related to the supply chain (i.e., refinery construction investments and biomass and biofuel transportation cost), the existing roadway users (i.e., travel delay and vehicle operating cost), and the pavement infrastructures (i.e., agency rehabilitation cost). The freight traffic is assumed substitutable and centrally controlled in a SO fashion in the upper level, while the background traffic follows the UE principle. Compared to those in Bai et al. (2011), the major difference in terms of constraints is the bi-level hierarchy (industry constraints in the upper level and UE constraints for the background traffic in the lower level problem). As discussed in Sect. 1.3, the bi-level DC-MPEC is extremely challenging. The solution method developed in Hajibabai et al. (2014) reformulates it into an equivalent single-level MINLP based on Karush-Kuhn-Tucker (KKT) conditions. Non-convexity is resolved by approximating the user cost by a piece-wise linear function. Please see Hajibabai et al. (2014) for more details.

3 Illinois Case Study

In both studies, real scale problems are solved applying the above models to Illinois State. We will summarizes their numerical results and key findings.

First, a set of experiments were conducted in Bai et al. (2011) applying the SO model to an Illinois network with 98 nodes and 374 links (including interstate highway and major local arterials). The nodes are mostly county centroids and highway intersections, and nodal biomass supply and ethanol consumption are aggregated at county level resolution. 20 candidate biorefinery locations are selected considering the access to major transportation facilities, water availability and proximity to biomass supply regions. Figure 2a illustrates the Illinois transportation

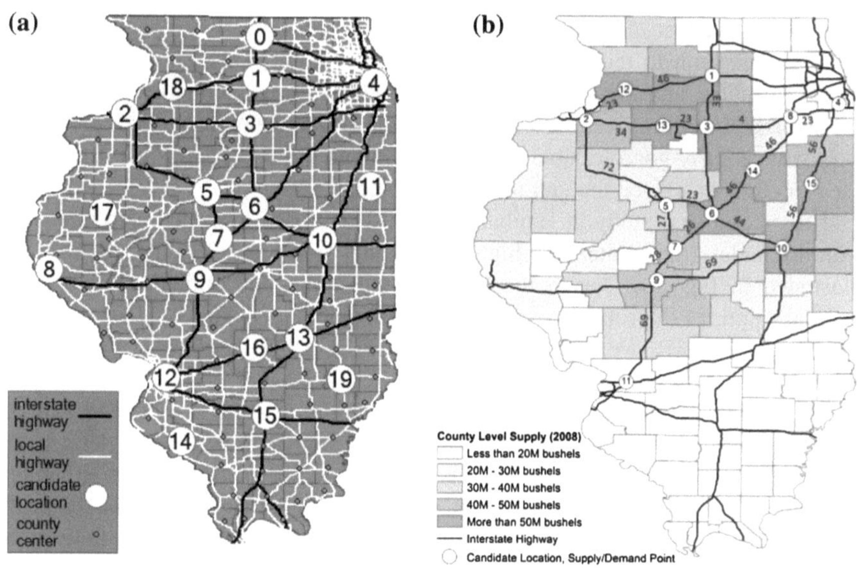

Fig. 2 Illinois network maps in Bai et al. (2011) and Hajibabai et al. (2014)

network in this paper. The details on data sources and assumptions will be omitted here, interested readers are referred to the paper for more information.

To study the impact of traffic congestion on the total system cost, the SO model considering congestion impact are compared with a benchmark scenario. The benchmark scenario (omitting the congestion effect) is solved using a regular fixed charge facility location model, where all biomass and bioethanol shipment traffic are simply assigned based on shortest-distance paths. The results revealed that congestion could lead to quite different design solutions. The benchmark design tends to build fewer refineries to save fixed cost, leading to much higher transportation cost for both the biofuel industry and the general public. While if congestion is considered in the decision making, more refinery locations will be selected to form a more dense and more scattered supply chain network with shorter shipping distance on average. Thus compared to the benchmark scenario, the SO model considering congestion is able to not only save transportation costs for the biofuel industry but also reduce total travel time for the background traffic. The congestion experienced by the shipment fleet and the public (i.e., background flow) are also different. Compared to the benchmark, considering congestion could save the industry by a much higher percent than the background travelers (in terms of transportation cost). The results also implies that the impact of biofuel traffic induced congestion on the general public is smaller than that on the biofuel industry itself.

In addition, the optimal system design and congestion impact is found to depend, to some extent, on the value of several important parameters, such as the time value, annual traffic growth rate and BPR parameter. A series of sensitivity analyses were further conducted to examine the impact of these parameters. The results show that when each of the time value (implying the weight on transportation cost), annual traffic growth rate (implying the background traffic demand), and BPR parameter (implying the effect of traffic volume on congestion) increases respectively, more refineries are built to reduce the average biomass/biofuel shipment distance, mitigate congestion, and thus save transportation cost. Computationally, the proposed LR and BB algorithms were shown effective in solving the model, where LR is able to provide a fairly good feasible solution with small optimality gaps and BB can be used to close or reduce the gap to near optimum.

The second case study in Hajibabai et al. (2014) used a much smaller network with 15 nodes (main interstate highway conjunctions) and 38 directed links, as illustrated in Fig. 2b. Due to the model complexity, it is difficult to find exact optimal solutions to problem instances larger than this size. The same set of data is used, and similarly we refer the readers to Hajibabai et al. (2014) for more information.

The optimal solutions to two scenarios are also compared, i.e., with and without considering pavement impacts. As expected, the joint optimization model considering pavement impacts is shown to reduce the total system cost and exhibit a more balanced and disperse flow pattern. Since the link cost function is likely to be convex, the results are consistent with the intuition that well-balanced flow saves cost for convex nonlinear networks. As such, the impacts on congestion and pavement deterioration are alleviated to some extent. It is also found that the high percentage cost reductions are from pavement related costs (i.e., agency cost and users vehicle operating cost). This is because most links have lower traffic volumes which reduces the average roughness level and extends the rehabilitation cycles. In short, the joint model balances the freight traffic assignment and optimizes the system pavement deterioration and rehabilitation. The effects are even more evident when traffic demand increases.

4 Summary and Managerial Insights

This chapter discusses current challenges and critical issues on the transportation infrastructure in bioenergy supply chain development. The literature on biofuel supply chain and biomass logistics is briefly reviewed. In particular, we analyze the impacts on highway traffic network and infrastructure. We also develop optimization models for an integrated sustainable biofuel supply chain design. The first model focuses on the traffic congestion impact of biomass and ethanol traffic and its incorporation into the traditional facility location problem. The second model extends the first model, accounting for the impact of biomass and ethanol traffic on the supporting transportation infrastructure. This integrated optimization model minimizes the total social cost jointly for the biofuel industry (refinery construction and biomass/biofuel transportation cost), the existing roadway users (public travel delay) and the pavement infrastructures (agency cost and vehicle operating cost).

Empirical case studies and a series of numerical analysis indicate that those impacts can be significant in light of infrastructure aging and freight demand growth. Expanding the highway capacity or using other modes of transportation are possible solutions. But they are neither immediately available nor practically easy (given the tremendous cost and land use restrictions). Without substantial investment on new infrastructure, integrated optimization of strategic supply chain design could facilitate better use of the existing infrastructure and help minimize such adverse consequences.

Currently, the biofuel industry gains profit from biofuel manufacturing but pays little directly for the pavement damage and extra congestion near its refineries. To lesson these negative externalities, regulations or incentives are needed for the industry to be willing to adopt sustainable supply chain design and management. For example, to more equitably re-assign these additional costs induced by the industry, tolls can be enforced in the network to influence drivers' route choice towards the optimal routings for shipping biomass and biofuel. It is also shown that biofuel companies' transportation cost can be lowered under the optimal supply chain design solutions. Biofuel companies could be incentivized to select optimal refinery location and shipment routes seeing the potential congestion-induced excessive transportation time and cost. In this sense, considering these externalities in biofuel supply chain design improves the supply chain efficiency.

We have also demonstrated that the social cost can be minimized if the externalities are included in the objective. The joint optimization models tend to result in well-balanced flow patterns, which have a two-fold effect on mitigating both traffic congestion and pavement damage. The alleviated congestion and pavement deterioration and its corresponding cost (i.e., delay, agency cost and users vehicle operating cost) accounts for a large portion of the reduction in overall cost. In the meantime, the savings are actually at the cost of building more refineries. There could be cases when transportation cost savings may not offset the additional fixed cost, so the biofuel industry ends up with losing profit. However, compared with the traditional design without considering the transportation externalities, the total cost reduction in optimal design shows the potential for Pareto improvement. In other words, these models could be used to make both the industry and the society better-off, if proper mechanisms or leverages (e.g., subsidies or taxes) are used. Some challenging but interesting topics for future research include how to design policies or market mechanisms to (1) appropriately reallocates benefits and costs among involved stakeholders, and (2) motives the biofuel industry to make sustainable decisions and to achieve the SO.

There are other important infrastructure and social-economic subsystems, which are essential for the sustainability of the biofuel industry, such as water supply for both irrigation and ethanol production, biomass crop production and land use, biomass and fuel markets of different substitute goods, and uncertainties from natural and social environments. The complex biofuel system needs to be carefully designed to ensure efficiency and economic viability, without undermining the community and future development. It is an extremely challenging problem to model the comprehensive biofuel system of subsystems involving complex interdependencies and different decision makers. Operations research methodologies will be valuable tools to provide scientific guidelines for policy makers, manufacturers and suppliers to embrace a prosperous and sustainable next-generation biofuel industry.

References

Bai Y, Hwang T, Kang S, Ouyang Y (2011) Biofuel refinery location and supply chain planning under traffic congestion. Transp Res Part B Methodol 45(1):162–175

Bar-Gera H (2002) Origin-based algorithm for the traffic assignment problem. Transport Sci 36(4):398–417

Bowling I (2010) Optimization of supply chain management and facility location selection for a biorefinery. Ph.D. thesis, Texas A&M University

BPR (1970) U.S. Bureau of Public Roads. Urban transportation planning. General information and introduction to system 360. Washington. D.C. (U.S. Department of Transportation)

Brown R, Orwig E, Nemeth J, Rocha C (2007) The economic potential for ethanol expansion in illinois

Chatti K, Zaabar I (2012) Estimating the effects of pavement condition on vehicle operating costs. NCHRP report 720. National Research Council. Washington, D.C. (Project 1-45)

Daskin M (1995) Network and discrete location: models, algorithms, and applications. Wiley, New York

Drezner Z (1995) Facility location: a survey of application and methods. Springer, Berlin

Fernandez J, Friesz T (1981) Influence of demand-quality interrelationships on optimal policies for stage construction of transportation facilities. Transp Sci 15(1):16–31
Frank M, Wolfe P (1956) An algorithm for quadratic programming. Naval Res Logist Q 3(1–2): 95–110
Friesz T, Fernandez J (1979) A model of optimal transport maintenance with demand responsiveness. Transp Res Part B Methodol 13(4):317–339
Gkritza K (2010) Iowas renewable energy and infrastructure impacts. Technical report. Center for Transportation Research and Education (CTRE), Iowa Highway Research Board (TR-593), and Iowa Department of Transportation (InTrans Project 08-334)
Gonzales D, Searcy E, Ekşioğlu S (2013) Cost analysis for high-volume and long-haul transportation of densified biomass feedstock. Transp Res Part A Policy Pract 49:48–61
Graves SC, Willems SP (2005) Optimizing the supply chain configuration for new products. Manage Sci 51(8):1165–1180
Hajibabai L, Bai Y, Ouyang Y (2014) Joint optimization of freight facility location and pavement infrastructure rehabilitation under network traffic equilibrium. Transp Res Part B (In press)
HCM (2000) Highway capacity manual (HCM). Transp Res Board
Housh M, Cai X, Ng T, McIsaac G, Ouyang Y, Khanna M, Sivapalan M, Jain A, Eckhoff S, Gasteyer S, Al-Qadi I, Bai Y, Yaeger M, Ma S, Song Y (2014) System of systems model for analysis of biofuel development. J Infrastruct Syst 0401–4050
Huang YH (1993) Pavement analysis and design, 2nd edn. Pearson, Prentice Hall, Upper Saddle River 07458
Jacobson J, Searcy E, Muth D, Wilkerson E, Sokhansanj S, Jenkins B, Tittman P, Hart Q, Nelson R (2009) Sustainable biomass supply systems. Techical report, Idaho National Laboratory
Jayakrishnan R, Tsai WK, Prashker JN, Rajadhyaksha S (1994) A faster path-based algorithm for traffic assignment. Transport Res Rec 1443:75–83
Kang S, Onal H, Ouyang Y, Scheffran J, Tursun D (2010) Optimizing the biofuels infrastructure: Transportation networks and biorefinery locations in illinois. In: Khanna M, Scheffran J, Zilberman D, Dinar A, Zilberman D (eds) Handbook of bioenergy economics and policy, vol. 33. Natural resource management and policy. Springer, New York, pp 151–173
Larsson T, Patriksson M (1992) Simplicial decomposition with disaggregated representation for the traffic assignment problem. Transport Sci 26(1):4–17
Mahmudi H, Flynn P (2006) Rail vs truck transport of biomass. Appl Biochem Biotechnol 129(1):88–103
Mapemba L (2005) Cost to deliver lignocellulosic biomass to a biorefinery. Ph.D. thesis, The Oklahoma State University. http://bioeconomyblog.blogspot.com/2007/06/dissertation-cost-to-deliver.html
Ouyang Y (2007) Pavement resurfacing planning on highway networks: a parametric policy iteration approach. ASCE J Infrastruct Syst 13(1):65–71
Ouyang Y, Madanat S (2004) Optimal scheduling of rehabilitation activities for multiple pavement facilities: exact and approximate solutions. Transp Res Part A 38(5):347–365
Ouyang Y, Madanat S (2006) An analytical solution for the finite-horizon pavement resurfacing planning problem. Transp Res Part B Methodol 40(9):767–778
Paterson WDO (1990) Quantifying the effectiveness of pavement maintenance and rehabilitation. In: Proceedings at the 6th Road Engineering Association of Asia and Australasia (REAAA) Conference, Kuala Lumpur, Malaysia
Searcy E, Flynn P, Ghafoori E, Kumar A (2007) The relative cost of biomass energy transport. Appl Biochem Biotechnol 137–140(1–12):639–652
Sheffi Y (1985) Urban transportation networks: equilibrium analysis with mathematical programming methods. Prentice Hall, Englewood
Sokhansanj S, Kumar A, Turhollow A (2006) Development and implementation of integrated biomass supply analysis and logistics model (ibsal). Biomass Bioenergy 30(10):838–847

Swenson D (2008) Ethanol's economic impact on rural communities. AgMRC Renew Energy Newsletter. http://www.agmrc.org/renewable_energy/renewable_energy/ethanols-economic-impact-on-rural-communities

Tan F, Thoresen T, Evans C (2012) Review of vehicle operating costs and road roughness: past, current, and future. In: 25th ARRB conference, Perth, Western Australia, Australia

Tatineni VC, Demetsky MJ (2005) Supply chain models for freight transportation planning. Technical report, Center for Transportation Studies at the University of Virginia (UVACTS-14-0-85)

Tembo G, Epplin F, Huhnke R (2003) Integrative investment appraisal of a lignocellulosic biomass-to-ethanol industry. J Agric Resour Econ 28(3):611–633

Vedenov D, Fuller S, Power G, Ahmedov Z, Vadali S, Burris M (2010) Bio-fuels energy policy and grain transportation flows: Implications for inland waterways and short sea shipping. Techical report, Research and Innovative Technology Administration, Department of Transportation

Biofuel Lifecycle Energy and Environmental Impacts: The Challenges of Co-product Allocation

Valerie M. Thomas, Dong Gu Choi and Dexin Luo

Abstract Calculations of energy, greenhouse gas emissions, and other environmental impacts from biofuel production often allocate impacts between biofuel and its co-products by calculating that co-products substitute for other products. We illustrate the issues of co-product allocation with a case study of corn-derived ethanol, and show that the choice of allocation procedure and parameters can significantly influence the results. This is an open issue in environmental lifecycle assessment methodology; there are research opportunities to determine co-product allocation values by using data that relate co-product utilization to land use or market changes.

1 Introduction

The energy and environmental impacts of biofuel production include impacts from growing and harvesting the feedstock, from producing the fertilizers, pesticides and herbicides, from transporting the feedstock to the processing facility, and from processing the feedstock into biofuel. Biofuel is produced in conjunction with

V.M. Thomas (✉)
School of Industrial and Systems Engineering, Georgia Institute of Technology, Atlanta, GA 30332, USA
e-mail: vthomas@isye.gatech.edu

V.M. Thomas
School of Public Policy, Georgia Institute of Technology, Atlanta, GA 30332, USA

D.G. Choi
Korea Institute of Energy Research, 152 Gajeong-ro, Yuseong-gu, Daejeon 305-343, Republic of Korea
e-mail: dgchoi@kier.re.kr

D. Luo
Linde AG, Seitnerstr. 70, 82049 Pullach, Germany
e-mail: luodexin@gmail.com

S.D. Eksioglu et al. (eds.), *Handbook of Bioenergy*,
Energy Systems, DOI 10.1007/978-3-319-20092-7_11

co-products; the energy and greenhouse gas implications of these co-products have received significant attention (Fargione et al. 2008; Searchinger et al. 2008; Shapouri et al. 1995, 2002; Wang et al. 1999; ISO 2006b). The land use effects of biofuel development have also received attention, because choice of land for biofuel crops has energy, greenhouse gas and other environmental and ecosystem implications, and because use of feedstocks for biofuel production can cause indirect land use change at other locations. As new processes for biofuel production are developed and as the net energy and greenhouse gas emissions of biofuel production become criteria for meeting policy goals, quantitative determination of the energy and greenhouse gas implications becomes increasingly important.

There are some biofuel production processes that produce only one product—ethanol production from blue-green algae, for example (Luo et al. 2010). For these single-product biofuel systems, optimization modeling can obviously proceed without taking co-products into account. For example, Luo et al. (2014) evaluate the optimal scale of ethanol production with the blue-green algae process. Also, in analyses focused on production costs, rather than on an environmental variable such as greenhouse gas emissions, the economic value of co-products can be addressed through standard approaches; the modeling of co-products can even be suppressed when the effects will be small and distract from the main analysis. For example Okwo and Thomas (2014) develop a game theoretic approach to characterize the land use change that may be induced by the development of cellulosic biofuel production; that analysis does not delve into co-product analysis.

However, in many biofuel systems there is more than one product, and environmental assessments will allocate between them. This chapter is aimed at optimization and simulation practitioners who will be modeling bioenergy systems. These practitioners may develop lifecycle assessments as part of the analysis, or they may use the results of lifecycle assessments as inputs to their models. In either case, it will be helpful for practitioners to understand the issues involved in co-product allocation: different methods of co-product allocation can strongly affect the results, and approaches have yet to be standardized and can be expected to change over time.

To demonstrate the consequences of different allocation choices, we draw on the existing literature on corn-derived ethanol as a case study. Production of ethanol from corn also produces co-products that can be used as animal feed supplements. There are two main processes for producing ethanol from corn: the dry mill process and the wet mill process. The co-product from the dry mill process is called dried distillers grains and solids (DDGS); the co-products from the wet mill process are corn gluten meal (CGM), corn gluten feed (CGF) and corn oil. As discussed in Fargione et al. (2008), co-products of ethanol have a positive economic value and are assumed to displace competing products that require energy to make. Increases in corn ethanol production will lead to more co-products that are assumed to displace products such as whole corn and soybean meal in animal feed, and the energy thereby saved is calculated as partly offsetting the energy required for ethanol production.

2 Formalism

Many processes produce more than one product. When a product and co-product are produced together, there is no physical basis for quantifying the energy or other inputs needed to make the product or co-product alone. Nevertheless, it is the physically undefined quantity, "energy to make biofuel", and the associated greenhouse gas emissions, that are key to public and political support for biofuel production. Determination of any of the quantities associated with the energy and environmental impacts of biofuel requires allocation of energy and impacts to co-products and to biofuel separately.

Even though the physical quantities are not well-defined, it is possible to write down an equation that expresses this concept of separate energy values for each co-product. In Eq. (1) α_i is the amount (e.g., kg) of co-product i that is produced for each volume (e.g., liter) of ethanol.

$$E_{ethanol} = e_{ethanol} + \Sigma_i \alpha_i e_i \tag{1}$$

We use the notation e to refer to energy allocation values and E to refer to measurable energy values. That is, $E_{ethanol}$ is the total energy input to making ethanol and all its co-products and is in principle measurable, whereas $e_{ethanol}$ is the energy to make a liter of ethanol alone and e_i is the energy to make a kilogram of co-product alone; these quantities might not be physically measured.

The standards for environmental lifecycle assessment are codified by the International Standards Organization (ISO). ISO 14044 (2006b) states that inputs and outputs should be allocated to the different co-products using process disaggregation, system expansion, or allocation on a basis such as mass or energy content. Process disaggregation means the separation of the modeling of two different processes; this is possible when the processes are distinct, for example when one facility contains two separate productions lines producing different productions. Process disaggregation is not possible for processes that inherently produce more than one product, for example the crushing of soybeans to make soy oil and soy meal. Allocation using mass or energy content or other basis means allocating the upstream environmental inventory proportional to the mass, energy, etc. For example, the products of a petroleum refinery, which include jet fuel, diesel fuel, gasoline, naphtha, asphalt and other products, might be allocated proportional to the energy content of each product, or according to the economic value of each product. While ISO 14044 does not advocate use of this type of proportional allocation, there are numerous respected lifecycle assessment practitioners who prefer and use this approach (Allen et al. 2009).

System expansion is the approach recommended in ISO 14044. In principle, system expansion means the expansion of the system boundary of the modeling and analysis to include all of the co-products. In practice this has often been implemented by what is called a displacement calculation, which means taking into account the effect of the co-products in displacing one or more other products in the economy. Generally, in a standard attributional lifecycle assessment, these calculations are made without consideration of resulting price and market changes, although these effects are included in what is called consequential lifecycle assessment. The worked example below illustrates a displacement calculation and the potential implications for evaluation of energy, greenhouse gas and other environmental metrics of biofuel production.

Specifically, in order to arrive at a quantitative value for $e_{ethanol}$, the displacement approach is to identify products for which the co-products can substitute, and to define the energy to produce these substitutes as the energy to produce the co-products. That is, the energy for each co-product *i* is defined as the energy to make a set of products *j* that can substitute for a unit of co-product *i*, as expressed in Eq. 2, with β_{ij} as the amount (kg) of each substitute product required to replace a unit of co-product *i*.

$$e_{ethanol} = E_{ethanol} - \Sigma_{i,j}\alpha_l\beta_{i,j}e_j \tag{2}$$

Equation (2) is written with the notation E_j, which indicates that the energy to produce the substitutes for the co-product *i* is directly measurable. However, as will be discussed below, often the energies to produce the substitute products are also not measurable quantities.

3 Illustrative Calculation

To illustrate the calculation of co-product energy allocation, we use published values for key parameters such as the amount of biomass needed to produce a liter of ethanol, the energy to convert corn to ethanol, and the amount of agricultural land needed to produce a given mass of corn (Table 1). The purpose here is to illustrate the methodology, which applies to production of other biofuels, other biofuel co-products, and other energy product systems.

Approximately 2.5 kg of corn are needed to produce a liter of ethanol; this requires energy inputs of approximately 13.1 MJ[1] or 14.6 MJ for the wet or dry mill processes, respectively (Shapouri et al. 2002). The wet mill process also produces about 0.64 kg of dry distillers grains and solids (DDGS); the dry mill process also produces about 0.12 kg of corn gluten meal (CGM), 0.51 kg of corn gluten feed (CGF), and 0.09 kg of corn oil (Arora et al. 2008).

$$\text{corn}_{2.5\,\text{kg}}(\text{wet mill}) \rightarrow \text{ethanol}_{1\,\text{liter}} + \text{DDGS}_{0.69\,\text{kg}} \tag{3}$$

$$\text{corn}_{2.5\,\text{kg}}(\text{dry mill}) \rightarrow \text{ethanol}_{1\,\text{liter}} + \text{CGM}_{12\,\text{kg}} + \text{CGF}_{51\,\text{kg}} + \text{cornoil}_{0.09\,\text{kg}} \tag{4}$$

[1] 1 MJ = 10^6 Joules.

Table 1 Ethanol co-products, the quantities of products they can replace, and the corresponding energy values

Production method	Co-product	Yield (kg/L ethanol)	Substitute product (per L ethanol)	Energy to make substitute (MJ/L ethanol)	
Dry mill	DDGS	0.64	0.69 kg Corn	1.37	4.39
			0.53 kg Soy meal	3.02	
Wet mill	CGM	0.12	0.18 kg Corn	0.36	1.77
			0.0027 kg Urea	0.07	
	CGF	0.51	0.51 kg Corn	1.01	
			0.0077 kg Urea	0.20	
	Corn oil	0.09	0.095 kg Soy oil	0.13	
Weighted average					3.52

Data from Arora et al. 2008. (L = liters)

Correspondingly, the energy input to each product can be expressed as a sum of the energy attributed to each output.

$$e_{ethanol}(\text{wet mill}) = E_{ethanol} - 0.69e_{DDGS} \tag{5}$$

$$e_{ethanol}(\text{dry mill}) = E_{ethanol} - 0.12e_{CGM} - 0.51e_{CGF} - 0.09e_{cornoil} \tag{6}$$

In the displacement approach to co-product allocation, each co-product is taken to substitute for another product, and the energy and other lifecycle values of that product are used for the co-product.

For example, DDGS can substitute for a combination of corn and soybean meal used as cattle feed. CGM and CGF substitute for corn and urea, which are used as a feed supplement for cattle, and corn oil can substitute for soy oil. These co-product substitutes are shown in Table 1.

Taking the weighted average of US production from Table 1, based on 2/3 wet mill and 1/3 dry mill (O'Brien and Wolverton 2010), the production of co-products per liter of ethanol can be expressed as follows, using co-product substitution:

$$\begin{aligned} \text{corn}_{2.5\,\text{kg}} \rightarrow\ & \text{ethanol}_{1\,\text{liter}} + \text{corn}_{0.69\,\text{kg}} + \text{soymeal}_{0.35\,\text{kg}} + \text{soyoil}_{0.031\,\text{kg}} \\ & + \text{urea}_{0.0034\,\text{kg}} \end{aligned} \tag{7}$$

and the energy to produce ethanol "alone" can, correspondingly, be defined as

$$e_{ethanol} = E_{ethanol} - 0.69E_{corn} - 0.35e_{soymeal} - 0.031e_{soyoil} - 0.0034E_{urea} \tag{8}$$

Equation (8) cannot be directly quantified because the substitute products soybean meal and soybean oil also have an energy allocation issue; soybean meal and soybean oil are co-products of each other. Equation (9) shows the mass balance of soybeans into soymeal and soy oil, with 4 % loss or waste.

$$soy_{1.27\,kg} \rightarrow soymeal_{1\,kg} + \ soy\ oil_{0.22\,kg} + \ loss_{0.05\,kg} \tag{9}$$

In a previous study (Wang et al. 1999), the energy to produce soymeal alone has been defined through a second substitution: barley has been taken as a substitute for soybean meal, with 1 kg of soymeal equivalent to 2.3 kg barley (Ahmed et al. 1994). The energy to grow and prepare barley as feed, 2.49 MJ/kg$_{barley}$, has been used as the energy value allocated to an equivalent amount soybean meal. Specifically,

$$e_{soymeal} = 2.3\ E_{barley} = 5.73\ \mathrm{MJ/kg}_{soymeal} \tag{10}$$

where E_{barley} is 2.49 MJ/kg$_{barley}$ (Arora et al. 2008). Correspondingly, the energy to make soybean oil can be defined through mass balance of Eq. (9), with the loss included as part of the soymeal, as

$$e_{soyoil} = 5.77\ E_{soy} - 4.77 e_{soymeal} = 5.77\ E_{soy} - 10.97\ E_{barley} \tag{11}$$

E_{soy} is the energy to grow, transport and process soy into soyoil and soybean meal, and is taken here to be 4.95 MJ/kg$_{soy}$ (Arora et al. 2008). As a result, the energy to make soyoil, e_{soyoil}, can be evaluated as 1.33 MJ/kg$_{soyoil}$. A range of other values can be generated through different substitution and allocation choices (Wong 2008). Substituting barley for soybean meal, Eq. (8) can be rewritten as

$$e_{ethanol} = E_{ethanol} - 0.69\ E_{corn} - \ 0.47\ E_{barley} - \ 0.18\ E_{soy} - 0.0034\ E_{urea} \tag{12}$$

Using this approach, the weighted average co-product credit for ethanol production is a total of 2.51 MJ/L$_{ethanol}$, and the energy to produce a liter of ethanol "alone", $e_{ethanol}$, is 19.82 − 2.51 = 17.3 MJ/L$_{ethanol}$.

In Wang et al. (1999) the energy allocation between soyoil and soymeal was done using a hybrid approach taking into account both energy and mass for the barley substitution, with the result reported as e^*_{soyoil} = 17.6 MJ/kg and $e^*_{soymeal}$ = 2.4 MJ/kg.

Using the hybrid approach for determining the energy values of soyoil "alone" and soymeal "alone", as used in Wang et al. (1999), the co-product credit is 2.63 MJ/kg$_{soyoil}$ and 3.31 MJ/kg$_{soymeal}$ for the dry and wet mill processes, respectively, for a weighted average co-product credit of 2.85 MJ/L$_{ethanol}$, and a net energy to produce a liter of ethanol "alone" of 20.2 − 2.85 = 17.35 MJ/L$_{ethanol}$.

4 Discussion of the Co-Product Substitution Approach

There are a number of issues associated with calculating the effect of co-products in this way. First, the co-product substitution approach leaves open the question of whether the substitution represents a real energy savings or whether it is only an accounting procedure. Although the production of co-products may indeed result in reduced production of the substitute products, economic supply and demand considerations would suggest partial rather than full substitution. Verifying co-product substitution does not entail simply verifying that the co-product is used for the same purpose as the identified substitute. Verification also entails determining that the production of the substitute is reduced on a one-to-one basis, or other quantified basis, with increased production of the co-product. This remains an open area for research.

Second, some co-products may substitute for more than one product, and those products may not have identical energy values. For example, corn oil can substitute for soy oil or canola oil; CGM can be used either as a cattle feed supplement or as an herbicide. A mix of 15 % urea and 85 % corn can substitute for an equal weight of soybean meal (Stanton and Whittier 2006). Thus the co-product allocations are not unique. As shown in the previous section, different assumptions about the relative energy content of soybean oil versus soybean meal can produce significantly different results.

Third, if the substitute product is an agricultural product, the energy allocation calculations imply a specific reduction in agricultural production of the specific crops for which the co-products are substitutes. These changes in crop production and land use will have environmental impacts that need to be addressed consistently throughout the energy and environmental analysis.

Fourth, the co-product allocation approach typically does not take into account direct feedstock displacement. That is, if corn that would have been used as feed or food is used instead to make ethanol, the energy and inputs to grow and harvest the corn would not increase over the base case (Fargione et al. 2008; Searchinger et al. 2008).

Some lifecycle assessment models use the displacement approach to co-production allocation; others use direct allocation proportional to mass, energy, economic value or other metric.

5 Toward a Verifiable Approach to Co-Product Allocation

The scale of ethanol production may become large enough to allow verification of co-product calculations. An alternative to assuming full substitution of a fixed set of substitutes is to include co-product subtraction as a variable. This may be considered a form of system expansion (ISO 2006a, b; Kim and Dale 2005).

In principal, a biofuel environmental assessment should determine the net energy and environmental impacts from producing biofuel, including the impacts from co-products that may displace other products, and including change in land use and agricultural products from production of both the main biofuel feedstock and the feedstocks for products that may be displaced by co-products. Some parts of the energy and environmental impacts of biofuel are direct and can be measured: these directly measurable parts include the energy to produce the inputs to biofuel feedstock production, the energy and impacts of growing the feedstock, and the energy to transport the feedstock and produce the full suite of biofuel and all its co-products. The parts of the energy and environmental impact that are more difficult to measure are the changes in production of products due to the sale of co-products, and the changes in agricultural production due to use of land or crops for biofuel feedstocks.

Equation (13) expresses the change in energy use

$$\Delta E_{ethanol} = E_{ethanol} - \sum_i L_i Y_i E_i + \Delta E_{non-agroc} \quad (13)$$

corresponding to a land use change of L_i times the sum over the affected crops and land uses. A simplified approach is to consider, for corn-derived ethanol, the changes in production of corn, the changes in production of the main co-product substitute which in this case may be soy, and the changes in use of noncrop land. The fraction of land displaced can be denoted L_{corn}, $L_{other\ crop}$, and $L_{noncrop}$, with corresponding yields Y. A combination of all three types of changes can be expected: use of corn to produce ethanol will plausibly result in some reduction in corn for other uses, some reduction in other crop production, and some increase in total agricultural acreage. These options are illustrated in Fig. 1.

The typical co-product substitution allocation is equivalent to assuming a specific amount of corn and soy land is taken out of production, equivalent to the amount of corn and soy that could be substituted by the ethanol co-products. Using the data in Table 2, these values are equivalent to assuming that 51 % of the corn for ethanol is grown on land that would otherwise not be planted in crops, 24 % is grown on land that would otherwise be planted in soy, and 25 % is grown on land that would otherwise produce corn for other purposes.

Land use change resulting from biofuel production can occur overseas as well as in the U.S. Here we explore the U.S. data, as a step toward understanding how the co-product and land use impacts of corn-derived ethanol may change over time. Between 1976 and 2006, the amount of U.S. land planted in corn and the amount of corn land harvested remained roughly constant, at 78 ± 6 million acres and 70 ± 6 million acres per year, respectively (Fig. 2). During this time, the yield of corn has risen steadily and approximately linearly, from 88 bushels per acre in 1976 to 149 bushels per acre in 2006, an increase of 69 %. As a result, total corn production rose from 6.3 billion bushels in 1976 to 10.5 billion bushels in 2006. Through 2006, there was no significant increase in U.S. land used to produce corn, despite the growth of corn ethanol production. The amount of corn used to make ethanol in the U.S. has increased from 0 in 1979 to 2.95 billion bushels in 2007, accounting for 22.5 % of corn production in 2007, and corresponding to 17.6 million acres of land growing corn for ethanol (USDA 2014).

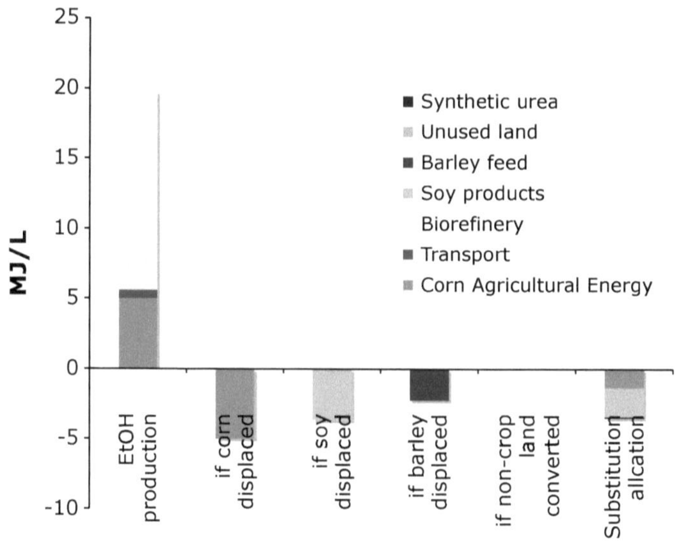

Fig. 1 Energy balance for corn ethanol, showing four orthogonal combined co-product and feedstock land use change modes, as well as a product substitution allocation

Table 2 Soybean and barley yield and energy requirements

Quantity	Value	Reference
Y_{soy} yield (kg/m^2) (2004–2008 weighted average)	0.28	USDA (2014)
Soy agriculture energy (MJ/kg)	1.8	Arora et al. (2008)
Soy processing energy (MJ/kg soy)	2.92	Arora et al. (2008)
Soy transport energy (MJ/kg)	0.23	Arora et al. (2008)
E_{soy} (agriculture + transport + processing)	4.95	
Ratio soybean to soybean meal (kg/kg)	1.27	Arora et al. (2008)
Ratio soybean to soyoil (kg/kg)	5.7	Arora et al. (2008)
E_{barley} (agriculture + transport + processing)	2.49	Arora et al. (2008)
e_{soyoil} total soy oil "energy" (MJ/kg oil)	1.33	Ahmed et al. (1994)
$e_{soymeal}$ total soy meal "energy" (MJ/kg meal)	5.73	Energy substitution method
e^*_{soyoil} total soy oil "energy" (MJ/kg oil)	17.6	Hybrid substitution method
$e^*_{soymeal}$ total soy meal "energy" (MJ/kg meal)	2.4	Hybrid substitution method
E_{urea}	26	Grabowski (2002)
Y_{barley} yield (kg/m^2)	0.34	

The co-product energy calculation implies a reduction of 6 billion kg of soy, a decrease of about 7 % over what would otherwise have been produced. However, from 1976 to 2006 the area of harvested soybean has risen from 49 million acres in 1976 to 75 million acres in 2006 (Fig. 2). Total production was 3.2 billion bushels in 2006, corresponding to 87 billion kg of soy. If it is corn that has been displaced

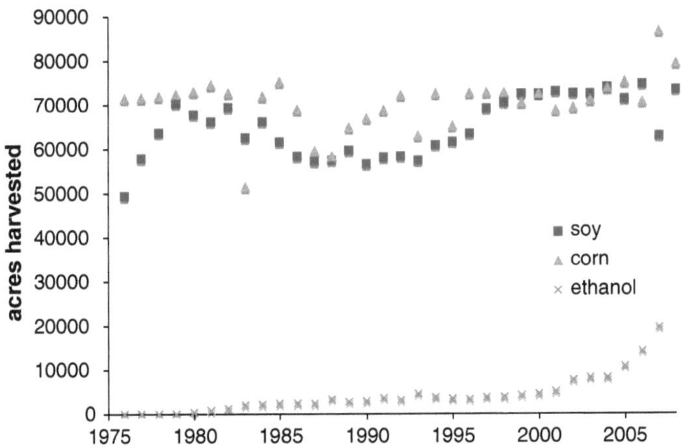

Fig. 2 U.S. acres harvested of corn and soy, and equivalent acres for ethanol production (data from USDA 2014)

rather than soy, the net energy to produce ethanol from corn in the U.S. has been 14.2 $MJ/L_{ethanol}$. As the amount of corn used to make ethanol increases, the land use pattern may change; if more land is used to grow corn, and less land is used for soy or current noncrop land is used to crop corn, then the net energy requirement for ethanol will increase. The 2007 and 2008 data suggest this kind of shift but several years of data would be needed for further evaluation.

Abayomi et al. (2011) have developed a statistical method for evaluating how use of biomass for energy affects other uses of the crop or land use generally. That type of statistical approach could be used to provide an empirical basis for co-product allocation calculations. Other market data could also be used to verify co-product allocation parameters (Hellweg and Canals 2014).

6 Conclusions

The lifecycle energy and greenhouse gas emissions are basic metrics for characterizing biofuels. For some biofuels and for some parts of the analysis, standard lifecycle assessment is well-defined and straightforward. However, when production of a biofuel or other product involves production of co-products, displacement of other products, or changes in the global economy over time, then the potential to uniquely determine the energy or greenhouse gas impact may not only be challenging but also become inherently subject to choices or convention.

There are two key messages. First, optimization and simulation practitioners, as well as lifecycle assessment practitioners, should be aware of the different approaches to co-product allocation, and understand the extent to which changes in the approach could change the results of the analysis. There are situations in which different choices of allocation or substitution procedures can result in substantially different energy and environmental results. Co-product allocation and other methodological issues are not yet a completely settled area of environmental lifecycle assessment, and the standards and conventions for the calculations are likely to continue to evolve.

Second, in the future it is possible that appropriate co-product allocation could be determined from data. Observation over time of changes in land use or changes in markets for co-products and their substitutes could provide evidence for how co-products influence markets and hence provide a basis for energy, greenhouse gas and other calculations. Collection or use of physical and market data to support allocation modeling is a promising open area for research.

References

Abayomi K, Luo D, Thomas VM (2011) Statistical evaluation of the effect of ethanol in US Corn Production: a flexible test for independence on a constrained sum. Int J Ecol Econ Stat 22 (S11):105–126. http://www.ceser.in/ceserp/index.php/ijees/article/view/1060

Ahmed I, Decker J, Morris D (1994) How much energy does it take to make a gallon of soydiesel? Institute for Local Self-Reliance

Allen DT, Allport C, Atkins K, Cooper JC, Dilmore RM, Draucker LC, Eichmann KE, Gillen JC, Gillette W, Griffin WM, Harrison WE, Hileman J, Ingham JR, Kimler FA, Levy A, Murphy CF, O'Donnell MJ, Pamplin D, Schivley G, Skone TK, Strank SM, Stratton RW, Taylor PH, Thomas VM, Wang M, Zidow T (2009) Framework and guidance for estimating greenhouse gas footprints of aviation fuels, U. S. Air Force AFRL-RZ-WP-TR-2009–2206. http://www2.isye.gatech.edu/~vthomas/Aviation_GHG_Framework.pdf

Arora S, Wu M, Wang M (2008) Update of distillers grains displacement ratios for corn ethanol life-cycle analysis. Argonne National Laboratory, Energy Systems Division

Fargione J, Hill J, Tilman D, Polasky S, Hawthorne P (2008) Land clearing and the biofuel carbon debt. Science 318:1235–1238

Grabowski MS (2002) Fossil energy use in the manufacture of corn ethanol, Colorado School of Mines, Prepared for the National Corn Growers Association

Hellweg S, Canals MI (2014) Emerging approaches, challenges and opportunities in life cycle assessment. Science 344:1109–1113

ISO (International Standards Organization) (2006a) Environmental management—life cycle assessment—principles and framework. ISO 14040

ISO (International Standards Organization) (2006b) Environmental management—life cycle assessment—requirements and guidelines. ISO 14044

Kim S, Dale BE (2005) Environmental aspects of ethanol derived from no-tilled corn grain: nonrenewable energy consumption and greenhouse gas emissions. Biomass Bioenergy 28:475–489

Luo D, Hu Z, Choi DG, Thomas VM, Realff MJ, Chance RR (2010) Lifecycle energy and greenhouse gas emissions for an ethanol production process based on blue-green algae. Environ Sci Technol 44(22):8670–8677. doi:10.1021/es1007577

Luo D, Realff MJ, Thomas VM (2014) Bioreactor-based fuel systems I: optimal production capacity considering start-up dynamics. Comput Chem Eng 71:141–153. doi:10.1016/j.compchemeng.2014.07.006

O'Brien D, Woolverton M (2010) Updated trends in U.S. wet and dry corn milling production, AgMRC renewable energy Newsletter, February 2010. http://www.agmrc.org/renewable_energy/ethanol/updated-trends-in-u-s-wet-and-dry-corn-milling-production/

Okwo A, Thomas VM (2014) Biomass feedstock contracts: role of land quality and yield variability in near term feasibility. Energy Econ 42:67–80. doi:10.1016/j.eneco.2013.11.004

Searchinger T, Heimlich R, Houghton RA, Dong F, Elobeid A, Fabiosa J, Tokgoz S, Hayes D, Yu T-H (2008) Use of U.S. croplands for biofuels increases greenhouse gases through emissions from land-use change. Science 318:1238–1240

Shapouri H, Duffield JA, Grabowski MS (1995) Estimating the net energy balance of corn ethanol *721*. U.S. Department of Agriculture, Agricultural Research Service

Shapouri H, Duffield JA, Wang M (2002) The energy balance of corn ethanol: an update 813, U.S. Department of Agriculture, Office of the Chief Economist

Stanton T, Whittier J (2006) Urea and NPN for cattle and sheep *1.608*. Colorado State University Extension

USDA (U. S. Department of Agriculture) (2014) National Agricultural Statistical Service

Wang M, Saricks C, Santini D (1999) Effects of fuel ethanol use of fuel cycle energy and greenhouse gas emissions *ANL/ESD-38*. Argonne National Laboratory, Energy Systems Division

Wong HM (2008) Life-cycle assessment of greenhouse gas emissions from alternative jet fuels. SM thesis, MIT

Life-Cycle Assessment of Bio-Fuel Production Using Syngas from Biomass

Nelson Andrés Granda-Marulanda, Mingzhou Jin and Fei Yu

Abstract This study conducts a life-cycle assessment (LCA) for a gasification process that converts wood biomass into various biofuels. The analysis is based on the mass balance and input/output of the two steps of the 10-ton biofuel production per day. By using OpenLCA, an open-source licensed LCA software package, and the NREL and databases, this LCA shows that the system producing the syngas from biomass has very similar emissions to natural gas and the gasification process is the one with the major impact potential.

1 Introduction

The total energy consumption in the world is expected to grow by 56 % from 2010 to 2040 (IEO 2013). The renewable and nuclear energy alone will have an increase of 2.5 % per year each, making them the fastest growing energy types (IEO 2013). However, fossil fuel is still expected to be the major supply of the world energy demand by 2040 with about 80 % of the energy market (IEO 2013). The Fourth Assessment Report of the Intergovernmental Panel on Climate Change (IPCC) sates that the fossil fuel consumption growth, together with the world's growing population, are the leading contributors to the rapid increase in greenhouse gas (GHG) emissions (Parry et al. 2007). The worldwide energy-related carbon dioxide (CO_2) emissions is expected to increase from about 31 billion metric tons in 2010 to 36 billion metric tons in 2020 and then to 45 billion metric tons in 2040 (IEO 2013).

N.A. Granda-Marulanda · M. Jin (✉)
Department of Industrial and Systems Engineering, The University of Tennessee at Knoxville, Knoxville, Tennessee 37996, USA
e-mail: jin@utk.edu

F. Yu
Agriculture and Biological Engineering, Mississippi State University, Mississippi, USA

S.D. Eksioglu et al. (eds.), *Handbook of Bioenergy*,
Energy Systems, DOI 10.1007/978-3-319-20092-7_12

The need of finding more sustainable fuel sources has become compelling in the world. The US government has aggressively urged the research community to find sustainable options to fossil fuels in the last decades. The Clean Air Act of 1990, the California Global Warming Solution Act of 2006, the Energy Independence and Security Act of 2007, and other laws are some example of the US Government support towards more sustainable fuel options (EISA 2007). One of the options to reduce greenhouse gas emissions and reduce the dependency on fossil fuels is the use of biofuels as an alternative, but biofuels may have emission and other indirect effects during their life cycles that create concerns on their sustainability (Petrou and Pappis 2009, Sheehan 2009). The *Energy Independence and Security Act of 2007* (EISA 2007) stated in Section 526 that "*No Federal agency shall enter into a contract for procurement of an alternative or synthetic fuel, including a fuel produced from nonconventional petroleum sources, for any mobility-related use, other than for research or testing, unless the contract specifies that the lifecycle greenhouse gas emissions associated with the production and combustion of the fuel supplied under the contract must, on an ongoing basis, be less than or equal to such emissions from the equivalent conventional fuel produced from conventional petroleum sources.*" In order to establish the environmental benefit of the use of biomass as a valid alternative, a life-cycle assessment (LCA) needs to be performed.

LCA has been extensively used to quantify the environmental burden of different processes. Luo et al. (2010) investigated the life cycle energy required to produce ethanol from Blue-green algae and their corresponding green gas emissions. Xu et al. (2014) recently performed an LCA on thirteen sewage sludge-treatment scenarios in China. Their results showed that a sewage sludge-treatment scenario with anaerobic digestion, dewatering, and incineration technologies was the most environmentally and economically sustainable method to treat sewage sludge. Rubin et al. (2014) applied the LCA methodology to evaluate and compare two processes for recovering copper from Printed Circuit Boards scrap. Their study showed that the process using acqua regia has better environmental performance. The review by Cabeza et al. (2014) presented several LCA studies in the building industry, including assessments of residential buildings, non-residential buildings, and civil engineering constructions. LCA has also been applied in the agriculture and farming industries. Haas et al. (2000) used the LCA methodology to assess the environmental impact of the grassland dairy farms in the southern region of Germany as an example to introduce a framework to use LCA in the agriculture industry. They argued that the impact categories and the functional units of classical LCA must be adapted to specific agricultural production processes when performing agricultural LCA. Thévenot et al. (2013) performed an LCA study of a poultry supply chain applying a cradle-to-slaughterhouse gate methodology. Finally, one of the most studied areas using the LCA methodology is the electric power industry, since this sector is one of the major contributors to GHG and plays a critical role in the world's sustainability. Jaramillo et al. (2007) showed a comparative analysis of the air emissions during the life cycle between coal, natural gas, liquefied natural gas and synthetic natural gas used to the generation of electricity. In the analysis Jaramillo et al. (2007) mentioned that the used of coal as fuel to generate electricity has the higher CO_2 emission and the use of natural gas the lower emissions. It was also discussed by Jaramillo et al. that the technology to generate synthetic natural gas need to overcome various technological challenges in order to be able to reduce the overall Life Cycle GHG emissions. In a more recent study Turconi et al. (2013) presented a review of numerous case studies involving the LCA of electricity generation based on hard coal, lignite, natural gas, oil, nuclear, biomass, hydroelectric, solar photovoltaic (PV) and/or wind. In the transportation area there are also many studies using LCA to evaluate various fuel types. Ally and Pryor (2007) performed an LCA study to evaluate the environmental burden of three different transportation fuels used in the bus transportation system of Western Australia, including diesel, natural gas and hydrogen fuel cell. Their analysis quantified the improvements that can be expected in future generations of fuel cell vehicles. Xue et al. (2012) performed an LCA study to quantify the environmental impacts of different biodiesel blend production in the Pennsylvania area. The study showed the fertilizer usage in the agricultural phase, soy oil extraction and refining, feedstock and fuel transportation, and fuel combustion in the usage phase are main contributors to biodiesel's life-cycle environmental impacts for all blends. All the above examples showed that LCA has been widely used to evaluate sustainability for various industries. Biofuel production is still in its development stage and a standard LCA is necessary to evaluate various pathways to produce biofuels from biomass.

This chapter will introduce how to implement an LCA on possible production processes of biofuel regarding their sustainability and present an application example. The application case is about a newly designed biofuel production process based on a gasification pathway using wood chips.

1.1 LCA Background and Methodology

LCA was originally created by the Society of Environmental Toxicology and Chemistry (SETAC) (Klöpffer 2006). In 1990 SETAC generated the first document in which the name of the method and a general structure, still valid today, were presented. Later on in 1997 the LCA methodology presented by SETAC was used as a basis to create the standard of ISO14040-43 called life-cycle assessment standards by the International Standard Organization (ISO 1997). Furthermore, ISO14041, ISO14042, and ISO14043 were developed to detail the standards at different LCA steps (Finkbeiner et al. 2006). In ISO14040 the term of the Life-cycle assessment is defined as "compilation and evaluation of the inputs, outputs and the potential environmental impacts of a product system throughout its life cycle". This definition means that the LCA methodology studies the environmental burdens and impacts of all stages in a product's life, from raw material acquisition to disposal. The stages also include product production or manufacturing, product usage, and product and material recycle. This all-stage analysis is also known as a "cradle-to-grave" approach. ISO14040 mentions that some of the general categories

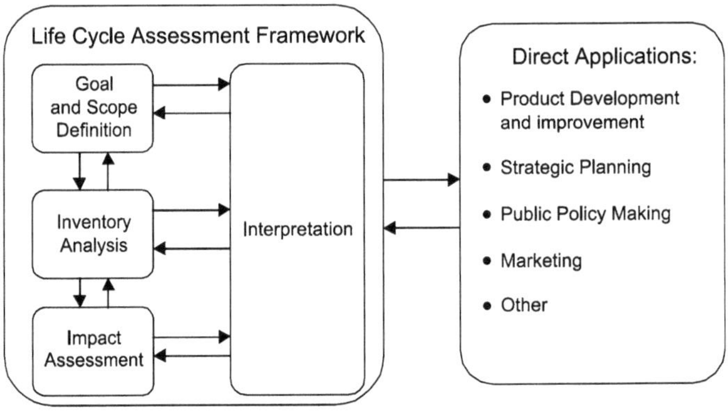

Fig. 1 Phases and applications of an LCA. (ISO 1997)

of environmental impacts to consider are resource use, human health, and ecological consequences.

Under ISO14040 the LCA framework was defined into four major steps; (1) Goal and scope definition, (2) Life cycle inventory analysis, (3) Life cycle inventory impact assessment, and (4) Life cycle interpretation. The relationship between these stages is illustrated in Fig. 1 and all four steps will be introduced in details below with examples from the literature.

1.2 Goal and Scope Definition

The first step of the LCA is to define the problem or intended application for the assessment, the reason to perform the assessment, and the intended audience of the study (ISO14040). The goal of an LCA may be descriptive and rather brief. For example, Rubin et al. (2014) defined the goal of their LCA study as "to evaluate and compare two electrochemical processes for recovering copper from PCB scrap: one using sulfuric acid and one using nitric and chloridric acid".

During the definition of the scope of an LCA study, ISO14040 suggests some or all of the following items be considered and clearly described.

- Functions of the product system or, in the case of comparative studies, the systems;
- Functional unit;
- Product system to be studied;
- System boundaries;
- Allocation procedures;
- Types of impact and methodology of impact assessment, and subsequent interpretation to be used;
- Data requirements;

- Assumptions, limitations, and initial data quality requirements; and
- Type of critical review (if any), such as type and format of the report required for the study.

Even though the scope definition is flexible and could be very different across applications and it is possible not to include all of the above items, two of these items are critical in defining an LCA scope. They are functional unit and system Boundaries and should be included in all LCA studies. These two items are introduced in more details as follows.

Functional Unit was defined by Rebitzer et al. (2004) as a quantitative description of the service performance (e.g., the needs to be fulfilled) of the investigated product systems. The functional unit is a very critical aspect of the scope of the LCA and is more important if the LCA is used to compare multiple products or systems. If two or more productions are involved in an LCA, the functional unit should be the same for all products in order to have a meaningful comparison. Functional units need to be carefully selected based on the goal of the LCA study. For example, Rubin et al. (2014) used the recovery of 102 g of copper as the functional unit in the comparison of two ways of recovering copper from PCB scrap. In a case of sewage treatment in China (Xu et al. 2014), the treatment of one ton of dry sludge was selected as the functional unit. In the study by Ally and Pryor (2007) to compare three different fuel types for the Bus system in Western Australia, the functional unit was defined based on vehicle kilometers.

System Boundaries are the restrictions that should be considered in order to collect data for an LCA study. These restrictions can be natural restrictions of the system under study, geographic restriction or time restrictions. The system boundaries determine which processes shall be included within the LCA (ISO 1997). Figure 2 presents the most common system boundaries used in performing an LCA.

In the study performed by Xu et al. (2014), the system boundaries of the study of sewage treatment were defined as thirteen different scenarios including different stages to treat sewage, such as raw materials and energy production, road transport, direct emissions, wastewater treatment, energy recovery from anaerobic digestion,

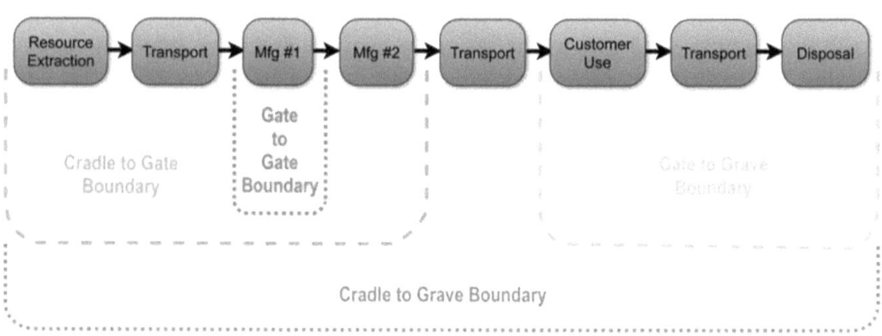

Fig. 2 Common boundaries for life cycle assessment. (Autodesk)

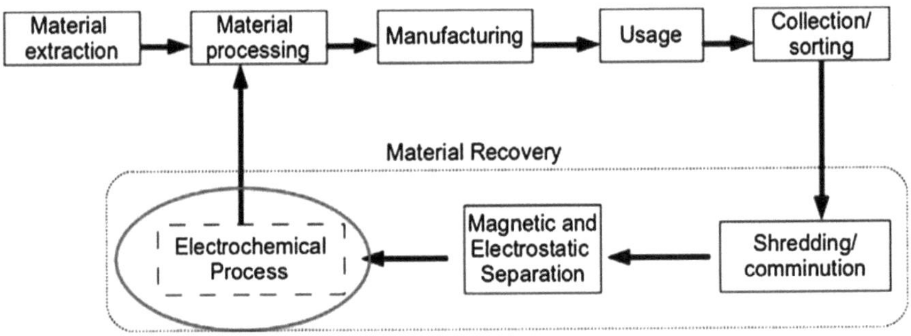

Fig. 3 Life cycle of the PCB production and the selected system boundaries (Rubin et al. 2014)

incineration, and landfill stages. In a case of poultry production (Thévenot et al. 2013), the system boundaries cover the supply chain of the poultry process the production of resources and waste treatment with their associated emissions. Another example of how to define system boundaries can be found in the study of the copper recovery from PCB scraps (Rubin et al. 2014) and is illustrated in Fig. 3. The system boundary was defined as the complete life of the PCB, from material extraction through material recycle, even though they performed the LCA to compare the two recycle strategies (i.e., the usage of different chemicals) only during the electrochemical process. It is important to note that the results of an LCA heavily depend on the selected system boundaries, even if the exactly same functional unit is studied.

1.3 *Life-Cycle Inventory (LCI) Analysis*

The inventory analysis is comprised of the data collection and calculation procedures to quantify the relevant inputs and outputs of a product system or systems. Some examples of these are raw materials, required energy, and any ancillary or physical inputs needed by the system. As outputs we have all the waste and emissions including all releases into the air, water and soil by the system and also by all by-products (ISO 1997). The inventory analysis process is typically iterative. During data collection, how to calculate the system inventory could be better understood so that more requirements and limitations may be identified. It is not unusual that some changes are necessary in the procedure of collecting data in order to meet the specified goal of the study. In some cases issues found during this phase may cause revisions of the goal or scope of the study (1997).

There is a large amount of data that has already been collected for common processes in different countries. These data sets are known as LCA Inventory databases. There are two types of access to these databases, public access and proprietary access. Some examples of publicly accessible databases are the U.S. Life Cycle Inventory (LCI) database administered by the National Renewable

Energy Laboratory (NREL 2012), the Greenhouse Gases, Regulated Emissions, and Energy Use in Transportation (GREET) Model administered by Argon National Labs, and the Building for Environmental and Economic Sustainability (BEES) tool administered by the National Institute of Standards and Technology (NIST). Some of the most commonly used proprietary databases are EcoInvent provided by the Swiss Centre for Life Cycle Inventories and PE International LCI database provided by PE International. PE International LCI database can only be used in GaBi software, which is a commercial LCA software package. All these databases follow the ISO 14040 procedure to collect and present data in a way so that they can be easily used by the LCA research community. Figure 4 illustrates a simple flow to collect and validate the data under ISO 14040.

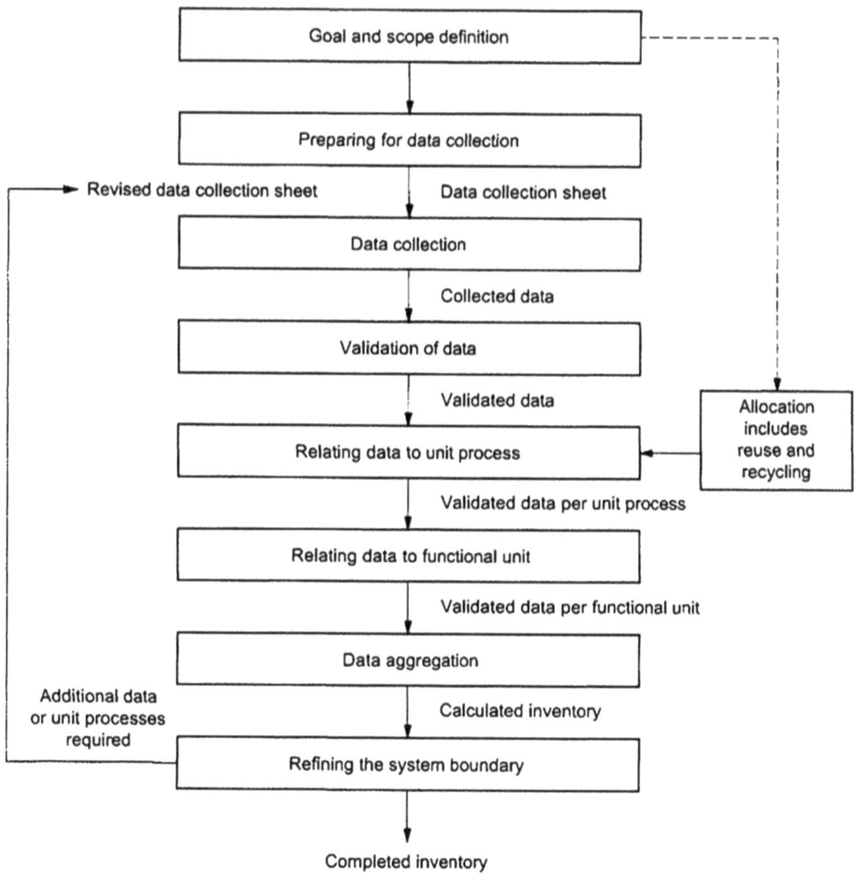

Fig. 4 Simplified procedure for LCI analysis. (1997)

1.4 Life Cycle Inventory Impact Assessment

All outputs or results of the LCI analysis are used as inputs in the life cycle inventory assessment (LCIA) stage of the LCA. All data is evaluated and associated with specific environmental impacts. The level of detail, choice of impacts to evaluate, and selected assessment methodologies can significantly vary based on the scope of the assessment (1997). Some of the common or baseline environmental impacts in LCA studies are depletion of abiotic resources, impacts of land use, land competition, climate change later known as Global Warming Potential index (GWP), stratospheric ozone depletion, human toxicity, aquatic and terrestrial eco-toxicity, photo-oxidant formation, acidification, and eutrophication. The same as the Inventory analysis stage, the inventory impact assessment stage is an iterative process of evaluating the goal and scope that need to be performed.

1.5 Interpretation of LCA

The interpretation stage of the LCA is where the results of the LCI analysis and impact assessment get combined to present the LCA findings to the desired audience. In order to reach conclusions and recommendations from this stage, the findings should be according with the goal and scope of the LCA study. Three major components are suggested in the interpretation stage: (1) identify major issues resulting from the LCI and LCIA, (2) evaluate if the study is complete based on the goal and scope and perform a sensitivity analysis, and (3) report conclusion, limitations and recommendations of the study. Based on the report of this stage the audience (mostly decision makers) can make better decisions in terms of selecting and designing the most environmentally beneficial product or process.

2 LCA Study on the Production of Biofuel from Wood Biomass Through Synthetic Natural Gas

2.1 Case Introduction

Approximately 9.19 % of the 81.85 quadrillion BTU of the annually consumed energy in the US was produced from the renewable sources in 2013 (EIA 2000). Biomass, in all types, has constituted about half of the renewable resources (i.e., 4.49 % of all energy production). Negative environmental impacts caused by fossil energy sources and consequently national energy security concerns have increased the motivation and interest in developing renewable energy markets. Therefore, biomass in general and specifically the dedicated energy crops have been on the center of attention as sustainable sources of energy (Heller et al. 2003). Numerous studies have been conducted on developing and improving various processes to produce biomass-based fuels as a replacement to regular fossil fuel. Two major types of conversion of biomass to fuels are thermochemical conversion and biochemical conversion. Numerous studies, propelled by the needs of the reduction of the fossil fuel consumption (EISA 2007), have been conducted to investigate various biomass-derived products through many conversion pathways. Some examples of fuel products and their conversion methods are bio-ethanol from fermentation, bio-methanol from gasification, bio-naphtha from flash pyrolysis and hydro treating, bio-oil from flash pyrolysis and liquefaction, methane from anaerobic digestion, hydrogen from gasification, and many other (Grassi and Bridgwater 1993).

Corresponding to those new conversion technology developments, LCA researches have been conducted to study their environmental impact and to assess their advantages and disadvantages from a systematic way. Those researches try to address various environmental questions for the system under study, such as the emitted substances of the system along the entire life cycle, the most accountable part of the system for such emissions, the rate of emitted substances, the energy balance of the system as a whole or by part, and the net carbon dioxide emissions of the entire system. Among numerous conversion technologies, the pathway based on biomass gasification seems promising because of its high thermal efficiency and potential of realizing neutral CO_2 emissions. There have been numerous attempts to generate renewable energy from biomass in the literature and all the efforts show progress in gasification technology and brought more efficiency while the systems are more environmental friendly and easy and convenient to implement and operate. In addition, to be adjacent to feedstock, countryside installation of gasification systems promotes rural economic development and may reduce costs, energy consumption, and emissions of biomass feedstock transportation.

This LCA case study is specifically conducted for a gasification process that converts wood biomass into various biofuels. The process has been under development at Mississippi State University during the last decade. Preliminary experimental studies feed data to this LCA study. Components of the facility include feedstock preparation, gasification, and syngas cleaning. The bio-gasifier is an atmospheric, downdraft, and fixed bed gasifier. The bio-gasification system was designed for producing syngas at a capacity of 60 nm^3/h (Kim et al. 2012). The gasification process at the downdraft gasifier is controlled by a host computer. The main parts of the system include the feeding system, the gasifier, the heat exchanger and the filters, as illustrated in Fig. 5. In order to keep track of the input and output substances and monitor the partial and overall performance of the system, thermocouples and pressure transducers have been installed, connected to, and controlled by the computer. During the gasification reaction and operations in the gasifier, the feeder automatically starts feeding new fuel into the gasifier in case the level of fuel in the gasifier drops below a pre-set point. All the energy that is required for the system to operate is provided by the wood chip consumption as the biomass and input fuel. No other external energy or fuel in any form is necessary to maintain the system operations after the system has been started with some initial energy. The produced gas then is cooled down from 500–700 to 110 centigrade

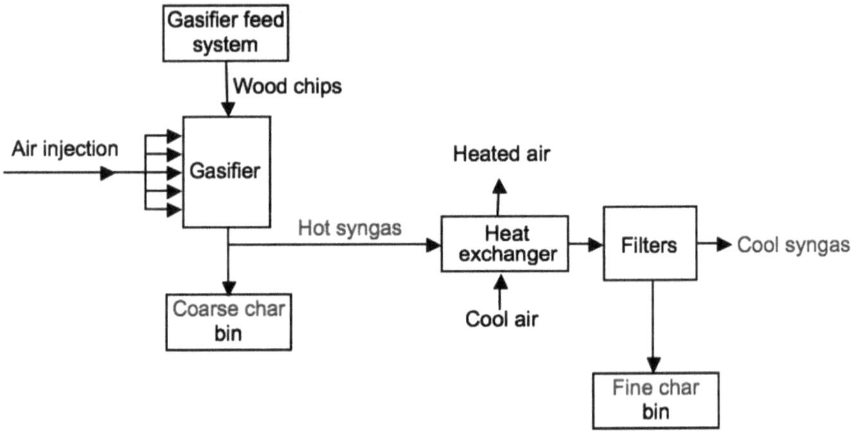

Fig. 5 Gasification process

degrees by the heat exchanger and through passing across the parallel bag filters where fine particles are removed from the gas. After bag filters the activated carbon filter absorbs the tar in the gas. The throughput syngas contains about 48 % N_2, 21 % CO, 18 % H_2, 10 % CO_2, 1.5 % CH_4, some water vapor, and a trace amount of other gases. The cool syngas is then compressed and stored in a storage tank (Wei 2005, Wei et al. 2009). The gasification process has been investigated using the wood chips of oak trees as raw materials and air as the oxidation gas. The wood chips contain about 9.5 % moisture content and temperature in the combustion is stable at 600–900 degrees centigrade.

Under the flow rate of producing gas at 65 nm^3/h in a stable process, the consumption of wood chips is 26–28 kg/h with the syngas yield of 2.3–2.4 nm^3/Kg wood. The carbon conversion rate from wood to carbon monoxide, carbon dioxide and methane is about 92–99 % and the hydrogen conversion rate from wood to H_2 and methane is 67–75 %. Figures 6 and 7 show the mass balance and inputs/outputs of the biofuel production and carbon balance during conversion respectively. The emission of 6.45 tons of carbon dioxide after conversion in Fig. 6 is the total emission. Most of this emission is created in the gasification process and is contained in the syngas before conversion.

According to the heating value of wood chips and produced syngas for the energy balance, the energy conversion rate from wood to syngas is 59–65 %. The wasted energy heat that is entrapped in the produced syngas is calculated according to the gas composition and gas temperature when getting in the heat exchanger and its rate is 10–12 %. The low heating value (LHV) of the output syngas is also 5.1–5.3 MJ/nm^3.

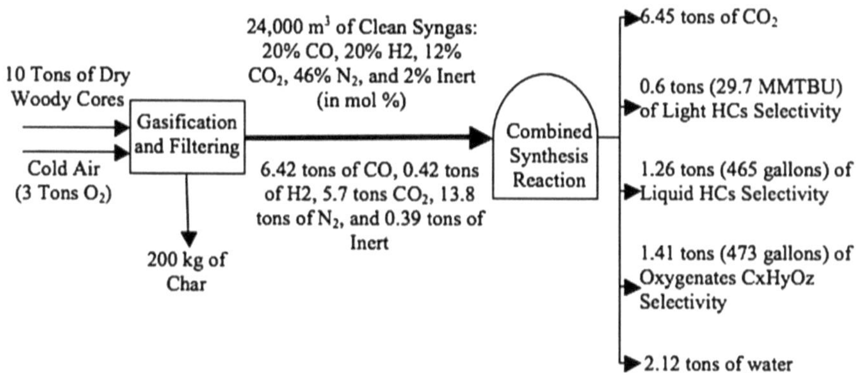

Fig. 6 The mass balance and input/output of the two steps of the 10-ton biofuel production per day

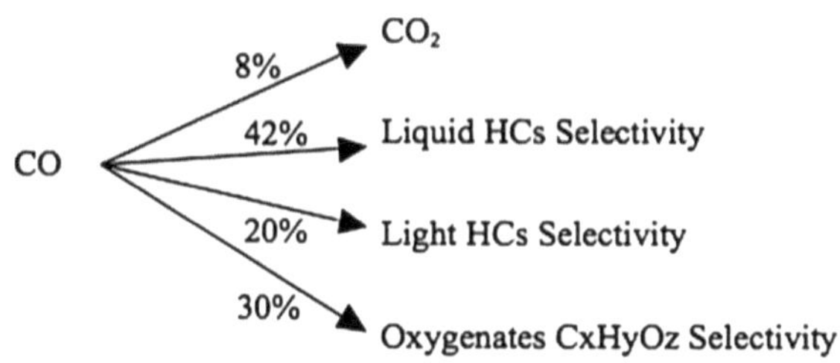

Fig. 7 Carbon balance during conversion

2.2 *Software Used in the Study*

A variety of software could be used to facilitate the large efforts in data calculation and data analysis during LCA studies. Some of them are commercially available and others are free or have open-source licenses. In this case study, we selected an open-source licensed software package called OpenLCA (GreenDelta 2013). Figure 8 demonstrates the basic processes to perform an LCA and where the software is mostly involved. The software is also very helpful in the impact assessment by calculating the impact characterization index using various methods based on the inventory results. Figure 9 shows how to setup OpenLCA and what external data are needed to complete the assessment.

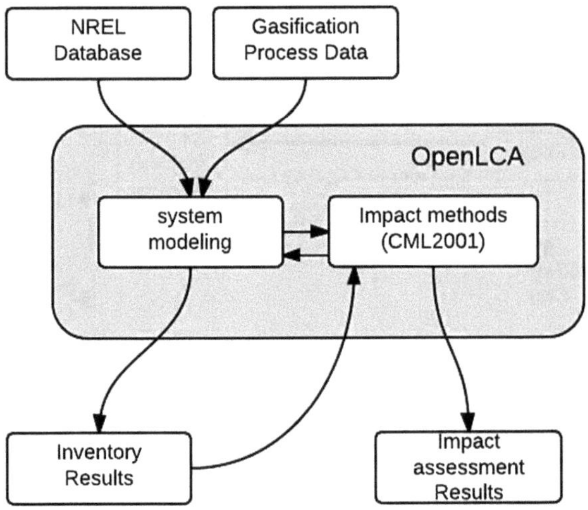

Fig. 8 Basic model flow (GreenDelta 2013)

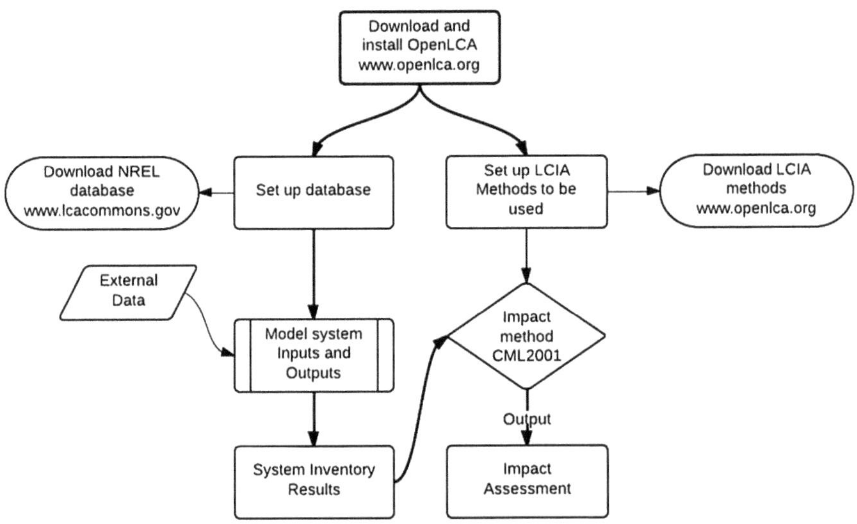

Fig. 9 OpenLCA setup and data flow (GreenDelta 2013)

2.3 *Case Goal and Scope*

The goal of this case study is to evaluate the environmental impact of the production of biofuel using woodchips through the synthesis gas pathway. A cradle-to-grave approach is done by separating the system in four subsystems,

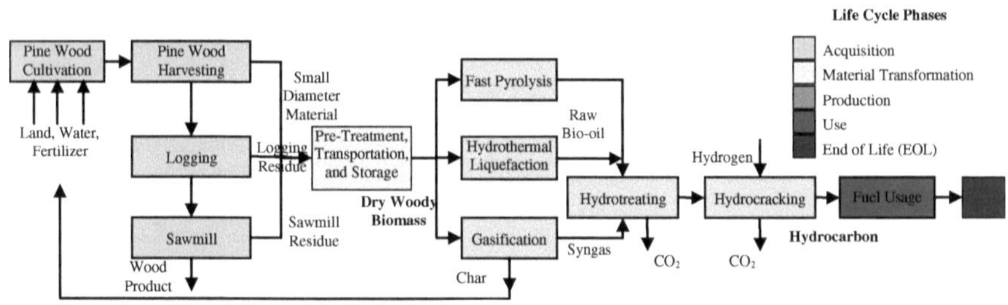

Fig. 10 Life cycle stages flow

Table 1 Data from experimental results obtained from the pilot gasifier

Data from plant	
Woodchips used LHV	18.7 MJ/Kg
Production Yield	2.4 Nm^3/kg
Syngas output LHV	5.8 MJ/Nm^3
Electric demand	3.5 Kw/h at 60 Nm^3/h

Table 2 Processes used from the NREL database

Processes from NREL
Dry rough lumber, at kiln, US SE
Wood chips, hardwood, green, at sawmill, SE
Transport, combination truck, average fuel mix
Electricity, diesel, at power plant

wood acquisition, wood transportation, production of syngas, and use of syngas. In order to compare these stages of the system, we select a functional unit of 1 MJ in order to evaluate different stages of the system and make comparisons. All the inventory data was collected based on the geographic boundary of the southeast region in the US. Figure 10 shows all stages of the system and other components. Tables 1 and 2 list the data used from experiments that have been conducted on the gasifier at Mississippi State University and the involved processes based on the NREL database (NREL 2012) in order to model the system in OpenLCA.

2.4 *Inventory Assessment for the Case*

The database provided by the NREL is used to model the system in OpenLCA. All the inputs and outputs related to the wood acquisition process are from the NREL inventory for softwood logs with bark, harvested at average intensity site, at mill, and in US southeast region. These inputs and outputs are for the acquisition amount of 1 kg of wood chips. For the second stage of the process, wood transportation, the calculation to obtain the inventory is based on the transport of 1 ton per mile of

Table 3 Inventory input results for the use stage of the LCA

Inventory input flows	Unit	Amount
Carbon dioxide, in air	Kg	4.73E-11
Coal, 26.4 MJ per kg, in ground	Kg	8.23E-15
Diesel, at refinery	$m^3 \times a$	4.68E-05
Diesel, combusted in industrial boiler	$m^3 \times a$	7.93E-10
Diesel, combusted in industrial equipment	$m^3 \times a$	2.22E-11
Electricity, biomass, at power plant	MJ	9.26E-06
Electricity, bituminous coal, at power plant	MJ	0.000328
Electricity, lignite coal, at power plant	MJ	9.84E-06
Electricity, Natural Gas, at power plant	MJ	5.76E-05
Electricity, nuclear, at power plant	MJ	0.00013
Electricity, onsite boiler, hardwood mill average, SE	MJ	4.19E-10
Electricity, residual fuel oil, at power plant	MJ	1.87E-05
Gas, natural, 46.8 MJ per kg, in ground	Kg	8.13E-14
Gasoline, combusted in equipment	$m^3 \times a$	5.98E-12
Heat, onsite boiler, hardwood mill average, SE	MJ	4.32E-12
Limestone, in ground	Kg	2.98E-12
Liquefied petroleum gas, combusted in industrial boiler	$m^3 \times a$	7.13E-15
Natural gas, combusted in industrial boiler	m^3	2.84E-12
Natural gas, combusted in industrial equipment	m^3	1.84E-13
Oil, crude, 42 MJ per kg, in ground	Kg	3.67E-13
Residual fuel oil, at refinery	$m^3 \times a$	3.79E-09
Residual fuel oil, combusted in industrial boiler	$m^3 \times a$	3.97E-16
Softwood logs with bark, harvested at average intensity site, at mill, US SE	m^3	7.32E-08
Transport, combination truck, diesel powered	$t \times km$	3.77E-10
Uranium, 2291 GJ per kg, in ground	Kg	1.23E-18
Water	$m^3 \times a$	2.27E-06
Water, process and cooling, surface	m^3	2.24E-17
Water, well, in ground	m^3	7.79E-09
Wood and wood waste, 3.9 MJ per kg, oven dry basis	Kg	2.59E-11
Wood, unspecified, standing/kg	Kg	0.416667

wood chips by diesel truck. The average distance traveled from the mill to the gasification plant was assumed to be 15 miles while the truck load was assumed to be 5 tons. For the gasification stage, the reference is 60 nm^3 of output gas, which is about the hourly yield of the pilot bio-gasifier at Mississippi State University. For the last stage of the process, the use of the bio-gas, the reference unit is our functional unit of 1 MJ. In this stage we used the output of all the other stages to complete the system analysis. Some of the data obtained from the NREL inventory database are listed in Tables 3 and 4, where the inputs and outputs of the system are presented.

Table 4 Inventory output results for use stage of the LCA

Inventory outputs flow	Unit	Amount
Acetaldehyde	Kg	1.44E-15
Acrolein	Kg	6.94E-15
Antimony	Kg	1.37E-17
Arsenic	Kg	3.07E-09
Bark	Kg	5.21E-07
Benzene	Kg	4.98E-10
Beryllium	Kg	6.46E-11
BOD5, Biological Oxygen Demand	Kg	3.37E-12
Cadmium	Kg	9.25E-10
Carbon dioxide, biogenic	Kg	3.38E-10
Carbon dioxide, fossil	Kg	0.137507
Carbon monoxide	Kg	1.04E-12
Carbon monoxide, fossil	Kg	0.00014
Chloride	Kg	1.38E-10
Chlorine	Kg	1.37E-15
Chromium	Kg	1.96E-09
Cobalt	Kg	1.4E-08
Copper, ion	Kg	6.9E-10
Dinitrogen monoxide	Kg	2.28E-06
Dioxins, measured as 2,3,7,8-tetrachlorodibenzo-p-dioxin	Kg	3.22E-14
Ethene, tetrachloro-	Kg	1.77E-10
Formaldehyde	Kg	7.67E-08
Gas_use	MJ	1
Hydrogen chloride	Kg	1.63E-06
Iron	Kg	6.9E-10
Lead	Kg	3.51E-09
Manganese	Kg	6.97E-09
Mercury	Kg	2.63E-10
Metals, unspecified	Kg	7.42E-14
Methane	Kg	3.64E-14
Methane, dichloro-, HCC-30	Kg	1.04E-08
Methane, fossil	Kg	1.42E-06
Methane, tetrachloro-, R-10	Kg	7.81E-17
Naphthalene	Kg	2.63E-09
Nickel	Kg	1.96E-07
Nitrogen oxides	Kg	0.000596
Nitrogen oxides	Kg	1.85E-14
NMVOC, non-methane volatile organic compounds, unspecified origin	Kg	1.2E-09
Oils, unspecified	Kg	1.03E-08
Particulates, >2.5, and <10um	Kg	1.18E-05

(continued)

Table 4 (continued)

Inventory outputs flow	Unit	Amount
Particulates, unspecified	Kg	1.54E-16
Particulates, unspecified	Kg	3.22E-08
Phenols, unspecified	Kg	7.84E-09
Radioactive species, unspecified	kBq	0.002101
Selenium	Kg	1.59E-09
Sulfur oxides	Kg	3.03E-05
Sulfur oxides	Kg	6.17E-16
Suspended solids, unspecified	Kg	2.07E-08
TOC, Total Organic Carbon	Kg	7.1E-15
VOC, volatile organic compounds	Kg	3.07E-05
VOC, volatile organic compounds	Kg	4.51E-17
Wood waste, unspecified, combusted in industrial boiler	Kg	0.008333

2.5 *Impact Assessment*

After modeling the system in OpenLCA and obtaining the inventory for each stage of the system, we characterized the impact that the outputs of each stage have. In order to have the impact assessment we need to select a method to generate the characterization of the data output of the inventory. In this LCA study we used the CML2001 (Hischier et al. 2010), which was developed by the Center of Environmental Science at Leiden University. This method considers a midpoint method. Table 5 lists the results of the four most commonly used impact categories based on CML2001 for this LCA study. Table 5 indicates that the gasification process is the stage of the whole system where the major impact is found.

OpenLCA also provides comparative results for the major contributors of the system. Figure 11 compares the gasification process against the other stages of the system in terms of GWP.

Table 5 Impact assessment results for all stages of the system

	Stages			
CML2001 impact	Acquisition	Transportation	Gasification	Use
Climate change—GWP 20a (Kg CO_2-Eq)	5.7970E-11	1.2919E-01	3.4345E+00	1.3849E-02
Acidification potential—generic (Kg SO_2-Eq)	8.3202E-11	5.9615E-04	2.6244E-03	4.1866E-04
Human toxicity—HTP infinite (Kg 1,4-DCB-Eq)	1.4432E-06	1.0340E-03	5.1362E-01	9.2107E-03
Photochemical oxidation (summer smog) (Kg formed ozone)	5.0523E-12	7.5346E-06	3.2733E-05	5.2841E-06

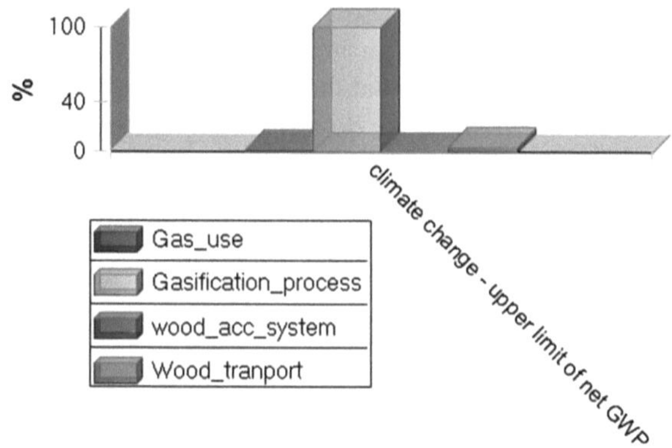

Fig. 11 Graphic output of OpenLCA compression analysis

We also compare the CO_2 emissions of the system against the productions of various fuels in Table 6 based on data from EIA (2014). The table shows that system producing the syngas from biomass has very similar emissions to natural gas. The natural gas has an emission of 53.1 kg of CO_2 per Million Btu which is equivalent to 0.01372 kg of CO_2 per MJ. In our analysis the use stage of the system generate approximately 0.01385 kg of CO_2 per MJ. Even though the emissions are a little higher if we compare the production of natural gas vs the environmental benefits of the overall system of syngas production the use of syngas as an alternative fuel is substantial. We need to take into consideration that the data of CO_2 emissions caused by the energy required in the process are sourced from diesel

Table 6 Carbon dioxide emissions coefficients by fuel type (EIA 2014)

Carbon dioxide (CO_2) factors:	Pounds CO_2 per unit of volume or mass	Kilograms CO_2 per unit of volume or mass	Pounds CO_2 per million Btu	Kilograms CO_2 per million Btu
For homes and businesses				
Propane	12.7/gallon	5.8/gallon	139.0	63.1
Butane	14.8/gallon	6.7/gallon	143.2	65.0
Butane/propane mix	13.7/gallon	6.2/gallon	141.1	64.0
Home heating and diesel fuel	22.4/gallon	10.2/gallon	161.3	73.2
Kerosene	21.5/gallon	9.8/gallon	159.4	72.3
Coal (all types)	4,631.5/short ton	2,100.8/short ton	210.2	95.3
Natural gas	*117.1/thousand feet^3*	*53.1/thousand feet^3*	*117.0*	*53.1*
Gasoline	19.6/gallon	8.9/gallon	157.2	71.3
Residual heating fuel (businesses only)	26/gallon	11.8/gallon	173.7	78.8

power production only. If we used energy produced by other means of power generation like wind or solar farms we could easily reduce the total emissions of the system, thus having a greater impact on the GWP. Also, there are other factors, like the acidification potential, human toxicity or photochemical oxidation, that need to be considered when we do a full comparison of the overall environmental impact.

2.6 Recommendations and Discussion

After the impact assessment, we can identify that the gasification process is the one with the major impact potential. Even though this is the stage of the system with most impact, its environmental impact is still below that of the normal production of other fossil fuels. We need to contemplate that the process to produce the syngas used in this study is pioneering and experimental; it is expected to have procedures optimized in the future. Those optimizations are expected to improve the process and reduce the overall environmental impacts. Furthermore, we believe that if this process is industrialized it can benefit from the economy of scale and therefore the overall emissions can be reduced. The analysis recommends more studies on the optimization of the gasification processes to enhance the environmental sustainability of the whole system since this is the more energy intensive process of the system. More data related to the source of the energy used in the system are recommended in order to have higher accountability for the emissions during the gasification process.

References

Ally J, Pryor T (2007) Life-cycle assessment of diesel, natural gas and hydrogen fuel cell bus transportation systems. J Power Sources 170(2):401–411

Autodesk. Autodesk systaibability workshop autodesk education community. http://sustainabilityworkshop.autodesk.com/products/lifecycle-assessment. Accessed 3 Oct 2014

Cabeza LF, Rincón L, Vilariño V, Pérez G, Castell A (2014) Life cycle assessment (Lca) and Life cycle energy analysis (Lcea) of buildings and the building sector: a review. Renew Sustain Energy Rev 29:394–416

EIA (2000) Annual energy review: U.S. energy information administration. http://www.eia.gov/totalenergy/data/annual/previous.cfm

EIA (2014) Carbon dioxide emissions coefficients: U.S. energy information administration. http://www.eia.gov/environment/emissions/co2_vol_mass.cfm

EISA (2007) H.R. 6–110th congress: energy independence and security act of 2007 Congress, 110th session. http://www.govtrack.us/congress/bills/110/hr6. Accessed 8 Mar 2013

Finkbeiner M, Inaba A, Tan RBH, Christiansen K, Klüppel H (2006) The new international standards for life cycle assessment: ISO 14040 and ISO 14044. Int J Life Cycle Assess 11 (2):80–85

Grassi G, Bridgwater AV (1993) The opportunities for electricity production from biomass by advanced thermal conversion technologies. Biomass Bioenergy 4(5):339–345

GreenDelta (2013) Openlca, 1.3.4. http://www.openlca.org/download_page

Haas G, Wetterich F, Geier U (2000) Life cycle assessment framework in agriculture on the farm level. Int J Life Cycle Assess 5(6):345–348
Heller MC, Keoleian GA, Volk TA (2003) Life cycle assessment of a willow bioenergy cropping system. Biomass Bioenergy 25(2):147–165
Hischier R, Weidema B, Althaus H-J, Bauer C, Doka G, Dones R, Frischknecht R, Hellweg S, Humbert S, Jungbluth N (2010) Implementation of life cycle impact assessment methods. Ecoinvent report
IEO (2013) International energy outlook 2013: U.S. energy information administration. http://www.eia.gov/forecasts/ieo/pdf/0484(2013).pdf
International Standard Organization (ISO) (1997) Iso 14040: environmental management—life cycle assessment—principles and framework
Jaramillo P, Michael Griffin W, Scott Matthews H (2007) Comparative life-cycle air emissions of coal, domestic natural gas, Lng, and Sng for electricity generation. Environ Sci Technol 41 (17):6290–6296
Kim H, Parajuli PB, Yu F, Columbus EP, Batchelor WD (2012) Economic evaluation of syngas production: model development and analysis. Trans ASABE 55(3):1033–1045
Klöpffer W (2006) The role of setac in the development of Lca. Int J Life Cycle Assess 11(1): 116–122
Luo D, Hu Z, Choi DG, Thomas VM, Realff MJ, Chance RR (2010) Life cycle energy and greenhouse gas emissions for an ethanol production process based on blue-green algae. Environ Sci Technol 44(22):8670–8677
NREL (2012) U.S. life cycle inventory database. https://www.lcacommons.gov/nrel/search. Accessed 19 Jan 2014
Parry ML, Canziani OF, Palutikof JP, van der Linden PJ, Hanson CE (2007) Ipcc, 2007: Climate Change 2007: Impacts, Adaptation and Vulnerability. Contribution of Working Group Ii to the Fourth Assessment Report of the Intergovernmental Panel on Climate Change. Cambridge University Press, Cambridge
Petrou EC, Pappis CP (2009) Biofuels: a survey on pros and cons. Energy Fuels 23(2):1055–1066
Rebitzer G, Ekvall T, Frischknecht R, Hunkeler D, Norris G, Rydberg T, Schmidt WP, Suh S, Weidema BP, Pennington DW (2004) Life cycle assessment: part 1: framework, goal and scope definition, inventory analysis, and applications. Environ Int 30(5):701–720
Rubin RS, Soares MA, de Castro D, Brandão VS, Ometto AR (2014) Utilization of life cycle assessment methodology to compare two strategies for recovery of copper from printed circuit board scrap. J Clean Prod 64:297–305
Sheehan JJ (2009) Biofuels and the conundrum of sustainability. Curr Opin Biotechnol 20(3): 318–324
Thévenot A, Aubin J, Tillard E, Vayssières J (2013) Accounting for farm diversity in life cycle assessment studies—the case of poultry production in a tropical island. J Clean Prod 57:280–292
Turconi R, Boldrin A, Astrup T (2013) Life Cycle Assessment (Lca) of electricity generation technologies: overview, comparability and limitations. Renew Sustain Energy Rev 28:555–565
Wei L (2005) Experimental study on the effects of operational parameters of a Downdraft Gasifier
Wei L, Pordesimo LO, Igathinathane C, Batchelor WD (2009) Process engineering evaluation of ethanol production from wood through bioprocessing and chemical catalysis. Biomass Bioenergy 33(2):255–266
Xu C, Chen W, Hong J (2014) Life-cycle environmental and economic assessment of sewage sludge treatment in China. J Clean Prod 67:79–87
Xue X, Collinge WO, Shrake SO, Bilec MM, Landis AE (2012) Regional life cycle assessment of soybean derived biodiesel for transportation fleets. Energy Policy 48:295–303

Physical and Economic Aspects to Assessing Woody Biomass Availability for Bioenergy Production and Related Supply Constraints

Donald L. Grebner, Robert K. Grala, Omkar Joshi and Gustavo Perez-Verdin

Abstract Questions related to energy supply, security, environmental sustainability, and possible alternative sources are of a growing concern due to population growth and increased energy demand. The main goal of this Chapter is to present different aspects related to the availability and recovery of woody biomass as a feedstock for bioenergy production in the southern United States. To facilitate growth of an emerging bioenergy industry it is important to identify feedstocks that are appropriate for bioenergy production. Common feedstocks include logging residues, small-diameter trees, mill waste, and urban wood waste. In addition, there are many factors affecting the estimation of woody biomass feedstock available for processing. They include the intensity and frequency of thinning operations, woody biomass accessibility and recovery, soil nutrient compensation as well as availability and type of existing forest inventory data. Other potential constraints that can affect utilization of woody biomass feedstocks for bioenergy include logging and transportation costs, landowner willingness to produce and sell woody biomass, feedstock storage issues, and mill processing capacity.

1 Introduction

Energy is critical to sustaining and growing any society because it is commonly used in the production of goods people use every day such as food, automobiles, gasoline, electronic equipment, clothes, etc. The list is almost limitless. With the Earth's human population currently being 7.2 billion and its expected increase to 10.8 billion by 2100 (Worldometers 2014), it is clear that demand for energy will be

D.L. Grebner (✉) · R.K. Grala
Mississippi State University, Starkville, USA
e-mail: don.grebner@msstate.edu

O. Joshi
Texas A&M Forest Service, College Station, Texas, USA

G. Perez-Verdin
Instituto Politécnico Nacional CIIDIR, Durango, Mexico

S.D. Eksioglu et al. (eds.), *Handbook of Bioenergy*,
Energy Systems, DOI 10.1007/978-3-319-20092-7_13

increasing. Consequently, questions related to energy sources, its future supply, security, environmental sustainability, and possible alternative energy sources are of a growing concern.

Since the advent of fire, humankind has used woody biomass from trees to generate a variety of products. The most important of which has been the use of wood as a heating source during periods of cold climatic conditions, for cooking foods, and providing light. Wood has been used in this role for thousands of years and is still used that way today in many geographic locations (Meiggs 1982; Westoby 1989; Williams 1989; Perlin 2005; Grebner et al. 2013). However, over time people have discovered and developed uses of other energy sources such as coal, oil, and natural gas as more effective substitutes to woody biomass. These fossil fuels are dense and generate more energy per unit volume than does woody biomass (Biomass Energy Centre 2011). Currently, fossil fuels account for about 83 % of total energy consumption in the United States (Perlack et al. 2005; U.S. Department of Energy 2011). However, unlike woody biomass these fossil fuels are not renewable implying that at some point this energy source will be exhausted or its future extraction might not be economically viable. In addition, a continuing use of fossil fuels for energy production raises concerns related to an increasing concentration of atmospheric carbon dioxide (CO_2) and its impact on climate (Field et al. 2004). Numerous efforts have been undertaken to both decrease reliance on foreign energy sources and mitigate CO_2 emissions by promoting domestic renewable energy sources such as woody biomass. One of such efforts is an initiative to replace 30 % of current petroleum consumption with biomass (Perlack et al. 2005; U.S. Department of Energy 2011).

Renewable energy sources account for 8 % of total energy consumption, of which biomass derived from forest and agricultural sectors constitutes about half of this amount (Perlack et al. 2005; U.S. Department of Energy 2011). Woody biomass is an attractive substitute because of its perceived renewability and lower CO_2 emissions. Trees are planted or are established naturally, grown, and can be harvested for a variety of products including bioenergy. This process can be started all over again after harvest. Thus, researchers have thought of woody biomass as a theoretically viable alternative to fossil fuels given this resources more limited supply and the politically unstable locations from which it is imported. This factor amongst others has stimulated much interest within forested regions of the United States with a goal to develop new economically viable industries that convert woody biomass into bioenergy and stimulate job creation, income generation, and rural development. Estimates derived by Perlack et al. (2005) and U.S. Department of Energy (2011) suggest that forest and agricultural sectors can supply a needed quantity of biomass to reach a 30 % petroleum substitution goal.

In addition, the distribution of fossil fuels is not uniform around the world. In many locations across the globe especially in the United States and China, there are large reserves of coal that will last centuries, but increases in proven petroleum reserves, given current technologies, come from revisions to existing reserves not new discoveries (EIA 2013, 2014). In addition, many of the largest petroleum reserves are located in politically unstable parts of the globe (IAGS 2014). During the 1970s, the OPEC oil embargo generated concern over the future availability of petroleum and stimulated much research on identifying alternative fuel sources (Solomon and Johnson 2009; Rockwood et al. 2012). This issue was revisited in the 2000s during new military conflicts in the Middle East (Solomon and Johnson 2009, 2012).

Many studies have been conducted to examine various aspects of utilizing woody biomass for bioenergy production. Some examples include Overend (1982), Börjesson (2000), Gallagher et al. (2006), Gan and Smith (2006; 2007), Petrolia et al. (2006), Polagye et al. (2007), Guo et al. (2007), Perez-Verdin et al. (2009, 2012), Gruchy et al. (2012), and Radhakrishnan et al. (2013). These studies have focused on various biological growth aspects of trees, soil nutrient compensation issues, the social aspects of growing trees for energy, the technological aspects of converting wood into viable liquid fuels, use of wood in combined heat and power facilities, harvesting and transportation, current policies, landowner willingness to provide woody feedstocks, as well as the economics of production and conversion processes.

The main goal of this Chapter is to present different aspects related to the availability of woody biomass as a feedstock for bioenergy production in the southern United States. The Chapter presents topics related to assessing the availability of woody biomass as a feedstock for bioenergy production by a facility producing energy directly or producing liquid biofuels. The Chapter starts with an identification of relevant feedstock sources and assessment methods for quantifying woody biomass feedstock physical availability. Then, woody biomass feedstocks are discussed in terms of their economic availability. This step requires an estimation of the costs of harvesting, transporting, and transforming the feedstock into a form useable for bioenergy production. This, in turn, requires an understanding of resource ownership patterns, its spatial distribution, and willingness of landowners to provide woody feedstock. Figure 1 shows an outline for assessing different woody biomass feedstocks.

2 Potential Woody Biomass Feedstock Sources

The first step in assessing potential availability of woody biomass feedstocks for an emerging bioenergy industry is to identify feedstocks that are appropriate for bioenergy production (Fig. 1). In the bioenergy industry, there are many such feedstocks that are derived from either the agricultural or forest sectors. In the agricultural sector, common feedstocks include corn and soybeans utilized in producing ethanol and bio-diesel, respectively (Worldwatch Institute 2007). In the forest sector, feedstocks generally are woody in nature and come in the form of logging residues, small-diameter trees, woody byproducts generated by wood processing facilities, and urban wood waste. Based on work by Perez-Verdin et al. (2009), four feedstock types are identified and described below.

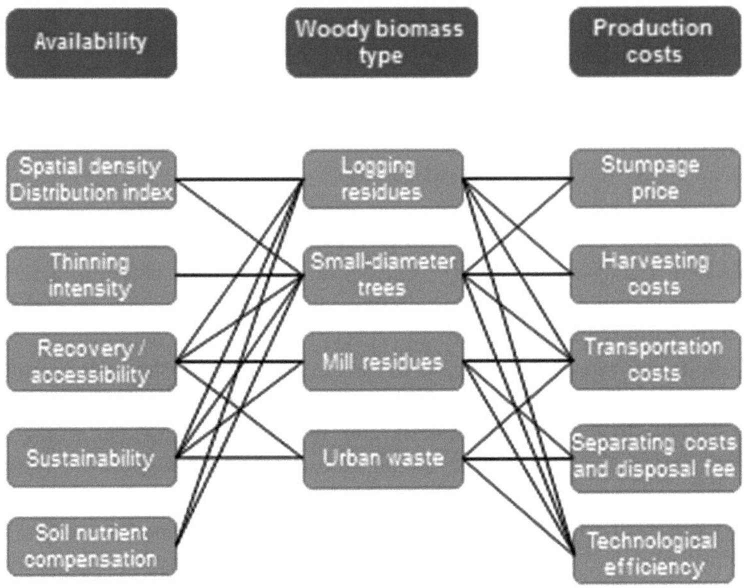

Fig. 1 An evaluation analysis outline for determining availability of woody biomass feedstock for bioenergy production

2.1 *Logging Residues*

Logging residues are typically thought as those parts of a tree that are left in a forest after conventional harvesting operations. They include such tree components as branches, tree tops, twigs, leaves, needles, stumps, and culled trees not suitable for conventional wood products.

Extraction of logging residues from a harvest site can be achieved by a variety of methods. Perez-Verdin et al. (2009) assumed logging residues were extracted by skidding a harvested tree with branches and foliage from the stump location to the landing. At the landing, the harvested tree would be processed into a traditional log form suitable for transportation, whereas the logging residues such as tree tops, branches and foliage were processed by a portable chipper, loaded on a chip truck, and hauled to a bioenergy facility. There are other proposed processing options, such as using a slash bundlers; however, research on using this method is relatively new in the United States and elsewhere (Rummer et al. 2004; Johansson et al. 2006; Karha and Vartiamaki 2006; Meadows and Gallagher 2009; Wilkerson and Perlack 2009; Patterson et al. 2010).

The amount of logging residues generated on the harvest site depends on several factors. First factor includes tree morphology with specific distinction between hardwood and coniferous tree species. Deciduous species tend to be branchier and thus during harvest generate more logging residues and provide more feedstock for bioenergy production. Second factor relates the wood products specifications, such as those of sawtimber and pulpwood, that can greatly impact the amount of tree tops left after harvest. Last factor relates to the amount of logging residues that need to be left on the harvest site to maintain soil nutrients and future site productivity, and depends on the initial site conditions as well as other factors.

2.2 *Small-Diameter Trees*

Small-diameter trees are typically woody stems that have a diameter at breast height (DBH) of less than a specified dimension qualifying it for a desired wood product. Typically, small-diameter trees are pulpwood trees or smaller. For instance, the South Carolina Forestry Commission (SCFC) defines several product classes (SCFC 2014). The SCFC defines pulpwood trees as typically having a DBH of 6 to 9 inches. Chip-n-saw trees, produce both pulpwood and small dimension lumber, and have DBH of 10–13 inches. Sawtimber trees, used for producing dimension lumber, usually have DBH of 14 inches and higher. Veneer trees, higher quality stems used for plywood, have DBH of 16 inches and higher. Small-diameter trees can be obtained through thinning activities and stand regeneration harvests. The amount of potentially extractable small-diameter trees depends on the initial seedling density when a site is either planted or regenerated naturally and the stand's mortality rate.

2.3 *Mill Waste*

Mill waste is obtained during the conversion process of transforming a log into a final wood product and it can be the byproduct of processing a log into either a primary or secondary wood product. It occurs at several points in the process; for instance, when sawing a log into a conventional lumber, waste consists of bark, slabs, shavings, and sawdust. It can also include pieces of wood that are sawn off when trimming boards to specified lengths. In secondary wood product processing, mill waste can include sawdust, shavings, or pieces of wood sawn off to meet product specifications for table legs or chairs as an example.

All wood processing facilities make decisions on how to utilize their byproducts. The Billion Ton Study by Perlack et al. (2005) indicated that 69 % of mills utilize their wood waste for internal energy production such as heat and electricity as well as to extract chemicals. Some mills sell their wood residues to secondary forest products manufacturers which utilize it in production of final products such as different types of boards. Other facilities sell bark as mulch to local yard care facilities, give it away, or dispose in landfills (Joshi et al. 2014). An update to Billion Tons Study indicated that only 1.5 % of residues generated by primary forest products manufacturers is unutilized (U.S. Department of Energy 2011).

2.4 Urban Wood Waste

Urban wood waste is derived from several sources. First, urban wood waste can consist of wood collected from households and construction sites and include the bole of the tree, branches, and foliage as well as a stump. This woody material may also come from landowners removing dead or dying trees on their properties, utility companies trimming back tree tops and branches to protect power lines, and land clearing activities implemented for the construction of new buildings. The second source of urban waste includes demolition wood. Demolition wood is a wood-based material recovered from homes and buildings when they are renovated or simply demolished (Florida Department of Environmental Protection 2014).

Urban wood waste is disposed of in various manners. Yard waste can be collected by local governments involved in a weekly garbage collection. Boles, limbs are simply left by the side of road for collection. Sometimes, landowners bring it to a local landfill. If a landowner has a large enough property, yard waste might be burned on site. Demolition wood can be collected during renovation or demolition projects by loading it into portable metal containers brought to the site. When the containers are full, they are hauled away for recycling or disposal in landfills.

3 Estimating Availability of Woody Biomass Feedstocks

3.1 Factors Affecting Woody Biomass Availability

3.1.1 Thinning Intensity and Frequency

Often when managing forested stands of trees on a tract of timber, thinning operations are often considered when there is a desire to boost the growth rate of crop trees or to promote specific wildlife habitat. When deciding when to thin the stand and how many trees to harvest in a thinning operation, a tool called a stand density index (SDI) is used. The SDI uses the number of trees and tree diameter information to measure the stocking level on a particular forest site. This is typically implemented at a forest stand level. SDI indices are developed for individual tree species. Different species groups can have different maximum SDI which is used to determine how much the forest stand should be thinned to achieve desired forest management objectives. The first SDI was developed by Reineke (1933) and has undergone many modifications.

Perez-Verdin et al. (2009) used a modified SDI at a county level in the state of Mississippi. In their work, stand types were identified as either pine, hardwood, or mixed for each county. The stand type in dominance was selected to be used with the modified SDI. This index considers stand structure attributes such as crown closure and mortality.

$$SDI_{ij} = N_{ij}\left(qmd_{ij}/25\right)^{c}$$

where the SDI is quantified for county (*i*) and species group (*j*), *N* is the average number of trees per unit area for county and species group, *qmd* is the quadratic mean diameter per county and species group, and *c* is a constant assumed to be 1.605. Perez-Verdin et al. (2009) used a maximum SDI of 1028 for hardwood species, 1242 for pine species, and 1176 for mixed species. The maximum SDI allows the user to identify a minimum site occupancy measure when crown closure occurs at 30 % and when self-thinning, also called mortality, occurs at 50 % of maximum SDI. With this information, thinning can occur when targeted counties have a number of trees above the minimum site occupancy threshold, but below the self-thinning ceiling. To estimate the amount of small-diameter biomass that can be removed and utilized as a bioenergy feedstock, the number of trees above the 30 % threshold needs to be divided by the total number of trees per unit area. This approach may be used several times within a stand assuming that growth conditions and landowner objectives are being met. Although Perez-Verdin et al. (2009) applied this tool to the county level, it is important to remember that it is only a generalized guide to the density of physical woody biomass stocks over a broad area. Maximum SDI indices have been developed for specific species which would modify the woody biomass stock estimates at the forest track level.

3.1.2 Woody Biomass Accessibility and Recovery

Another important issue that affects the availability of woody biomass is the actual accessibility and recovery of the resource. In forest stands, accessibility typically refers to whether harvesting crews can drive to a site with their equipment, extract the resource, and transport it off site. Often, wet weather or periods of high rain fall may make many forested sites difficult to access during specific periods during the year. Another factor that affects accessibility is topography. Although states like Mississippi are relatively flat, forested areas in mountainous regions may be difficult to harvest due to steep slopes. In many cases, this difficulty can be overcome with the use of a high lead cable logging design but this is often a more expensive option which is not justifiable for low valued woody biomass feedstocks. Other access limitations pertain to sensitive sites due to threatened and endangered species and stream presence as well as vulnerability to soil erosion.

Recovery is also applicable to woody biomass produced as a byproduct from the processing of raw material to solid wood products. Often these feedstocks are referenced as a potential feedstock because they are often used for energy production on site. In many cases, wood processing facilities have contracts with outside companies to dispose of their waste. An example includes hardwood mulch which can be found in local agricultural co-ops or national retailers. Other potential issue with recovering this residue is the size of woody material. Recovering and using saw dust is potentially less costly than having to chip slabs and cut off portions of sawn boards.

Recovery of urban wood waste is another important consideration when conducting woody biomass feedstock assessments. Although this is an important potential resource, there are several issues that assessments need to account for. First, few if any urban waste disposal facilities extract their wood waste for bioenergy production. Therefore, there is no system developed for extracting and separating this resource efficiently. In many cases, wood waste is intermixed with metals and concrete when it is disposed of in a landfill. Second, in many cases wood waste from sources other than yard waste and trimmings may contain chemicals from paints and preservatives that may inhibit bioenergy production or make extraction more difficult and expensive (Perlack et al. 2005; U.S. Department of Energy 2011).

3.1.3 Soil Nutrient Compensation

An important issue in extracting woody biomass from forest lands for bioenergy production is maintaining the productive capacity of the forest to provide a multitude of resources over time. A key element in maintaining the productive capacity of the site is maintaining sufficient soil nutrient levels. During the life of a forest stand, trees take up nutrients from the soil to sustain their lives and grow in height and diameter. Prior to leaf fall, trees will translocate nutrients from its foliage before it is shed from the tree. Foliage shed from the tree decomposes, becoming part of the organic matter, and returns nutrients to the soil. However, when trees are harvested and moved off site, some nutrients tied up in the wood fiber are also removed from the site. If trees are harvested during the dormant season, more nutrients will be extracted from a forest than if it is done during the growing season. Annually, some nutrients are added to forested sites through the process of atmospheric mineral deposition (Kramer and Kozlowski 1979; Smith 1990). Over time, this accumulation can be substantial.

The Billion Ton Study conducted by Perlack et al. (2005) and later updated by U.S. Department of Energy (2011) suggested leaving a substantial portion of logging residues on site such as nutrient rich foliage and tree crowns to maintain site productivity. The exact amount of logging residues to leave on site to maintain site productivity may depend on factors such as soil type, climate, and topography. For example, Roberts et al. (2005) suggested that there is not much evidence that leaving a portion of logging residues on good sites will greatly affect the inherent site productivity across a typical forest stand rotation although best management practices that leave logging residues on site can help in soil erosion control. However, if forest stand management rotations were to resemble that of short rotation woody crops, then there might be an issue of nutrient drain from the system that can't be compensated through atmospheric nutrient deposition. This would require the addition of fertilizers to maintain adequate growth which would affect the financial returns of these types of woody biomass production systems.

3.2 Existing Data

3.2.1 Woody Biomass Inventory Collection Procedures

Collecting forest inventory data is a key component for assessing forest stocks necessary for achieving private and public forest management goals. Forest inventory can also be utilized to determine availability of woody feedstocks suitable for bioenergy purposes. Historically, forest inventory analysis has relied on aerial photography to identify stands of trees homogenous in terms of species composition. This stand classification system was then used to develop a random stratification process for assigning sampling plots throughout the forest. Typically, foresters would visit the forest and navigate to pre-defined sample plots or points to collect standard metrics such as tree DBH, height, species, and age for estimating the amount of standing timber volume in various tree stands and across a forest. Currently, some form of satellite imagery is used along with GPS units which also serve as data collection and retrieval devices. Although these procedures are commonly used on either small or large landowner properties for numerous objectives, they can also be used at state, regional, and national levels to collect data pertinent to developing state and national forest inventories and policies. The United States Forest Service manages such a system on a national scale and there are also state-level assessment programs developed to meet specific needs.

3.2.2 National and State-Managed Forest Inventory Systems

Quantity of woody biomass that can be used as a feedstock for bioenergy production is directly related to volume of standing timber. While a majority of a merchantable tree can be used for higher value wood products, unused portions can be utilized for bioenergy purposes. Information on quantities of woody biomass potentially available for extraction and utilization for bioenergy purposes can be derived from existing forest inventories which have been developed at national and state levels.

At the national level, the United States Forest Service created the Forest Inventory and Analysis (FIA) National Program in 1930 which was implemented in the Southern United States in 1933 (Reams and van Deusen 1999; United States Forest Service 2014a). The FIA program relies on systematic sampling plots laid out by state. The plots are periodically surveyed to measure changes in selected forest attributes such as tree species, DBH, height, age, density, basal area, regeneration, etc. The FIA inventory is conducted on a state level and involves a re-measurement period of 10 years (Reams and van Deusen 1999). Sample plot survey information can be used to develop inventory estimates for large forest areas as well as to predict growth rates for specific tree species. The plot number varies per state. For example, in Mississippi there are 3,190 plots across 82 counties which gives an average density of approximately 40 sample plots per county. FIA inventory data is available through numerous reports and facts sheets in print and online. FIA also offers access to various data sets and tools that allow users to process inventory data and develop customized output tables and maps through an online interface and offline. FIA tools can be used to estimate current and future availability of woody biomass. For example, Forest Inventory Data Online (FIDO) provides both FIA standard and custom reports that can be filtered by geographic region, type of report, and survey year as well as land and tree attributes (Fig. 2). FIA DataMart provides access to raw data that can be used to derive various stand characteristics. Timber Products Output (TPO) tool allows to derive timber output tables that can be organized by geographic area, year, and table type.

Another source of publically available inventory information can be derived from state-level inventory assessment programs. For instance, Mississippi created the Mississippi Institute for Forest Inventory Program in the early 2000s (MIFI 2010). The goal of this program was to provide a more precise estimation of state level inventory information compared to the FIA program. The goal of the initial pilot program was to provide county forest inventories at a sampling error of ±10 % with a 95 % level of confidence across the state (Parker et al. 2005). Many of the same types of forest attributes measured on the national level are measured at the state level. In contrast to FIA plot distribution within the state, MIFI plots are randomly allocated and there can be up to 100 plots per county (Parker et al. 2005). MIFI offers a Dynamic Reporter that is freely available from their website (Figs. 3). It divides the state into five regions: North, Central, Southeast, Southwest, and Delta. Each region is intended to be resampled every five years. The volume estimates are reported by species and product class for the entire state as well as at the county level. A unique feature is that the Dynamic Report Generator Software can be used to draw polygons to represent potential woody feedstock supply and procurement zones, and to provide all inventory statistics for this defined region. The Dynamic Reporter also provides FIA plot information for comparative

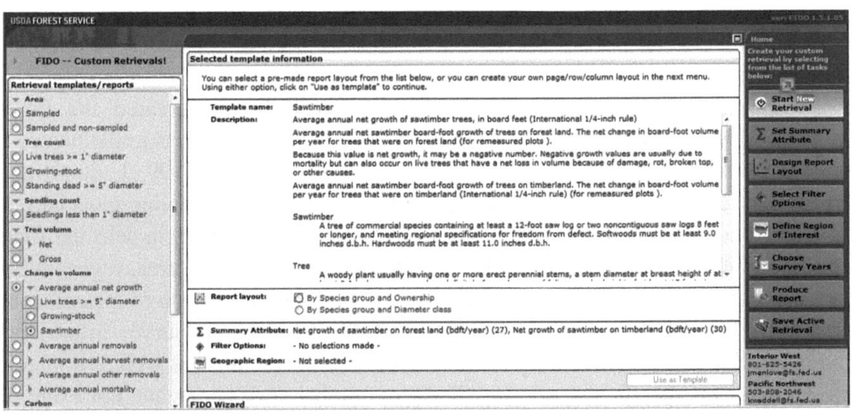

Fig. 2 Forest Inventory Data Online (FIDO) interface for developing customized timber inventory tables (*Source* USDA Forest Service 2014b)

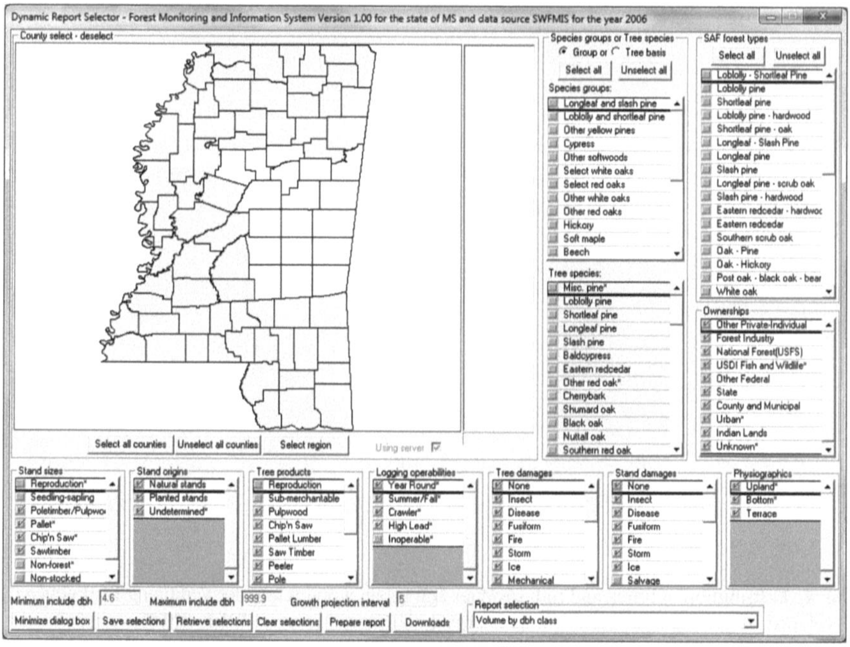

Fig. 3 Interface window of the Dynamic Report Generator Software provided by Mississippi Institute for Forest Inventory (*Source* MIFI 2014)

purposes. Woody biomass feedstock assessment work by Perez-Verdin et al. (2009) initially used MIFI estimates for some regions with MIFI data, whereas for the other regions not yet measured by MIFI they used FIA data. Since publication date, all regions in Mississippi have been measured at least once and forest inventory data is publically available.

3.2.3 Other Sources of Information on Woody Biomass Feedstock Availability

The National Renewable Energy Laboratory of the United States Department of Energy provides information on various sources of renewable energy, including woody biomass, via BioFuels and BioPower Atlases (Fig. 4). The atlases provide information on biomass quantities available for biofuel and power production that can be displayed visually in the form of maps. Each map layer is linked with quantitative information that can be displayed as a table. In addition to biomass information, the atlases provide location of power plants and bioenergy facilities and information on emissions of certain pollutants.

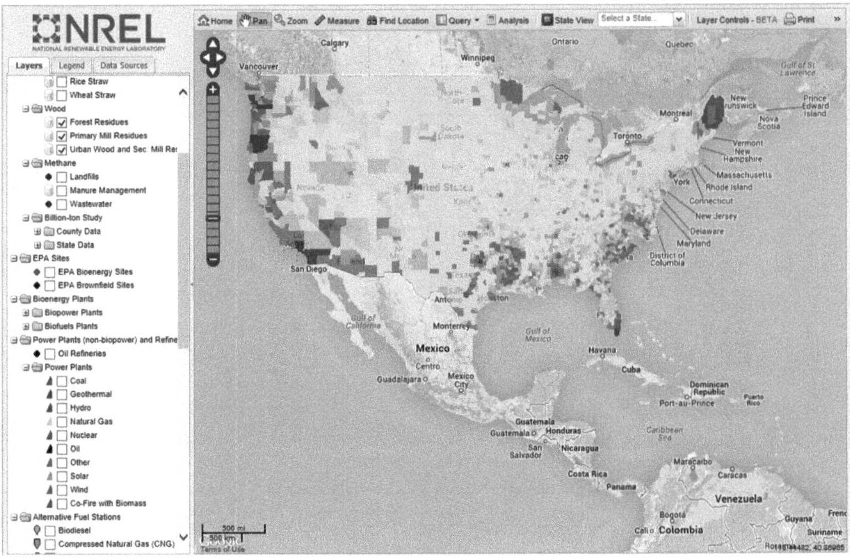

Fig. 4 Distribution of forest and mill residues and urban wood waste derived using BioFuels Atlas developed by the National Renewable Energy Laboratory (*Source* NREL 2014)

3.3 Geographic Information Systems

Geographic Information Systems (GIS) are commonly used in a wide variety of fields including forestry. In terms of woody biomass feedstocks, GIS can be used at various stages of conducting a forest inventory and estimating physical availability as well technical and economic availability of woody biomass feedstocks. In the assessment of standing inventory of woody biomass, feedstock is often stratified into stands that have a homogenous species composition such as pine, hardwood, and mixed species. This information can be used as a basis for designing a forest inventory. Satellite imagery can be used in conjunction with GIS to develop these inventories and illustrate their spatial distribution. In addition, GIS can be used to estimate supply of woody biomass at various stages of the supply chain, within specified geographical areas or in relation to a center-point or a location of a processing facility utilizing existing transportation networks and transportation cost information. Furthermore, GIS also can be utilized to estimate demand for woody biomass by individual and groups of existing and future bioenergy facilities (Figs. 5 and 6).

At the supply stage, GIS can be used to determine best locations for dedicated short-rotation woody crop plantations or best locations for planned bioenergy facilities. At the harvest level, a geospatial analysis can be applied to identify location of scheduled harvests and determine most cost effective routes for hauling woody biomass feedstocks (Fig. 7).

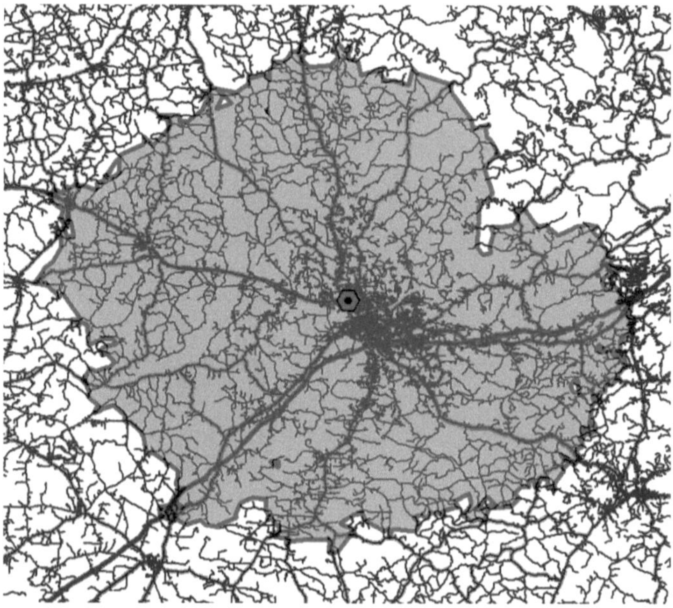

Fig. 5 A procurement zone for a woody biomass feedstock generated with a 30 mile hauling distance for a wood products manufacturer in Mississippi

GIS can also be used to determine woody feedstock supply based on harvest, existing transportation networks, and specified hauling distances. The quantity of potentially recoverable woody biomass can be quantified based on proportion of county area covered with a specific hauling distance. Such approach was used by Gan and Smith (2006, 2007) to identify hauling distance, bioenergy facility processing capacity, and determine availability of logging residues for electricity production. In both cases they used a straight line distance and tortuosity factor to account for transportation network density.[1] However, Langholtz et al. (2006) indicated that such approach might not be sufficient because it doesn't account for transportation infrastructure quality. To account for this limitation, they conducted a spatial analysis to determine supply of woody feedstock based on the actual haul time rather than straight radius distance.

3.4 Surveys

Estimating the physical stock of existing woody biomass is only part of the assessment process because physical availability does not mean that the feedstock

[1]The tortuosity factor accounts for terrain effects. Grebner et al. (2010) defines it as the ratio of actual travel distance to line of sight distance.

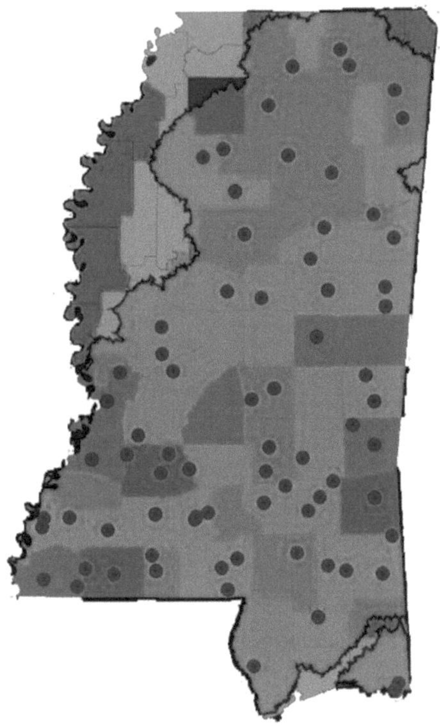

Fig. 6 A state-wide procurement zone for woody biomass feedstock established with a 30 mile hauling distance from primary and secondary forest products manufacturers in Mississippi

will be extracted and used for bioenergy purposes. An important aspect of the assessment is to understand how forest landowners will behave given a variety of factors including conditions in the local and global markets and whether they will be willing to supply woody biomass to the market. A key tool for collecting this information is the use of survey instruments that can be mailed, conducted by telephone, offered online, or used in face-to-face interviews. Forest landowners can vary greatly in motivations for managing their forests. This can depend also on whether they are public or private owners of the resource. Industrial forest landowners, broadly defined here to also include Timberland Investment Management Organizations (TIMOs) and Real Estate Investment Trusts (REITs), typically have a profit maximization motive (Bettinger et al. 2009; Grebner et al. 2013). Non-industrial private forest (NIPF) landowner objectives for managing their forests can vary greatly from some individuals interested in profit maximization to others concerned with wildlife habitat management, aesthetics, preservation, recreation, and other non-consumptive uses (Kluender and Walkingstick 2000; Majumdar et al. 2007; Grebner et al. 2013). Public forest landowners have a multiple-use perspective which requires them to often manage forests for many different objectives (Grebner et al. 2013). For instance, public forests can be managed to produce not only timber but also provide wildlife habitat and recreational opportunities.

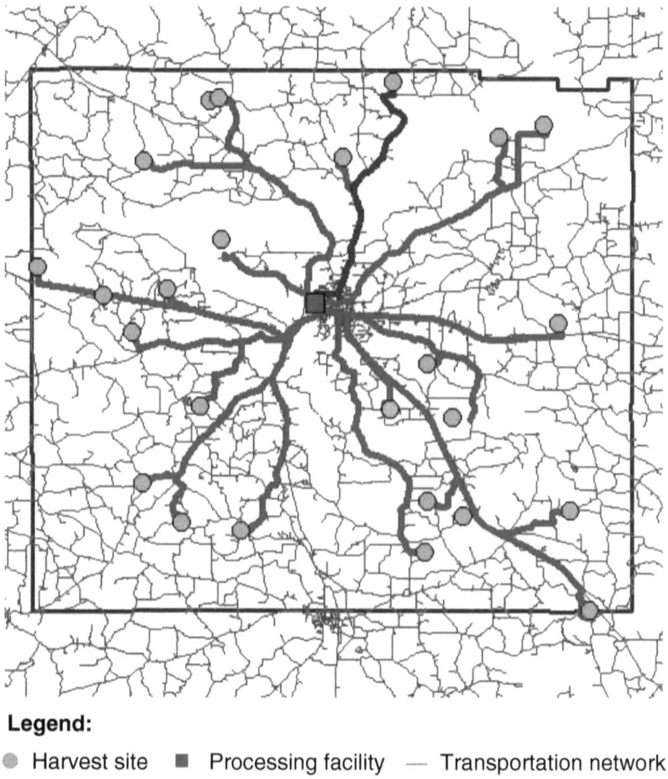

Fig. 7 Map illustrating the shortest distances for hauling woody biomass feedstocks from harvest sites to a processing facility using existing transportation network in Neshoba County, Mississippi

The design of these surveys and the questions they ask can vary widely. Studies conducted by Gruchy et al. (2012) and Joshi et al. (2013a, b) were based on a survey instrument with the goal of examining different aspects of landowner behavior with respect to providing woody biomass feedstocks. Gruchy et al. (2012) examined NIPF landowner behavior using a multinomial logit model to assess their preference for woody biomass for bioenergy production scenarios compared to standard forest harvesting regimes. Although these landowners preferred bioenergy production scenarios because of the extra revenue generated, many landowners were also willing to forego additional revenue and incur a loss to reduce the level of logging residues on site which can be utilized for bioenergy production. Joshi et al. (2013a, b) employed a choice experiment to evaluate NIPF landowner willingness to participate in bioenergy production. Most of the surveyed landowners preferred the management scenarios with bioenergy but did not want negative environmental impacts on their properties.

Surveys can also be used to assess availability of other woody biomass feedstock sources. One such study by Garrard and Leightley (2005) focused on a subregion within Mississippi to assess the potential availability of wood waste generated by wood processing facilities and the cost of disposal. Their study employed on-site interviews to assess woody residue stocks and waste disposal costs. Joshi et al. (2014) employed a survey instrument to estimate the available volume of wood waste from primary and secondary forest products industries in Mississippi. Their goal was to implement the Garrard and Leightley (2005) study format for the entire state. Garrard and Leightley (2005) results showed greater potential volumes of wood waste available for use than initially projected by Perez-Verdin et al. (2009). Part of the discrepancy may be due to the different sources of data used in the earlier study. Work by Carter (2010) and G.C. and Potter-Witter (2011) are other examples of state-wide assessments utilizing surveys to assess the availability of wood waste for bioenergy uses.

4 Existing Constraints to the Use of Woody Biomass Feedstocks

4.1 *Logging Costs*

An important constraint when evaluating the availability of woody biomass as a bioenergy feedstock is the cost of extracting the targeted biomass resource from a forest and transporting it to a bioenergy processing facility. In estimating the cost of this process, it is important to realize that the stumpage prices paid to the landowner for biomass are residual prices representing a difference between a delivered price at the processing facility and the cost of harvesting and transporting the feedstock to the facility gate (Rideout and Hesseln 1997; Grebner et al. 2005). Harvest and transportation costs play a critical role in the economic availability of woody biomass for bioenergy production. Ashton et al. (2007a) indicated that harvest costs can constitute up to 50 % of the total biomass delivered cost, whereas Perez-Verdin et al. (2009) and Grebner et al. (2010) suggested that at least 40 % of the feedstock cost is due to transporting the material from the forest to the processing facility.

When considering woody biomass sourced from logging residues, such as branches, tree tops, and foliage and as well as small-diameter trees, thought has to be given to how this material is to be moved out of the forest. Harvesting woody biomass for bioenergy purposes is affected by numerous factors such as harvested tract size, species composition, harvesting equipment, fuel cost, volumes of wood biomass to be harvested as well as forest restrictions and ordinances (Ashton et al. 2007b). Harvest of woody biomass for bioenergy purposes alone might not be economically feasible and typically has to be integrated with a harvest of traditional wood products such as sawlogs and pulpwood (Ashton and Jackson 2007). The major obstacle with extracting logging residues is the fact that it is a coarse material which results in high per unit delivery cost and typically requires compaction to make transportation economically viable (Gonzalez et al. 2011). Various approaches have been undertaken to lower costs associated with extraction of woody biomass from a harvest site. One approach is to include a wood chipper to convert logging residues into fuel chips on a harvest site. For example, Perez-Verdin et al. (2009) and Grebner et al. (2010) assumed that whole trees were harvested and hauled to the landing. At the landing, the tree tops, branches, and foliage were separated from the main bole and chipped. The chips were then blown into a trailer and transported to a bioenergy processing facility. This assumption differs from commonly adopted harvesting systems in which branches and tree tops are typically cut off and left in the forest. Bringing trees to the landing and producing wood chips on site helps to produce a denser material and improves the cost efficiency of transporting woody biomass feedstock (Gonzalez et al. 2011). Ashton and Jackson (2007) suggested using smaller and multifunction harvesting equipment to lower harvest costs; however, they also indicated that such equipment might not be sufficient to harvest larger volumes or meet safety standards.

When evaluating other potential woody biomass feedstocks such as mill residues and urban wood waste the process is different. In the case of the woody residues at mills, they are a byproduct of the converting logs into wood products such as 2” × 4”s, plywood, oriented strand boards (OSB), etc. During this transformation, the byproducts are generated. This material has costs pertaining to its separation, disposal, or transportation to end user. Commonly, this feedstock is considered superior to logging residues because it is considered clean, might have a more consistent form, and have low moisture content (Perlack et al. 2005; U.S. Department of Energy 2011). Clean mill residues include byproducts such as sawdust, shavings, and miscellaneous scraps that are not mixed with bark and dirt. Clean mill residues are less expensive to process with biofuel technologies or reutilize.

Urban wood waste is another feedstock that can serve as a bioenergy feedstock; however its use is associated with several challenges as there are many technical questions regarding how to estimate the quantity of this resource. Unlike woody biomass from wood processing mills, urban wood waste is considered dirty because often it is mixed with a variety of chemicals such as paints, stains, and oils as well as construction and demolition debris. Apart from fixed and incremental costs attributable to transporting this feedstock to a specific location, there are also costs associated with separating and chipping it. In Mississippi, Perez-Verdin et al. (2009) and Grebner et al. (2010) estimated that urban wood waste was a more expensive alternative to mill wood residues but a potentially less costly option than collecting logging residues and small-diameter trees.

4.2 *Landowner Willingness to Participate*

In the southern United States, the majority of forest land and woody biomass are owned by NIPF landowners. Therefore, the final woody biomass availability depends on landowner willingness to provide this feedstock. NIPF landowners have various and often multiple forest ownership objectives. Some of them are profit driven such timber production, whereas others might include non-market objectives such as game and non-game wildlife habitat, carbon sequestration, aesthetics, and recreation (Kluender and Walkingstick. 2000. In addition, the willingness of NIPF landowners to provide woody biomass might be affected by their knowledge of the woody biomass resource and their perception of bioenergy markets.

The larger landowners typically view their forest properties as an investment and are generally more active learners of their resource and potential forest management alternatives that might increase return on their investment. Smaller landowners, in particular NIPF landowners, are much more diversified in not only their objectives, but in what they know about their forested properties. Some of NIPF landowners are active managers, whereas others are passive owners. In some cases, owners who inherited their forest land are absentee owners and might have a limited knowledge of their property characteristics (Bullard et al. 2002). The level of involvement and knowledge levels on potential uses of woody biomass can have a substantial impact on whether they are willing to participate in bioenergy markets and provide woody biomass to bioenergy processing facilities.

4.3 Storage Issues

In bioenergy research, feedstock storage in an important issue that needs to be considered. When producing corn-based ethanol, corn has to be harvested from the field every year and then stored until it is used in the fermentation process. Storing the corn is a cost of the bioenergy processing facility and the farmer has to harvest the corn regardless of price. The woody biomass resource is different. If forest landowners are not satisfied with an offered price, they can decide to postpone harvest of their stand and let trees continue to grow. This delayed harvest can occur for several years while the trees grow in volume and can possibly reach a higher valued product class. Even though unharvested trees continue to grow in volume, landowners do face risks of forest damage due to wildfires, windstorms such as tornados and hurricanes, and pests as well as timber price fluctuations (Haight et al. 1995; Bright and Price 2000). Many of these risks depend on geographical location as well as other factors such as weather conditions and implemented forest management practices.

4.4 Mill Capacity

Mill wood residues are produced by primary and secondary wood products manufacturers as well as by pulp and paper manufacturing facilities. However, from a woody biomass feedstock perspective, much of these resides are already utilized on-site for heating, electricity generation, and extraction of chemicals. Perlack et al.

(2005) and U.S. Department of Energy (2011) indicated that the pulp and paper industry meets 60 % of its energy requirements in this manner. However, a state-level study by Joshi et al. (2014) suggested that even though approximately 69 % of the mill residues in Mississippi are utilized for on-site energy generation, substantial quantities of mill residues that are disposed of or sold to other vendors can be available for bioenergy purposes if a satisfactory price is offered. Few studies have closely examined the future potential of this feedstock and the behavior of mill owners and managers toward an increased utilization of mill residues.

5 Concluding Remarks

Evaluating the availability of woody biomass for bioenergy production is an important aspect of any wood-based bioenergy feasibility study. Woody biomass has been used as an energy source since the earliest human societies. Since that time, the use of wood has evolved into a diverse set of products used for human consumption. However, society is revisiting the use of woody biomass for energy in new forms because of concerns related to climate change as well as political instability in oil producing regions of the world. The first step in evaluating the availability of woody biomass is conducting an inventory of the existing physical stocks of woody biomass. While there are numerous methods for collecting this information, it is important to recognize that presence of woody biomass physical stocks is not synonymous with practical availability of woody biomass for bioenergy production due to technical, economic, and social constraints.

Estimating the physical stocks of woody biomass is only part of any assessment. Understanding resource ownership patterns and owner behavior is critical to evaluate woody biomass feedstock economic availability. Forest landowners have different landownership goals depending on whether they are public or private owners or if they are large or small owners. Often forest landowners have multiple objectives that are not dependent on the financial returns from timber production. Landowner ownership goals can have a very substantial impact on types and quantities of woody biomass feedstock available to bioenergy facilities.

Other aspects of woody biomass utilization that should be taken into consideration are the costs of logging and transporting woody biomass feedstock from the forest to the bioenergy facility. Currently, existing harvesting systems can accommodate the technical conversion of woody biomass into forms useable for bioenergy production. However, given the limited markets for this feedstock, it is difficult to estimate the price for this material other than estimating the costs of producing it and using these estimates as a proxy prices.

A key concept learned from conducting woody biomass feedstock assessments is that there are many approaches available. The approach and techniques used can vary depending on the available data. The general types of feedstock do not vary; however, estimating feedstock physical and economic availability can be challenging.

Any assessment critically depends on the availability of quality data. Quality data can depend on how and at what scale it is collected. An aim for this Chapter was to highlight some of the key factors that are important in completing any assessment of woody biomass used for bioenergy production. Not taking these factors into consideration can lead to an assessment process that is flawed. In addition, this Chapter can serve as a vehicle for communicating with professionals in agriculture, engineering, and public policy, as well as venture capitalists and lay persons.

There are many possible future research directions related to the use of woody biomass for bioenergy. Continuing to evaluate landowner willingness to provide woody biomass is critical. Especially, as more research explores woody biomass's role as an ecosystem service. Another potential research direction should include the development of new production technologies that improve the harvesting and transport of woody biomass. Given the lower value of this feedstock, it is important to develop lower cost extraction techniques as well as reduce the transportation costs of delivering this material to the processing facility. Although this Chapter is not focused on resource extraction technologies, exploring lower costs of production can have important impacts on NIPF landowner willingness to provide woody biomass. Other possible research ideas include examining the potential impacts of insects and diseases on dedicated tree plantations grown to meet bioenergy feedstock needs. This will be especially important with changing climatic conditions.

References

Ashton S, Baker S, Jackson B, Schroeder R (2007b) Conventional biomass harvesting systems. In: Hubbard W, Biles L, Mayfield C, Ashton S (eds) Sustainable forestry for bioenergy and bio-based products: Trainers curriculum notebook. Southern Forest Research Partnership, Inc, Athens, GA, p 316

Ashton S, Jackson B (2007) Small-scale woody biomass harvesting systems. In: Hubbard W, Biles L, Mayfield C, Ashton S (eds) Sustainable forestry for bioenergy and bio-based products: Trainers curriculum notebook. Southern Forest Research Partnership, Inc, Athens, GA

Ashton S, Jackson B, Schroeder R (2007a) Cost factors in harvesting woody biomass. In: Hubbard W, Biles L, Mayfield C, Ashton S (eds) Sustainable forestry for bioenergy and bio-based products: Trainers curriculum notebook. Southern Forest Research Partnership, Inc, Athens, GA

Bettinger P, Boston K, Siry J, Grebner DL (2009) Forest management and planning. Academic Press, New York, p 331

Biomass Energy Centre (2011) Typical calorific values of fuels. http://www.biomassenergycentre.org.uk/portal/page?_pageid=75,20041&_dad=portal&_schema=PORTAL. Accessed 11 Aug 2014

Börjesson P (2000) Economic valuation of the environmental impact of logging residue recovery and nutrient compensation. Biomass Bioenergy 19:137–152

Bright G, Price C (2000) Valuing forest land under hazards to crop survival. Forestry 73(4):361–370

Bullard SH, Gunter JE, Doolittle ML, Arano KG (2002) Discount rates for nonindustrial private forest landowners in Mississippi: how high a hurdle? South J Appl Forest 26(1):26–31

Carter JM (2010) North Carolina biomass availability and pellet production from mixed forest understory for bioenergy industry. MSc thesis, North Carolina State University, Raleigh, p 108

Field CB, Raupach MR, Victoria R (2004) The global carbon cycle: Integrating humans, climate, and the natural world, Chapter 1. In: Field CB, Raupach MR (eds) The global carbon cycle: integrating humans, climate, and the natural world. Island Press, Washington DC, United States, p 526

Florida Department of Environmental Protection (2014) Definition of construction and demolition debris. http://www.dep.state.fl.us/waste/categories/recycling/cd/candddefinition.htm. Accessed 15 Aug 2014

Gallagher T, Shaffer B, Rummer B (2006) An economic analysis of hardwood fiber production on dryland irrigated sites in the US Southeast. Biomass Bioenergy 30:794–802

Gan J, Smith CT (2006) Availability of logging residues and potential for electricity production and carbon displacement in the USA'. Biomass Bioenergy 30:1011–1020

Gan J, Smith CT (2007) Co-benefits of utilizing logging residues for bioenergy production: the case for East Texas, USA. Biomass Bioenergy 31:623–630

Garrard AW, Leightley L (2005) Characterizing wood waste from wood products companies in North Mississippi. Mississippi State University, Forest and Wildlife Research Center, Research Report

Gonzalez R, Phillips R, Saloni D, Jameel H, Abt R, Pirraglia A, Wright J (2011) Biomass to energy in the southern United States: Supply chain and delivered costs. BioResources 6 (3):2954–2976

Shivan GC, Potter-Witter K (2011) Status of Michigan's forest products industry in the changing market environment. For Prod J 61(1):77–82

Grebner DL, Grace LA, Stuart W, Gilliland DP (2005) A practical framework for evaluating hauling costs. Int J For Eng 16:115–128

Grebner DL, Perez-Verdin G, Sun C, Munn IA, Schultz EB, Matney TG (2010) Woody biomass feedstock availability, production costs and implications for bioenergy conversion in Mississippi, Chapter 12. In: Solomon BD, Luzadis VA (eds.) Renewable energy from forest resources in the United States. Routledge, London, p 330

Grebner DL, Bettinger P, Siry J (2013) Introduction to forestry and natural resources. Academic Press, New York, p 496

Gruchy SR, Grebner DL, Munn IA, Joshi O, Hussain A (2012) An assessment of nonindustrial private forest landowner willingness to harvest woody biomass in support of bioenergy production in Mississippi: a contingent rating approach. For Policy Econ 12(2):140–145

Guo Z, Sun C, Grebner DL (2007) Utilization of forest derived biomass for energy production in the U.S.A.: status, challenges, and public policies. Int Forest Rev 9:748–758

Haight RG, Smith WD, Straka TJ (1995) Hurricanes and the economics of Loblolly Pine plantations. Forest Science 41(4):675–688

Institute for the Analysis of Global Security (IAGS) (2004) The geopolitics of oil. http://www.iags.org/geopolitics.html. Accessed 11 Aug 2014

Johansson J, Liss J-E, Gullberg T, Bjorheden R (2006) Transport and handling of forest energy bundles: advantages and problems. Biomass Bioenergy 30:334–341

Joshi O, Grebner DL, Munn IA, Grado SC, Grala RK, Hussain A (2014) Factors influencing utilization of woody biomass from wood processing facilities in Mississippi. For Prod J 64 (1/2):64–71

Joshi O, Grebner DL, Munn IA, Hussain A, Gruchy SR (2013a) Understanding landowner preferences for woody biomass harvesting: a choice experiment-based approach. For Sci 59 (5):549–558

Joshi O, Grebner DL, Hussain A, Grado SC (2013b) Landowner Knowledge and willingness to supply woody biomass for wood-based bioenergy: sample selection approach. J For Econ 19 (2013):97–109

Karha K, Vartiamaki T (2006) Productivity and costs of slash bundling in Nordic conditions. Biomass Bioenergy 30:1043–1052

Kluender RA, Walkingstick TL (2000) Rethinking how non-industrial landowners view their lands. South J Appl Forest 24(3):150–158

Kramer PJ, Kozlowski TT (1979) Physiology of woody plants. Academic Press, San Diego, California, p 811
Langholtz M, Carter DR, Marsik M, Schroeder R (2006) Measuring the economics of biofuel availability: ArcGIS network analyst extension provides more comprehensive assessment, vol 9, no 4. ArcUser October-December 2006, p 3
Majumdar I, Teeter L, Butler B (2007) Characterizing family forest owners: a cluster analysis approach. For Sci 54(2):176–184
Meadows S, Gallagher T (2009) Optimizing the use of a John Deere bundling unit in a southern logging system. In: 2009 council on forest engineering (COFE) conference proceedings: "Environmentally Sound Forest Operations". Lake Tahoe, Nevade, 15–18 June 2009
Meiggs R (1982) Trees and timber in the ancient Mediterranean world. Oxford University Press, Oxford, p 553
Mississippi Institute for Forest Inventory (MIFI) (2010) Home. http://www.mifi.ms.gov/. Accessed 7 Feb 2014
Mississippi Institute for Forest Inventory (MIFI) (2014) MIFI software. http://www.mifi.ms.gov/. Accessed 21 July 2014
National Renewable Energy Laboratory (NREL) (2014) Bioenergy atlas. http://maps.nrel.gov/bioenergyatlas. Accessed 1 Aug 2014
Overend RP (1982) The average haul distance and transportation work factor for biomass delivered to a central plant. Biomass 2:75–79
Parker RC, Glass PA, Londo HA, Evans DL, Belli KL, Matney TG, Schultz EB (2005) Mississippi's forest inventory pilot program: Use of computer and spatial technologies in large area inventories. In: Forest and Wildlife Research Center, Bulletin FO 274. Mississippi State University, p 43
Patterson DW, Hartley JI, Pelkki MH, Steele PH (2010) Effects of 9 months of weather exposure on slash bundles in the mid-south. For Prod J 60(3):221–225
Perez-Verdin G, Grebner DL, Sun C, Munn IA, Schultz EB, Matney TG (2009) Woody biomass availability for bioethanol conversion in Mississippi. Biomass Bioenergy 33(3):492–503
Perez-Verdin G, Navar-Chaidez JJ, Grebner DL, Soto Alvarez CE (2012) Disponibilidad y costos de producción de biomasaforestalcomomateria prima para la conversión de bioetanol. For Syst 21(3):526–537
Perlack RD, Wright LL, Turhollow AF, Graham RL, Stokes BJ, Erbach DC (2005) Biomass as feedstock for a bioenergy and bioproducts industry: the technical feasibility of a billion-ton annual supply, DOE/GO-102005-2135. Prepared by Oak Ridge National Laboratory for the US Department of Energy and US Department of Agriculture, Washington, DC
Perlin J (2005) A forest journey: the story of wood and civilization. The Countryman Press, Woodstock, Vermont, United States, p 463
Petrolia DR (2006) The economics of harvesting and transporting hardwood forest residue for conversion to fuel ethanol: a case study for Minnesota, Staff Paper P06-15. Department of Applied Economics, University of Minnesota, Minneapolis
Polagye BL, Hodgson KT, Malte PC (2007) An economic analysis of bio-energy options using thinnings from overstocked forests. Biomass Bioenergy 31:105–125
Radhakrishnan S, Paz JO, Yu F, Eksioglu S, Grebner DL (2013) Assessment of potential capacity increases at combined heat and power facilities based on available corn stover and forest logging residue. Energies 6(9):4418–4428
Reams GA, Van Deusen PC (1999) The southern annual forest inventory system. J Agric Bio Environ Stat 4(4):346–360
Reineke LH (1933) Perfecting a stand density index for even-aged forests. J Agric Res 46:627–638
Rideout DB, Hesseln H (1997) Principles of forest and environmental economics. Resource & Environmental Management, LLC
Roberts, SD, Harrington CA, Terry TA (2005). Harvest residue and competing vegetation affect soil moisture, soil temperature, N availability, and Douglas-fir seedling growth. For Ecol Manage 205:333–350

Rockwood DL, Kirst M, Isebrands JG, Zhu JY (2012) Forest trees, Chapter 15. In: Kole C, Joshi CP, Shonnard DR (eds.) Handbook of bioenergy crop plants. CRC Press, Boca Raton, Florida, United States, p 824

Rummer B, Len D, O'Brien O (2004) Forest residues bundling project: new technology for residue removal. Prepared by Forest Service Forest Operations Research Unit, Southern Research Station, Auburn, Alabama

Smith WH (1990) Air pollution and forests: interaction between air contaminants and forest ecosystems. Springer-Verlag, New York, Inc., p 618

Solomon BD, Johnson NH (2009) Introduction, Chapter 1. In: Solomon BD and Luzadis VA (eds) Renewable energy from forest resources in the United States. Routledge, London, p 330

Solomon BD, Johnson NH (2012) Public policies, economics, public perceptions, and the future of bioenergy crops, Chapter 12. In: Kole C, Joshi CP, Shonnard DR (eds.) Handbook of bioenergy crop plants. CRC Press, Boca Raton, Florida, United States, p 824

South Carolina Forestry Commission (SCFC) (2014) Understanding timber as a commodity. http://www.state.sc.us/forest/lecom.htm. Accessed 11 Aug 2014

US Energy Information Administration (EIA) (2013) International energy outlook 2013. http://www.eia.gov/forecasts/ieo/pdf/0484(2013).pdf. Accessed 11 Aug 2014

US Energy Information Administration (EIA) (2014) Coal explained: how much coal is left?. http://www.eia.gov/energyexplained/index.cfm?page=coal_reserves. Accessed 26 Aug 2014

USDA Forest Service (2014a) We are the nation's forest census. http://www.fia.fs.fed.us/. Accessed 7 Feb 2014

USDA Forest Service (2014b) Forest inventory data online (FIDO). http://apps.fs.fed.us/fia/fido/index.html. Accessed 16 July 2014

U.S. Department of Energy (2011) U.S. Billion-ton update: Biomass supply for a bioenergy and bioproducts industry. R.D. Perlack and B.J. Stokes (Leads), ORNL/TM-2011/224. Oak Ridge National Laboratory, Oak Ridge, Tennessee, p 227

Westoby J (1989) Introduction to world forestry. Basil Blackwell. Oxford, United Kingdom, p 228

Wilkerson EG, Perlack RD (2009) Resource assessment, economics and technology for collection and harvesting, Chapter 4. In: Solomon BD, Luzadis VA (eds.) Renewable energy from forest resources in the United States. Routledge, London, p 330

Williams M (1989) Americans and their forests: a historical geography. Cambridge University Press, United Kingdom, p 599

Worldometers (2014) Curent world population. http://www.worldometers.info/world-population/. Accessed 18 Feb 2014

Worldwatch Institute (2007) Biofuels for transportation: global potential and implications for sustainable energy and agriculture. Earthscan, London, United Kingdom, p 452

The Role of Biofuels in Achieving a National Energy Independence Plan

Soheil Shayegh

Abstract Biofuel can play a crucial role in achieving energy independence goals. The success story of the Brazil's ethanol program started in 1976 and during the past four decades it has gained remarkable achievements in terms of economic development and energy independence. The aim of this study is to show how Brazil has achieved its goal of energy independence by implementing and updating its renewable fuel policies over the past 40 years. We adopt an evolutionary policy framework to study the emergence of ethanol as an alternative fuel in Brazil. We propose a dynamic framework for technology diffusion policies that can explain why Brazil chose ethanol as an alternative fuel and how this policy affected the country's energy market. We argue that the success of the Brazilian ethanol program is due to its flexibility and adaptability over the past four decades. In addition to this, we note that other domestic and international forces such as oil price and domestic politics have been largely responsible for the success or failure of these policies. Lastly, we discuss the key lessons from Brazil which can help the policy makers in other countries in devising a sustainable energy policy in the future.

1 Introduction

Energy production and consumption are significant contributors to economic growth of nations from the most developed countries in Europe and North America to the least developed parts of sub Saharan Africa. Our economic dependence on energy is so that in some places like the United States, foreign oil dependence represents a serious national security threat (Deutch et al. 2006). The consumption of fossil fuels at its current increasing rate imposes a serious threat to the future of our planet. Climate studies show a rapid increase in carbon emissions after the

S. Shayegh (✉)
School of Industrial and Systems Engineering, Georgia Institute of Technology, Atlanta, Georgia
e-mail: soheilsh@gatech.edu

S.D. Eksioglu et al. (eds.), *Handbook of Bioenergy*, Energy Systems, DOI 10.1007/978-3-319-20092-7_14

recent economic crisis in 2008–2009 (Peters et al. 2012). Fossil fuel dependency can be reduced or limited by providing viable alternatives which provides the same services and benefits to different sectors including power and transportation. Policies that aim to reduce the cost of renewable energies or make them accessible by larger population will nevertheless induce voluntary adjustments in energy usage which is more effective than policies to raise the cost of current fossil fuel consumption in order to limit its consumption. Although emissions trading schemes and carbon tax policies have received substantial attention among more developed countries, over the last few decades few developing countries have initiated different policy paths to reduce their oil dependency. The alternative policies may include market mechanisms for introducing new energy fuels or non-market mechanism through providing incentives for producers and consumers of these new energy resources. The ethanol fuel program (Proalcool) in Brazil is a successful example of creating new market for an emerging technology which has been supported by the national and regional levels of policy making domain in the country. In the next few sections we first propose an evolutionary policy framework for studying the biofuel production program as a mean to achieve the national energy independence. Then we provide a general overview of the biofuel program in Brazil. The aim of this study is to show how a country like Brazil has achieved its goal of energy independence by implementing and updating its renewable fuel policies over the past 40 years.

2 Evolutionary Policy Framework

For the purposes of this paper, we adopt an evolutionary policy framework to study the emergence of biofuels as an alternative to fossil fuel economy. We also propose a dynamic framework for technology diffusion policies that can explain why a certain country like Brazil has chosen ethanol as an alternative fuel and how this policy affected the country's energy market. We argue that the success of the Brazilian ethanol program is due to its flexibility and adaptability over the past four decades. In addition to this, we also take a look at the domestic and international forces that caused these policies to succeed or fail. Lastly, we discuss the technology evolution and its impact on policy.

We start by looking at what may cause a government to intervene in energy market and how such intervention will be received by the private sector. It is generally believed that market mechanisms are the most efficient way to coordinate resources and distribute wealth in the society. However, some studies have justified government intervention in the market in limited cases, such as:

- When the market fails to account for externalities,
- When market mechanisms fail to respond quickly to economic shocks or social changes, or

- When there are not enough players in the market or the market is losing its competitiveness

In these cases, *limited* government intervention is justified or even prescribed to guarantee that market functionality is not undermined. This intervention can happen through two categories of policies (Edquist 2002):

1. Non-market mechanisms: new legislations, regulations and standards such as tax incentives, emission control acts, etc.
2. Market mechanisms: improving competitiveness via anti-trust measures or creating new markets in the area of innovation.

Before any intervention, the government should make sure that:

- The market mechanism cannot solve the problems and achieve its goals
- Government intervention has the ability to solve these problems

Government intervention can be temporary (in response to a shock or new innovation) or permanent (in the case of health and environment standards). However, if a government decides to intervene in the market, its policies should be flexible enough for revisions in the future as the market evolves.

Figure 1 depicts this concept in the framework of Brazilian ethanol program. Each component of this graph is explained in more detail below, but it is worth noting that the ethanol program was first introduced to the Brazil energy market under the military dictatorship of the 1970s. In that era, any intervention and introduction of new policies did not need to be justified by the government, as the ethanol program was established by presidential decree and without gaining approval from the parliament. However, to make this program successful in the years to come, the Brazilian government deployed several measures (Hira and de Oliveira 2009):

- It established a market for ethanol and supported the industry during its infant phase and throughout market crisis;
- It invested in infrastructure and long-term research and development programs to increase efficiency and reduce the cost of technology; and
- It allowed market mechanisms to determine the direction of investment in ethanol industry after it enters its maturity phase.

Studies have shown systemic problems such as a lack of stable institutions to stimulate renewables, and a poor alignment of these institutions with practices in other sectors and regional/local institutions hamper the development of renewable energy technologies (Negro et al. 2012). Without intervention from government, transformation in the energy-economic system cannot happen. Government intervention is necessary to address the combination of environmental and knowledge externalities and to overcome locked-in situations (Veugelers 2012).

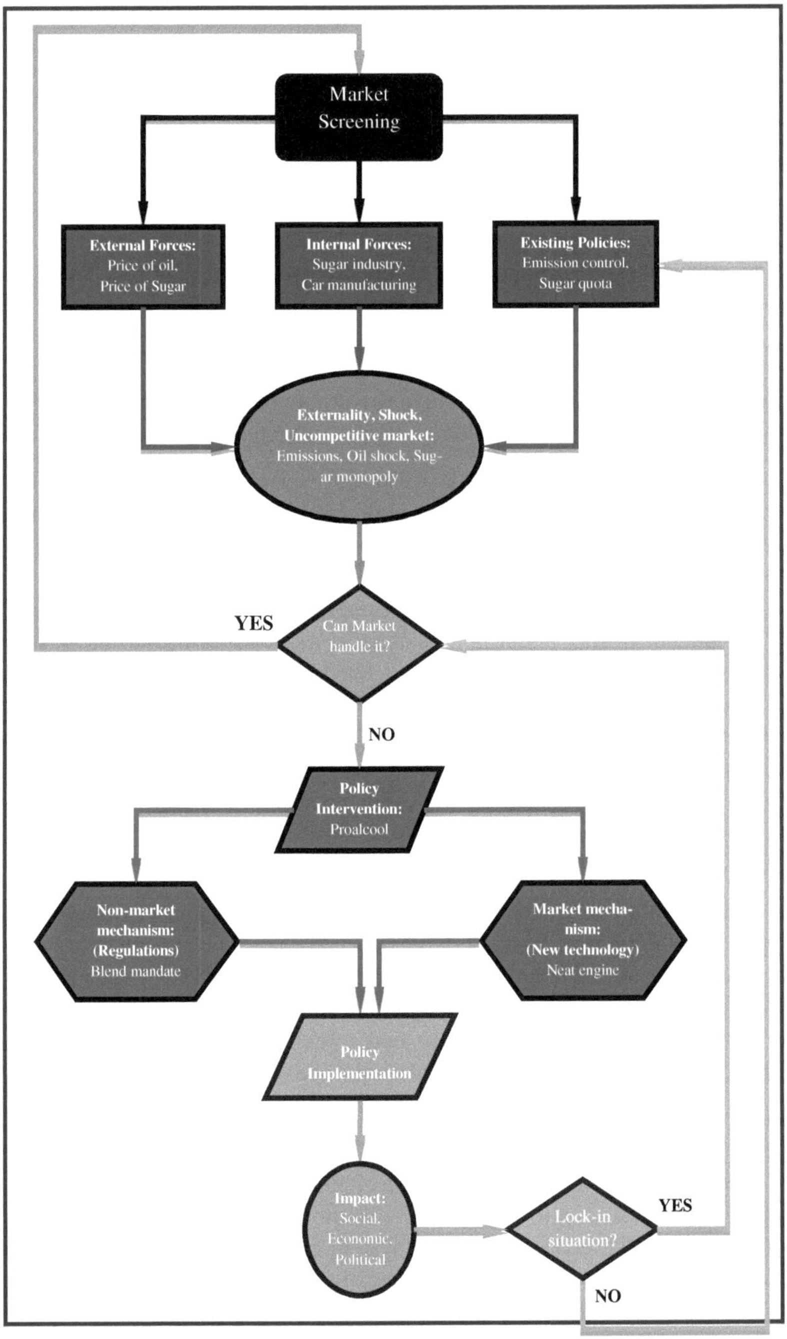

Fig. 1 Evolutionary policy framework for biofuel production programs

3 The Case of Brazilian Ethanol Program

Brazil is the largest country in South America. Its area is comparable to the United States, and its population is about two-thirds of the United States (Fig. 2). Brazil obtained its independence from Portugal in 1822, and since then it has experienced different political systems, ranging from a monarchy in the early years of independence to a presidential republic and eventually in its 1988 constitution a federal republic. Brazil was ruled by military government from 1964 to 1985, and since then it has transformed into a democracy with one of the fastest growing economies in the world.

The GDP of the country was estimated to be 2.324 Trillion USD in 2011 (about one seventh of the US economy) which places Brazil among the world's top eight economies (The World Factbook 2012). In addition to becoming an economic powerhouse in recent years, Brazil is a leading country in combating climate change. On December 29, 2009, the Brazilian Parliament adopted a law that instituted the National Climate Change Policy of Brazil and set a voluntary national greenhouse gas (GHG) reduction target by 2020.

As the largest tropical country, Brazil hosts one of the world's longest rivers, the Amazon, and significant portions of the country in the north and northwest are comprised of the Amazon Rainforest and Amazon River Basin. As Fig. 3 depicts deforestation has been a major environmental issue in these parts of the country for the last several decades. Agriculture and livestock activities are mainly responsible for this trend. These two sectors contribute to roughly 25 % of national GDP, and, at the same time, make land-use change the main source of GHG emissions in the

Fig. 2 Location of Brazil in South America

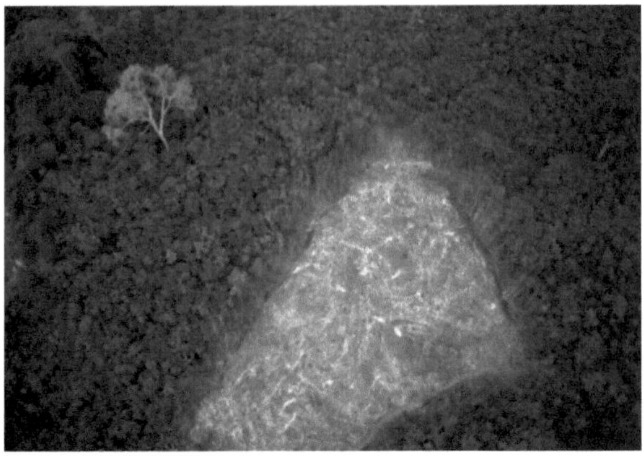

Fig. 3 Deforestation in the Brazilian Amazon. Prado (2011)

country (de Gouvello and World Bank 2010). However, and of particular interest to this paper, Brazil has one of the lowest per capita fossil-fuel-based GHG emissions in the world, thanks to the wide spread adaption of renewable sources of energy for electricity production and fuels. We will study the case of development of renewable fuel for vehicles in this paper.

3.1 Internal Factors

The sugarcane industry in Brazil has a long history and has played an important role in economic development and job creation. In 1933, the Institute of Sugar and Alcohol (IAA) was established as a national agency to oversee the sugarcane industry. In 1941, the IAA was charged with determining the production level for all growers, large or small, independent or mill owners (Cordonnier 2008/2009). However, and despite the official policy to protect smaller independent farmers, due to social and economic changes in the country, smaller producers were in decline over the next few decades. Given this brief history, by the early 1970s, the government in Brazil was well aware of the importance of having a stimulus program to jump start the sugarcane and ethanol industries in as part of an effort to uplift the national economy.

3.2 External Factors

There are two external factors that played a key role in introducing a new ethanol policy in the early 1970s.

First, the increase in the price of sugar in international markets made the Brazilian government temporarily suspend the quota system of 1941. This led to the

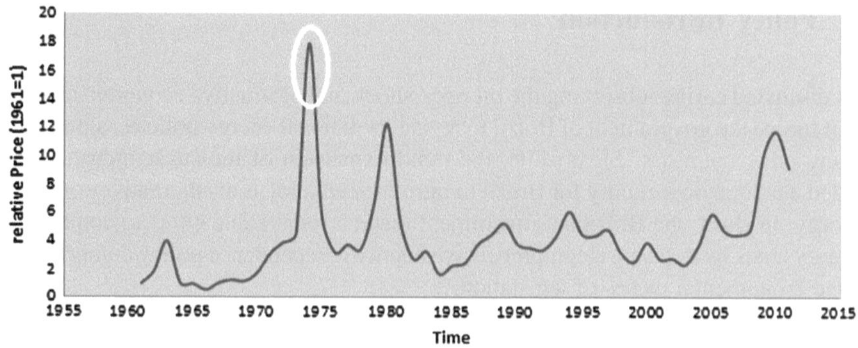

Fig. 4 Historical changes of sugar price. *Source of data* WIKIPOSIT

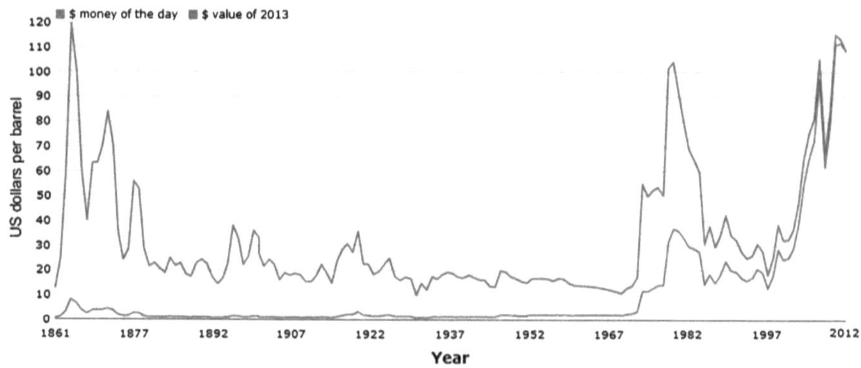

Fig. 5 Historic changes of crude oil price. Penny Arcade Forums (2012)

fast growth of large-scale sugar producers with better equipment and higher yields, allowing them to subsequently take over small independent farms and sugar mills. Figure 4 shows the historic change in price of sugar relative to the price in 1961.

Second, prior to 1975, Brazil, like many other countries, was heavily dependent on foreign oil to supply its growing domestic demand for energy. The oil crisis in 1973 and the jump in the price of oil had a significant impact on consumers around the world, forcing many oil-importing countries to look for alternative ways of supplying energy. Figure 5 shows the unprecedented magnitude of this price shock.

4 Policy Introduction

As discussed earlier, observing the oil price shock and destructive economic impacts of it forced the government of Brazil to revisit its national energy policies, along with many other countries. Meanwhile, the rapid expansion of the sugar industry provided a unique opportunity for Brazil to introduce ethanol as an alternative source of energy. In short, the Brazilian government made a remarkable effort to combat the energy crisis by devising a comprehensive energy independence policy comprised of three monumental pieces of legislation.

First, a significant effort was made to explore new sources of domestic petroleum, mainly under the deep waters of the continental shelf. After 30 years of implementation, this policy successfully helped the country to become energy independent. Figure 6 shows how this policy reached its goal over time by reducing the crude oil import and increasing the production of domestic oil, eventually leading to the exportation of this commodity in recent decades.

Second, Brazil turned to hydro-power to produce much of the country's electricity demand. The third policy was to establish a nationwide sugarcane program (The National Alcohol Program, Proalcool) based on a system of government subsidies and tax rebates to sugar cane producers and distilleries. This policy will be examined in more details throughout this paper, and we will look at how it changed the entire sugar industry and vehicle fuel economy in Brazil over the next 4 decades.

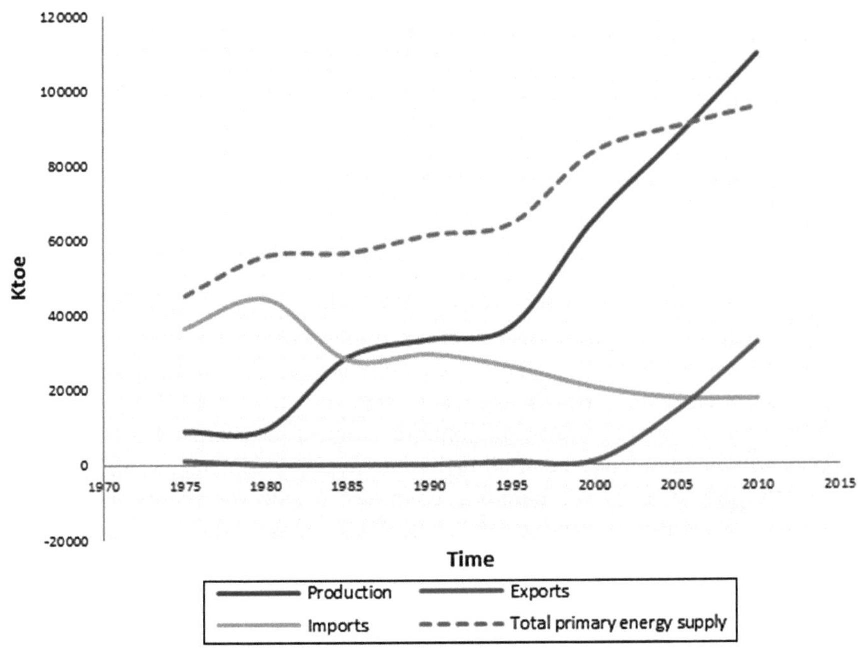

Fig. 6 Crude oil supply in Brazil 1975–2010. *Source of data* International Energy Agency

4.1 *Phase I: 1975–1979*

The government mandate of 1975 on ethanol was a landmark policy in the energy sector. It set specific goals for the ethanol industry and devise clear mechanism to assist the industry in reaching these goals. First, it demanded an increase in production of ethanol through (Colares 2008):

- An increase in agricultural productivity;
- The renovation and expansion of existing distilleries;
- The construction of new production facilities; and
- The construction of warehousing units.

Second, it assured the producers that the investments and expenditures related to the program would be funded by the government banking system. Lastly, in order to ensure the sale of ethanol for vehicle fuel, the National Petrol Council (CNP) was designated to establish a distribution program among the petrol distributing companies, which would receive the product at a price to be determined by the CNP.

Furthermore, the state-owned oil company, Petrobras, was required to purchase a guaranteed amount of ethanol, and more importantly the price ethanol was set at 59 % of the price of gasoline to make it more attractive to consumers (Goldemberg et al. 2004).

Proalcool implementation had two primary components in regard to road transportation fleets: a compulsory mandate to force all vehicles to use a 10 % blend of ethanol and gasoline, and a voluntarily mandate to use 100 % ethanol in newly designed vehicles. Later in 1979, the government increased the blend mandate to 20 %, which stimulated demand for ethanol even further (Cordonnier 2008).

4.1.1 Technology Selection: Bioethanol Crop

When a government decides to intervene in the market system to address an issue, its intervention is usually specific and selective, as it must solve a specific problem. In the case of Brazil, the government was selective in choosing two plants: cassava (manioc), a plant native to South America, and sugarcane. However, after the government intervention, it was up to the market to pick the winner. As shown in Table 1, cassava seems a superior alternative; however, surprisingly enough, the energy market favored sugarcane over cassava. The reason was that the infrastructure necessary for massive production of sugarcane was already in place, and the sugarcane industry was enjoying a boom due to the high international price of sugar. The industry was run by a small but politically powerful group of large producers who had control over 60 % of sugarcane production. However, the magnitude of the Proalcool mandate was significant even compared to the large scale of the industry. In the year before Proalcool was initiated, 130 ethanol

Table 1 Comparison of ethanol yield from Cassava and Sugarcane. Kuiper et al. (2007)

	Cassava	Sugarcane
Availability	Tolerant to extreme stress conditions, grown very easily in all areas of the country	Concentrated in two main areas in the country, the Northeast and the Center-South.
Yield (tonne/ha)	40	70
Conversion rate to ethanol (L/tonne)	150	70
Overall ethanol yield (Kg/ha)	6000	4900

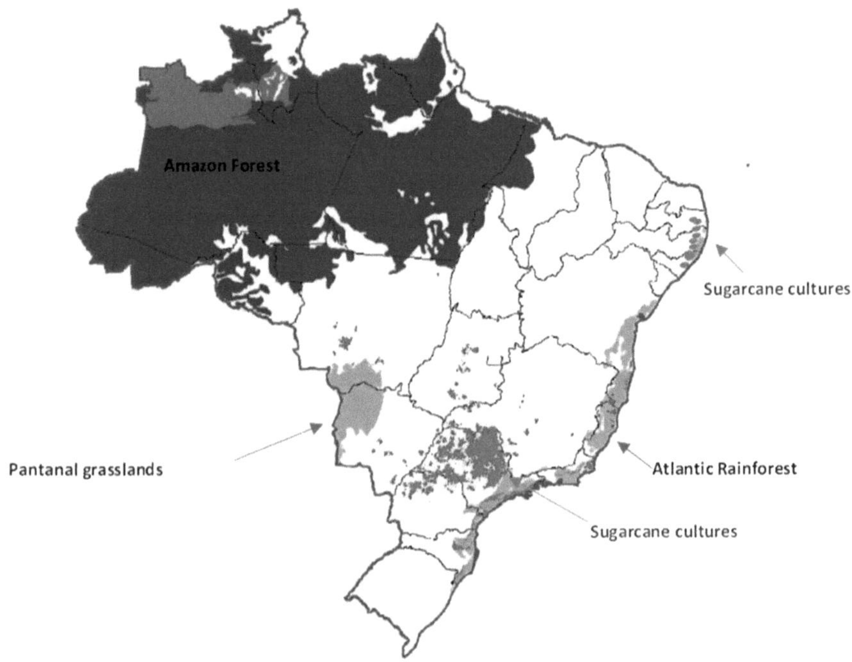

Fig. 7 Map of sugarcane crops in Brazil. Goldemberg (2008)

distilleries were operational, producing 625 million liters of ethanol from sugarcane, while the mandate was to produce 3.5 billion liters of ethanol by 1980. As Fig. 7 shows, the areas suitable for sugarcane production are concentrated mainly in two geographic regions of Brazil: the Northeast and the Center-South.

4.1.2 Technology Selection: Production Capacity

In the early years after Proalcool, ethanol was produced nearly exclusively by existing sugar mills that had expanded their capacity by adding annexed distilleries. Since the sugar industry as whole was run by a small number of large producers who controlled both sugarcane farms and sugar mills, this strategy allowed them to reduce their initial investments for ethanol production by adding distilleries to existing sugar facilities (Goldemberg 2006). However, as the goal of the program became more far-reaching, this strategy showed its shortcomings and eventually was abandoned. As it is shown in Fig. 4, the price of sugar dropped dramatically in 1979. This also forced many large sugar producers to switch from sugar production to ethanol distilling.

4.1.3 Technology Selection: Neat Ethanol Engines

Even before the implementation of the Proalcool, Brazil was hosting subsidiaries of some of the biggest car manufacturers, including Fiat, Volkswagen, GM, and Ford. These auto manufacturers were hit hard by waves of oil crises in the 1970s and once the Brazilian government approached them about designing a car that could run on an ethanol based engine (Neat ethanol engine), they had both financial interest and genuine motivation to support this initiative. The first car with an ethanol engine was manufactured and introduced to the market in 1979. Meanwhile, the Brazilian car producers modified their car engines to meet the compulsory mandate of a 10 % hydrous ethanol blend (Goldemberg 2008).

4.2 Phase II: 1979–1984

The second phase of the Proalcool can be characterized by rapid market growth and more regulations to support the use of ethanol in vehicles.

4.2.1 Demand Side: Market Expansion

At the beginning of this period, the future of ethanol vehicles was still not promising. Consumers' trust was low and the conversion from gasoline engines to ethanol engines was slow and inefficient. The government intervened again and worked with the auto manufacturers to improve the quality of ethanol engines. To make these goals achievable, the government implemented new regulations to ensure that ethanol would always be competitive with fossil fuel in the fuel market. First, the tax on fossil fuel increased, and then the pump price of ethanol was regulated to be no more than 64.5 % higher than the price of gasoline. Furthermore, following a government-sponsored protocol issued in 1979, all major auto makers in Brazil were required to phase in ethanol-consuming cars and phase out gasoline cars. Later, the government lowered the price cap on ethanol to be no more than 59 % of gasoline in for the hope of stimulating demand. To prepare for the demand to take off, the government also decreased the ethanol-gasoline blend mandate to 12 % to have sufficient ethanol in the market. After a successful introduction of the Neat ethanol engines in the first phase, the government decided to set an ambitious target for this program: producing 250,000 Neat ethanol vehicles by 1980, 300,000 by 1981 and 350,000 by 1982. In order to help raise awareness of the new vehicle, priority for these vehicles was given to governmental fleets. The government also mounted a comprehensive information campaign about the vehicles. These efforts were extremely successful and the Neat ethanol market share of new cars sold in Brazil jumped to 90 % in early 1980s. By 1985, two million of the totals of ten million cars were fueled completely by ethanol, and the rest were fueled by gasohol (blend of gasoline and ethanol). The ethanol-gasoline blend was set again at 20 % by the end of this period.

4.2.2 Supply Side: Emergence of New Players

Observing the market expansion, the government lifted the ethanol mandate to 10.7 billion liters by 1985 and poured generous subsidies and incentives on the ethanol industry. The second phase of Proalcool coincided with the second wave of oil crisis, which increased the price of gasoline and made ethanol production even more attractive. During this phase, the price of sugar, which had spiked in 1980, fell to a record low in 1985. This helped the sugar industry to stay in business of producing ethanol in annexed distilleries. However, growing ethanol demand and attractive government subsidies introduced new players to the supply side of the market. Instead of traditional combined sugar and ethanol production facilities, these distillation facilities were solely dedicated to producing ethanol. Seeking for ways of reducing the risk of investment and diversifying their portfolio, these facilities established partnerships with electricity suppliers to use bagasse (a byproduct of making ethanol from sugarcane) as a fuel for power plants (Soccol et al. 2005).

4.3 Phase III: 1985–1990

The military dictatorship in Brazil was abolished and a transition period to a new democracy started in the country. This sociopolitical change was accompanied by an economic recession, which endangered many new technological firms in the ethanol industry. As the global price of oil continued to fall, the price of sugar started to rise again.

4.3.1 Economy: Transition to Free Market

During this phase, the country was suffering from economic stagnation, inflation, and crisis. Many governmental programs and subsidies to the ethanol industry were cut and public investment in this industry was limited. As government involvement in the ethanol industry was shrinking, the sugar industry faced new challenges of competitiveness and efficiency to survive. Many inefficient companies were dependent on the government subsidies that had been phased out, which caused a shortage of ethanol by the end of 1980s. Subsequently, the sale of Neat ethanol vehicles dropped to only 11 % of total automobile sales in the mid-1990s. Consumers switched back to gasoline cars in the absence of a government mandate for buying Neat ethanol vehicles. Overall, these challenges made the ethanol industry focus on the efficiency and flexibility rather than market expansion and new generation capacity. The efficiency of ethanol production has improved over the last 40 years and starting from 1985 significant increases in agricultural yields was achieved. High productivity coupled with economics of scale in large production facilities led to a reduction in production costs. Figure 8 shows the learning curve for ethanol production.

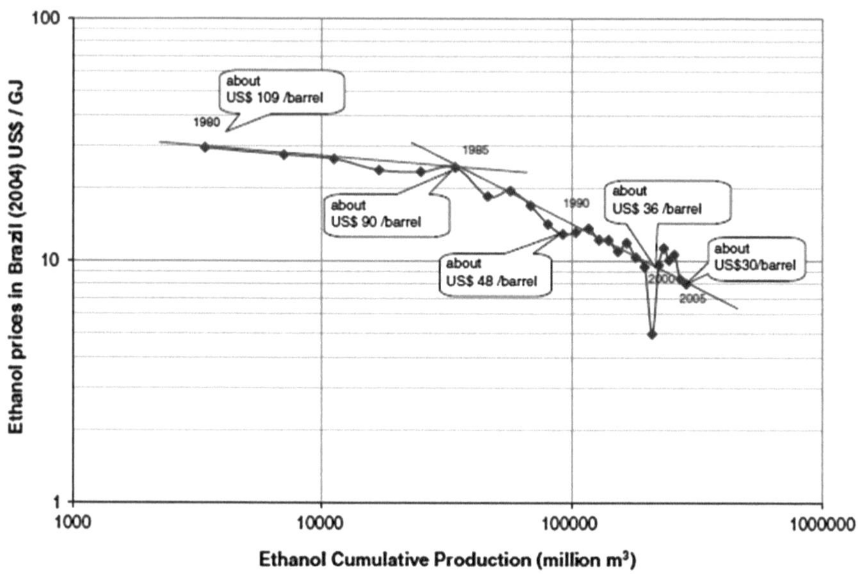

Fig. 8 Ethanol learning curve. Goldemberg (2006)

4.4 Phase IV: 1991-Now

The last and current phase of the program started with new democratic government and constitution in place in Brazil. Brazil faced social and political instability during the early years of this era but finally managed to control its inflation and stabilize its economy.

4.4.1 Policy Revision: Flexibility and Improvement

In 1993, the country was heavily dependent on foreign oil to supply its growing domestic demand of gasoline. New challenges, such as air pollution and GHG emissions, have surfaced in the international domain, and the ethanol industry, which was once seen as a symbol of national pride, was on the brink of collapse.

Due to these considerations, in 1993, a federal decree was passed to mandate the 22 % blend of ethanol-gasoline. This act gave new hope to the ethanol industry and saved it from bankruptcy. Later, when the global price of oil increased in early years of 20th century, the government set the blend mandate to a range of 20–25 % for gasoline cars, giving an opportunity to oil companies to adjust their blend with the fluctuation in the price of gasoline and ethanol. It also introduced a very innovative program to produce flexible-fuel vehicles (FFVs). These vehicles are equipped with sensors which detect the level of ethanol in the fuel and can adjust the engine settings accordingly. FFVs can run on any blend of ethanol from E0 (gasoline only) to E100 (Neat ethanol). With the price of gasoline increasing in recent years, FFVs are an excellent choice for consumers who seek cheaper fuel. Figure 9 shows the comparison between the prices of ethanol and gasoline in international market. In 2004, FFVs were introduced to the market and gained about 17 % market share compared to 3 % of Neat ethanol vehicles. They currently hold about 90 % of the market share and have proven to be a reliable choice for consumers (Brown and Sovacool 2011). The increase in the Brazilian FFVs is a major driving factor of long-run ethanol demand (de Freitas and Kaneko 2011). Economic studies of ethanol fuel substitution for gasoline shows that due to its higher price elasticity, ethanol consumers have more options when price goes up, while fossil fuel consumers are more vulnerable to price increases (Santos 2013).

As is clear from the figure above, the current price of ethanol has a downward trend and has fallen below the gasoline price since 2004. This is a clear indicator that ethanol is competitive now and that the market can be deregulated from subsidies and price caps on ethanol.

Other observations should be made about the Neat ethanol vehicles versus FFVs. This is a good example of a lock-in situation that is quite common in innovation policy practices. While the idea of a Neat ethanol vehicle was very innovative was considered a game changer in the auto industry during the early 1980s, it proved to be an unreliable option for consumers facing the shortage of ethanol. However, the Brazilian government showed more commitment to its energy independence policy

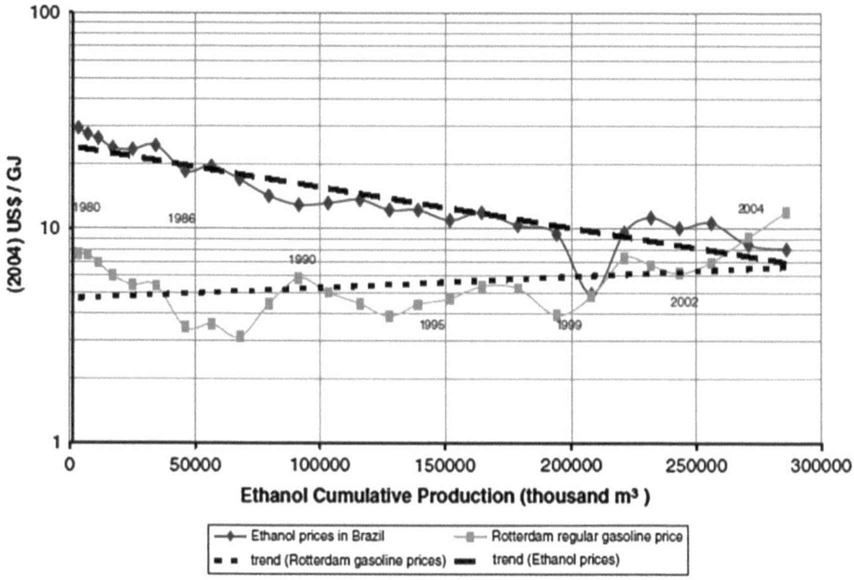

Fig. 9 Ethanol production in Brazil and comparison between prices of ethanol and gasoline in international markets. Goldemberg (2006)

rather than a particular tool or technology to achieve that policy. Therefore, faced with a lock-in situation over the Neat ethanol vehicle technology, Brazil decided to pursue a different path and eventually came up with an impressive alternative of flexible fuel vehicles.

Following the creation of the Inter-ministerial Work Group by Brazil's Federal Government in 2003, this commission recommended that: (i) the biodiesel use should not be mandatory, (ii) the biodiesel policy should be technology-neutral, and (iii) it should benefit the poor regions of the country the most (Pousa et al. 2007).

Currently about 57.6 % of national production of ethanol is concentrated in the State of São Paulo, with Minas Gerais being the second largest national producer with an 8.7 % share (Pereira et al. 2012). Analytical studies show that the domestic demand and price of ethanol continues to rise in Brazil as FFVs continue to dominate the new car sells (Du and Carriquiry 2013).

5 Lessons for Achieving a National Energy Independence Plan

In this paper, we showed the path that Brazil has paved since 1975 to reach energy independency in the road transportation sector. We focused on the case of the National Alcohol Program (Proalcool) and showed how this policy evolved during

4 phases of social, economic and technological changes. We observed and identified key factors that helped shape the policy in each phase by emphasizing both internal factors (economy, politics, and technology) as well as external factors (global price of oil and sugar). The key structure we sought to identify is when and under what circumstances government intervention in the market is justified. We noticed that during:

- The Oil Shock of 1973,
- The lock-in situation over Neat ethanol vehicle technology, and
- The fluctuating price of sugar,

the Brazilian government intervened in the market. In each case, the government employed two types of mechanisms:

- Non-market mechanisms: regulating the fuel market by imposing an ever changing ethanol blend mandate, and
- Market mechanisms: creating new markets for Neat ethanol vehicles in the 1980s and Flexible Fuel Vehicles in recent years

In any case, once the government decides to intervene in the market, each mechanism should be carefully chosen so as to avoid favoring a part of the market and cause an unfair advantage in other parts of the market. Policies should be designed for a limited range of impact and they must be evaluated both during and after they are employed. Another important observation is the persistence and courage which existed among high levels of the government and was required to pursue ambitious plans for energy independence. This point becomes even more interesting when we notice that the Proalcool started under the rule of the military dictatorship and survived to achieve its goals 40 years later under a democratically elected civilian government. Good policies invariably appeal to the public and build upon their social capital for support. As the case of energy independent policy in Brazil demonstrates, economic motivation is a strong driver in designing a robust independent policy. However, observing the emergence and decline of renewable energy policies in countries like the US clearly shows that the sole reliance on external factors (high oil prices) cannot provide a long term motivation for switching to a sustainable energy policy. The public sector has to facilitate an endogenous supporting mechanism at home before taking any step toward reducing its reliance to foreign energy resources to achieve the long-term energy independence. As proalcool relied on the traditionally strong sugarcane industry in the Brazil, other crops like corn can be considered as alternatives for developing a biofuel industry in other countries (Prasad et al. 2007). However, any dramatic increase in corn-based ethanol production may cause a competition for the limited agricultural land needed for food and feed production. Therefore, other potential sources for low cost ethanol production should be considered. These alternatives ranges from lignocellulosic materials such as crop residues, grasses, sawdust, and wood chips to solid animal waste and industrial wastes (Sun and Cheng 2002).

Furthermore, the Brazil's ethanol program shows a successful campaign for energy independence requires a high level of integration and adaptation in both supply and demand sides of the energy market. The introduction of Proalcool in Brazil created at least three new industries which enabled Brazil to impact the world energy markets: the distilling equipment manufacturing, ethanol engines manufacturing, and alcohol-based chemical industry (Levinson 1987). Flexibility should be considered as another important aspect of an effective policy. The remarkable coming back of the ethanol program would not be possible if the car manufacturers were not finally able to introduce the flex-fuel cars to the market. Adaptivity, flexibility, and responsiveness are the core characteristics of the future sustainable energy system whether in Brazil or in other countries. The success of Brazilian ethanol program has provided new hopes for displacing about 5 % of projected global gasoline use in 2025 by expansion of sugarcane-derived ethanol (de Cerqueira Leite et al. 2009).

Environmental considerations have to be taken into account in any biofuel expansion program. In Brazil, most biofuel plantations would replace rangeland areas and therefore direct land-use changes from such expansion will have a small impact on carbon emissions. However, indirect land-use changes, could offset the carbon savings from biofuels (Lapola et al. 2010). As air quality studies of the larger cities of São Paulo and Rio de Janeiro have pointed out, ethanol fuels are responsible for the increase in CH_3CHO and NO_x emissions that are important contributors to photochemical air pollution and ozone (O_3) formation (Anderson 2009).

Finally, it should be noted that the biofuel program can be effective only as part of a more comprehensive energy independence plan which may include the increase in domestic oil production and energy efficiency (Potter 2008). For example while ethanol satisfies only about twenty percent of Brazil's transportation fuel supply, large increase in domestic oil production accounts for the remaining transportation fuel needs (Fichera and Kueter 2006).

References

Anderson LG (2009) Ethanol fuel use in Brazil: air quality impacts. Energy Environ Sci 2 (10):1015–1037

Brown MA, Sovacool BK (2011) Climate change and global energy security: technology and policy options. MIT Press, Cambridge

Colares JF (2008) A brief history of Brazilian biofuels legislation. SSRN ELibrary, Jun 2008

Cordonnier VM (2008/2009) Ethanol's roots: how Brazilian legislation created the international ethanol boom. William Mary Environ Law Policy Rev 33:287

de Cerqueira Leite RC, Verde Leal MRL, Barbosa Cortez LA, Griffin WM, Gaya Scandiffio MI (2009) Can Brazil replace 5 % of the 2025 gasoline world demand with ethanol? Energy 34 (5):655–661

de Freitas LC, Kaneko S (2011) Ethanol demand in Brazil: regional approach. Energy Policy 39 (5):2289–2298

de Gouvello C, World Bank (2010) Brazil low-carbon country case study. World Bank, Washington

Deutch J, Schlesinger JR, Victor DG (2006) National security consequences of U.S. oil dependency, Nov 2006
Did peak oil cause the financialization of the US economy? Penny Arcade Forums. http://forums.penny-arcade.com/discussion/153588/did-peak-oil-cause-the-financialization-of-the-us-economy. Accessed 06 Oct 2012]
Du X, Carriquiry MA (2013) Flex-fuel vehicle adoption and dynamics of ethanol prices: lessons from Brazil. Energy Policy 59:507–512
Edquist C (2002) Innovation policy—a systemic approach. In: Archibugi D, Lundvall B (eds) The globalizing learning economy, Oxford, Oxford University Press
Fichera J, Kueter J (2006) Considering Brazil's energy independence. The Marshall Institute Policy Outlook series
Goldemberg J (2006) The ethanol program in Brazil. Environ Res Lett 1(1):014008
Goldemberg J (2008) The Brazilian biofuels industry. Biotechnol Biofuels 1(1):6
Goldemberg J, Coelho ST, Lucon O (2004) How adequate policies can push renewables. Energy Policy 32(9):1141–1146
Hira A, de Oliveira LG (2009) No substitute for oil? How Brazil developed its ethanol industry. Energy Policy 37(6):2450–2456
Kuiper L, Ekmekci B, Hamelinck C, Hettinga W, Meyer S, Koop K (2007) Bio-ethanol from cassava. ECOFYS, PBIONL062937
Lapola DM, Schaldach R, Alcamo J, Bondeau A, Koch J, Koelking C, Priess JA (2010) Indirect land-use changes can overcome carbon savings from biofuels in Brazil. Proc Natl Acad Sci 107 (8):3388–3393
Levinson M (1987) Alcohol fuels revisited: the costs and benefits of energy independence in Brazil. J Dev Areas 21:243–258
Negro SO, Alkemade F, Hekkert MP (2012) Why does renewable energy diffuse so slowly? A review of innovation system problems. Renew Sustain Energy Rev 16(6):3836–3846
Pereira MG, Camacho CF, Freitas MAV, da Silva NF (2012) The renewable energy market in Brazil: current status and potential. Renew Sustain Energy Rev 16(6):3786–3802
Peters GP, Marland G, Le Quéré C, Boden T, Canadell JG, Raupach MR (2012) Rapid growth in CO_2 emissions after the 2008–2009 global financial crisis. Nat Clim Change 2(1):2–4
Potter NI (2008) How Brazil achieved energy independence and the lessons the United States should learn from Brazil's experience. Wash U Glob. Stud Rev 7:331
Pousa GPAG, Santos ALF, Suarez PAZ (2007) History and policy of biodiesel in Brazil. Energy Policy 35(11):5393–5398
Prado P (2011) Disastrous forestry bill offers Brazil's president a critical test. Conservation International
Prasad S, Singh A, Joshi HC (2007) Ethanol as an alternative fuel from agricultural, industrial and urban residues. Resour Conserv Recycl 50(1):1–39
Santos GF (2013) Fuel demand in Brazil in a dynamic panel data approach. Energy Econ. 36:229–240
Soccol CR, Vandenberghe LPS, Costa B, Woiciechowski AL, de Carvalho JC, Medeiros ABP, Francisco AM, Bonomi LJ (2005) Brazilian biofuel program: an overview. J Sci Ind Res 64 (11):897–904
Sun Y, Cheng J (2002) Hydrolysis of lignocellulosic materials for ethanol production: a review. Bioresour Technol 83(1):1–11
The World Factbook (2012) Central Intelligence Agency. http://www.cia.gov/library/publications/the-world-factbook/geos/br.html Accessed 07 Oct 2012
Veugelers R (2012) Which policy instruments to induce clean innovating? Res Policy 41 (10):1770–1778

Index

S.D. Eksioglu et al. (eds.), *Handbook of Bioenergy*, Energy Systems, DOI 10.1007/978-3-319-20092-7

MIX
Papier aus verantwortungsvollen Quellen
Paper from responsible sources
FSC® C105338

If you have any concerns about our products, you can contact us on
ProductSafety@springernature.com

In case Publisher is established outside the EU, the EU authorized representative is:
Springer Nature Customer Service Center GmbH
Europaplatz 3, 69115 Heidelberg, Germany

Printed by Libri Plureos GmbH
in Hamburg, Germany